# Life, Work, and Rebellion in the Coal Fields

DAVID ALAN CORBIN

# Life, Work, and Rebellion in the Coal Fields

## The Southern West Virginia Miners 1880–1922



UNIVERSITY OF ILLINOIS PRESS
Urbana Chicago London

LIBRARY OF CONGRESS CATALOGING IN PUBLICATION DATA

Corbin, David.
Life, work, and rebellion in the coal fields.

(The Working class in American history)
Bibliography: p.
Includes index.
1. Coal-miners—West Virginia—History. 2. Labor disputes—West Virginia—History. 3. Trade-unions—Coal-miners—West Virginia—History. I. Title.
II. Series: Working class in American history.
HD8039.M62U61775 331.7'622334'.09754 80-25493
ISBN 0-252-00850-2 (cloth)
ISBN 0-252-00895-2 (paper)

DEDICATION

To my parents, Raymond and Dorothy Corbin,
for providing me with their love, their support,
and their deep sense of social justice.

# Contents

# Acknowledgments

The research and writing of this book was possible only with the cooperation of many people. I especially thank the hundreds of men and women of southern West Virginia who shared with me their time and experiences. I pray that I have not betrayed their trust; God knows they have been exploited enough.

For special help in the research, I thank Mrs. Mary Jenkins, West Virginia State Archives and History; Ms. Brenda Beasley and Drs. David R. Kepley and Sara Jackson, National Archives; Dr. Jonathan Grossman, Mr. Judson McClaury, and Mr. Henry Guzda, U.S. Department of Labor; Mr. Dale Lawson and Ms. Martha Spence, formerly of the United Mine Workers of America; Mr. James Jones, Mr. Thorold Funk, and Mr. Cecil Lacey, West Virginia Vocational Rehabilitation; Professor Fred Barkey, West Virginia College of Graduate Studies; Professor Richard Farrell, University of Maryland; Mr. Richard Hadsell, U.S. Department of Agriculture; Mr. Nathan Einhorn, Library of Congress; and Raymond, Dorothy, Ronnie, and Dee Corbin, family.

The preparation for publication profited from the help of others. Again, I extend my appreciation to Mr. Nathan Einhorn, whose readings of the drafts, handwritten as well as typed, made the job easier for the following readers. I sincerely thank Professors Ira Berlin and J. H. Laslett for carefully and tediously reading the manuscript and providing valuable criticism. Anyone who has worked with Professor David Grimsted knows well his humor and patience; these characteristics will be remembered and appreciated, as will his intellect and ability to improve my manuscript. Great pleasure came from working with Professor David Montgomery in preparation for publication; the manuscript and I both profited from his reading and advice, which was insightful and critical, and, as usual, constructive. My deep appreciation goes to Ms. Carol Boyer, who typed, corrected, and retyped; I am forever in her debt. It was my good fortune and privilege to work with Ms. Susan L. Patterson of the University of Illinois Press during the final stages of publication; her friendliness and cooperation were exceeded only by her thoroughness and professionalism. My wife, Candace H. Beckett, has my loving gratitude for her patience and encouragement, and for being there.

# Introduction

From 1912 to 1922, the southern West Virginia coal fields (now Districts 17 and 29 of the United Mine Workers of America [UMWA]) witnessed some of the most dramatic and violent events in American labor industry. These episodes included the bloody and protracted Paint Creek–Cabin Creek strike of 1912–13 that saw a labor-management dispute turn into civil war; a shootout in the streets between the citizens of Matewan and the representatives of law and order in the coal fields that cost the lives of eleven men; a ninety-mile armed march of 20,000 coal miners bent on the overthrow of two county governments; and the formation of a dual union in opposition to the UMWA. These events resulted in an untold number of deaths, indictments of over 550 coal miners for insurrection and treason, and four declarations of martial law. They shook the foundations of the largest trade union in the United States and rocked the nation's sense of honor and decency.

The violent behavior of the southern West Virginia miners was discussed intensely in their time. Radicals throughout the country publicized the miners' actions as evidence of class warfare and the beginnings of the socialist revolution in the United States.[1] Liberal, muckraking journalists and trade union leaders painted a more moderate picture. Interested primarily in exposing the exploitative and brutally anti-union policies of the southern West Virginia coal operators, the journalists and labor leaders explained the miners' actions as nothing more than attempts, albeit violent, to organize an area that the president of the American Federation of Labor, Samuel Gompers, called "the last remains of industrial autocracy in America."[2] It was this description that was written into the history books by the Wisconsin institutional labor historians who depended upon the writings and statements of the muckrakers and union officials for their interpretation of these events.[3]

Events in the late 1960s and early 1970s have again stimulated a current, and consequently an historical, interest in Appalachian coal mining. On the national level, the energy crises publicized the nation's dependence upon coal as an energy-producing resource, while the sensational murder of Joseph (Jock) Yablonski and trial of W. A. (Tony)

Boyle recalled the coal industry's bloody past. On the regional level, the Appalachian movement has been exposing the political corruption and economic colonialism that have historically plagued the region, and it has bitterly assailed the "John-Boy Walton" image of Appalachians as a people who have passively, if not happily, accepted that corruption and colonialism as well as poverty and exploitation. But events in southern West Virginia have again particularly caught the nation's attention as the coal miners in the area have staged massive wildcat strikes, boldly participated in the textbook protest in Kanawha County, and, against overwhelming obstacles, successfully carried out the Black Lung Rebellion and the Miners for Democracy movement, demonstrating that contemporary workers' militancy is not the figment in the minds of radical intellectuals, but a reality.

These national, regional, and local happenings initiated a wave of scholarly studies on the history of Appalachia, the coal industry, the UMWA, mine workers' leaders, and, of course, coal strikes—especially those bloody and dramatic ones that occurred in southern West Virginia between 1912 and 1922 against the anti-union coal operators in the area. Although separated by fifty years, a period marked by improved approaches toward the study of history and new concepts in the study of labor history, these recent studies of the coal strikes in southern West Virginia add little insight into, or knowledge about, those strikes and acts of violence.[4] They claim that the miners' violent behavior stemmed from their mountaineer tradition of "feuding," "gun-totin'," and "moonshining." This erroneous, pejorative reasoning fails to recognize the decades of economic growth and social change in southern West Virginia prior to the outbreaks of strikes and violence. It displays an ignorance of the fact that about two-thirds of the area's work force during this period came from outside the region, while two to three decades of industrialized life and work separated the native miners (the other one-third of the work force) from their mountain culture. This reasoning provides no better insight than the historical analysis of the U.S. Coal Commission in 1922, which explained that the violence had occurred in southern West Virginia because the area was "exclusively peopled by mountaineers" who "were a law unto themselves." Ignoring the sociocultural change prompted by the introduction of coal mining into the area, the Commission claimed: "Local traditions [still] exert a dominating influence and account very largely for the outbreaks of violence. Much of the violence had nothing to do with the coal industry but had to do with the nature and racial characteristics of the people. . . . The primitive conditions of life of this people can scarcely be paralleled anywhere."[5]

The failure to analyze more fully the southern West Virginia miners' seemingly bizarre deeds is an outrage and an injustice to the men and women who felt compelled to take those actions. Just as important is that our ignorance, or misunderstanding, of these events leaves a glaring void in American labor historiography. Until we understand and appreciate more completely the violent behavior of the southern West Virginia miners, we as social and labor historians cannot—and will not—fully comprehend the evolution of the American labor movement.[6]

Nearly fifty years ago, the Wisconsin school of labor historians, most prominently including John R. Commons, Selig Perlman, and Philip Taft, set the standards for the study of American labor history. Believing that the successes and failures, hopes and disappointments, and aspirations or lack of them of American workers could be understood by studying trade unions, labor leaders, and strikes (an approach that earned them the label of institutional labor historians), the Wisconsin school depended almost exclusively upon institutional sources (e.g., trade union journals and papers, the writings and speeches of labor officials) for their studies and interpretations.[7] A recent critic, Herbert Gutman, described this approach as a "narrow economic analysis" that encouraged labor historians "to spin a cocoon around American workers, isolating them from their own particular subculture and from the larger national culture. . . . [It caused] the study of American working-class history to grow more and more constricted and become more detached from larger developments in American social and cultural history."[8] Within the boundaries of their limited data and perception, the institutional labor historians concluded that American workers have wisely eschewed the radical ideologies of outside intellectuals and have chosen instead to follow the conservative, pragmatic ways of their union officials, who sought limited objectives on which compromises were possible. Content with their dependent and subordinate relationship to management, American labor unions, according to this interpretation, became only another powerful interest group in America's pluralistic structure, demanding a larger piece of the capitalist pie. Any strides toward a working-class consciousness or a formation of a labor party were slowed by America's democratic, political traditions, the opportunities for social mobility, and most important, the ethnic, racial, and religious diversities within America's polyglot work force.

Except for the polemical writings of a few radicals, the conclusions of the Wisconsin institutional labor historians, until recently, remained unchallenged. The consensus historians of the 1950s, who maintained that American history was void of class struggle, promoted their interpretations as they insisted that a "harmony of interests" has characterized labor-management relations in the United States.[9]

The first major wave of attacks upon the approach and conclusions of the Wisconsin institutional historians came with the social crises of the 1960s. Determined to discover—or rediscover—America's radical past, historians began reinterpreting the American labor movement, for here the historian could find social conflict. Some labor historians strained to find evidence of solidarity and militancy among American workers. Others emphasized, but more often romanticized, the importance of America's radical labor parties and leaders.[10] Still others reevaluated the trade union movement itself and the policies of American labor leaders and claimed that, rather than being pragmatic bureaucrats, many trade union officials were adroit thinkers, attempting to adapt radical European and Marxist labor thought to American conditions and values.[11]

Following the ideas and leads of several outstanding European social and labor historians, Gutman has provided probably the most valuable historical cross-examination of the American labor movement.[12] He has rendered an important service to the study of American labor history by challenging the institutional approach and stressing the need for historians to study the worker and his culture as well as officials and institutions. Gutman and his students have found that the worker of the Gilded Age was not as passive and flexible as had been proclaimed, but that he protested, often militantly, against the new socioeconomic order. While Gutman successfully emphasized the nineteenth-century workers' encounters with encroaching industrialism, few have carried his approach into the twentieth century to study the workers' plight and behavior under a matured, industrialized, capitalistic economy. The absence of working-class resistance, or at least our lack of knowledge of it, to capitalistic America has caused even a self-proclaimed Marxist scholar (although not a labor historian) to remark that with the exception of the antebellum South "we need to face the fact that . . . identification between bourgeois and general interests has existed throughout American history"[13]—a consensus interpretation from a Marxist!

The most serious efforts to rescue the twentieth-century worker from the conclusions of the institutional labor historians come from writers who have pointed out that, rather than representing the rank-and-file workers, unions and union leaders have been unresponsive to their desires and have often opposed their aggressive, sometimes collectivist interests.[14] In the best and most provocative of these books, *Strike!*, Jeremy Brecher argues that because of the nature and structure of the collective-bargaining process (e.g., the contract) and as a matter of self-preservation, union officials have exercised a moderating influence and control upon rank-and-file militancy. Discipline rather than democracy has characterized American trade unions. If this thesis is true, it contradicts the conclusions of the institutional labor historians who based their interpretations upon union documents. To understand the worker we must study the worker—not his institution nor his elite. Despite Brecher's emphasis on the militancy of rank-and-file workers, he does not quite claim that American labor has possessed, except during mass strikes, a sense of working-class consciousness. In fact, he denies the existence of class feelings and behavior in American history. Brecher concedes that social mobility, ethnic and racial conflicts, political traditions, and beliefs in American exceptionalism have inhibited class feelings and actions in American society.[15] With this conclusion we have returned to the Wisconsin institutional historians.

Brecher's failure to find class consciousness and class behavior in American labor history is not surprising, considering his approach. He, like several other labor historians, has extracted the strikes and labor violence from their sociocultural background. E. P. Thompson recently warned that "if we stop history at a given point, then there are no classes but simply a multitude of experiences. . . . We cannot understand class unless we see it as a social and cultural formation, arising from the processes which can only be studied as they work themselves out over a considerable historical period."[16] Similarly, Erik Erikson has assailed "students of society and culture . . . [who] blithely continue to ignore the simple fact that society consists of generations in the process of developing from children into adults . . . destined to absorb the historical changes of their lifetime and to continue to make history for their descendants."[17] Lifting strikes out of their historical context limits our understanding as much as does studying the union without studying the worker.

Working-class culture involves more than workers' solidarity during a labor-management dispute. It is not, Gutman points out, the "crude social history that defines class by mere occupation (and/or income) and culture as some kind of magical mix between ethnic and religious affiliations."[18] Rather, the working class, as Thompson declares, is a historical phenomenon and relationship that unifies a "number of desperate and seemingly unconnected events, both in the raw material of experience and in consciousness." According to Thompson, class occurs "when some men, as a result of common experiences (inherited or shared), feel and articulate the identity of their interests as between themselves, and as against other men whose interests are different from and usually opposed to theirs." It is "defined by men as they lived their own history, and in the end, this is the only definition."[19]

Following the leads and heeding the warnings of Gutman, Thompson, and Erikson, I here suggest the hypothesis that the strikes, outbreaks of violence, and formation of a dual union in the southern West Virginia coal fields were neither primitive nor sporadic occurrences. They were collective and militant acts of aggression, interconnected and conditioned by decades (beginning in the 1890s) of social change, economic exploitation and oppression, political corruption, and tyranny. They were products of decades of evolving attitudes toward and about work, life, existence, unionism, employers, law and order, social and economic justice, the state, the community, and the meaning of America. They were mature responses of fully sane and industrialized workers to conditions they understood and hated and wanted to change. They can be understood as extensions of a world view, a working-class culture, that was weaned and schooled in the company town (1890–1910), achieved a regional youth and identity during and following the Paint Creek–Cabin Creek strike (1912–16), and reached maturity during World War I (1917–18) and adulthood in 1919, when the southern West Virginia miners stood ready to match their will against a hostile world. If their culture, and the attitudes and actions of the miners prompted by that culture, seem strange, awkward, and primitive, it must be remembered that, as Thompson wrote of the Luddites, "they lived through those times of acute social disturbance, and we did not. Their aspirations were valid in terms of their own experience; and, if they were casualties of history, they remain, condemned in their own lives, as casualties."[20]

## NOTES

1. For the Paint Creek–Cabin Creek strike, see Fred Barkey, "The Socialist Party in West Virginia from 1898 to 1920" (Ph.D. diss., University of Pittsburgh, 1971), chap. 4; for the Mingo County strike, see "Stand by the Miners of Mingo," printed and circulated by the Executive Committee of the Communist Party (copy in author's collection), original in District 17 Correspondence Files, United Mine Workers of America Archives, UMWA Headquarters, Washington, D.C.

2. Typical of the muckraking accounts are: Winthrop Lane, *Civil War in West Virginia* (New York, 1921); John Owens, "Gunmen in West Virginia," *New Republic* 28 (Sept. 21, 1921): 90–92; Arthur Warner, "West Virginia—Industrialism Gone Mad," *Nation* 113 (Oct. 5, 1921): 372–73; McAllister Coleman, "A Week in West Virginia," *Survey* 53 (Feb. 1, 1925): 532–34; Arthur Gleason, "Company-Owned Americans," *Nation* 110 (June 12, 1920): 794–95.

3. See, for example, Selig Perlman and Philip Taft, *History of Labor in the United States, 1896–1932*, 4 (New York, 1935), 330–35, 479–82.

4. Howard Lee, *Bloodletting in Appalachia* (Morgantown, W.Va., 1969); Daniel Jordan, "The Mingo War," in *Essays in Southern Labor History*, ed. Merl Reed and Gary Fink (Westport, Conn., 1977); Maurer Maurer and Calvin Senning, "Billy Mitchell, the Air Service and the Mingo War," *Airpower Historian* 12 (Apr. 1965): 37–43; Cabell Phillips, "The West Virginia Mine War," *American Heritage* 25 (Aug. 1974): 58–61, 90–94.

5. U.S. Coal Commission, *Report of the United States Coal Commission*,

Senate Document no. 195, 68th Cong., 2nd sess., 5 vols. (Washington, D.C., 1925).

6. For example, in surveys and syntheses of the history of the American labor movement, radical as well as conservative and institutional labor historians have misunderstood and misinterpreted the labor-management conflicts in southern West Virginia. See Jeremy Brecher, *Strike!* (San Francisco, 1972); Sidney Lens, *The Labor Wars* (Garden City, N.Y., 1972); and Perlman and Taft, *History of Labor.* My disagreements with their interpretations will be discussed point by point in this book.

7. See especially Selig Perlman, *A Theory of the Labor Movement* (New York, 1949).

8. Herbert Gutman, "Work, Culture, and Society in Industrializing America, 1815–1919," *American Historical Review* 78 (June 1973): 536.

9. See, for example, Edward Kirkland, *Industry Comes of Age* (New York, 1961).

10. See, for example, James Weinstein, *The Decline of American Socialism* (New York, 1969); Bernard Brammel, "Eugene V. Debs: Blue-Denim Spokesman," *North Dakota Quarterly* 43 (Spring 1972): 12–28; Ronald Radosh, ed., *Debs* (Englewood Cliffs, N.J., 1971), 1–7; Melvyn Dubofsky, *We Shall Be All: A History of the Industrial Workers of the World* (Chicago, 1969); Patrick Renshaw, *The Wobblies* (Garden City, N.Y., 1968). For criticism of the tendency to romanticize the leftist past, see William Preston, "Shall This Be All? U.S. Historians versus William D. Haywood et al.," *Labor History* 11 (Summer 1971): 435–53.

11. Stuart Kaufman, *Samuel Gompers and the Origins of the American Federation of Labor, 1848–1896* (Westport, Conn., 1973), and William M. Dick, *Labor and Socialism in America: The Gompers Era* (Port Washington, N.Y., 1972).

12. Gutman, "Work, Culture, and Society in Industrializing America," 531–88; Gutman, "Protestantism and the American Labor Movement: The Christian Spirit in the Gilded Age," *American Historical Review* 71 (Oct. 1966): 74–101; Gutman, "The Workers' Search for Power: Labor in the Gilded Age," in *The Gilded Age: A Reappraisal*, ed. H. Wayne Morgan (Syracuse, 1963), 31–54.

13. Eugene Genovese, "On Antonio Gramsci," *Studies on the Left* 7 (Mar.-Apr. 1967): 301.

14. See, for example, Brecher, *Strike!*, and Stanley Aronowitz, *False Promises: The Shaping of American Working Class Consciousness* (New York, 1973). For a worker's view supporting this point, see Len De Caux, *Labor Radical: From the Wobblies to CIO* (Boston, 1970).

15. Brecher, *Strike!*, 260–63.

16. E. P. Thompson, *The Making of the English Working Class* (New York, 1963), 11.

17. Erik Erikson, *Identity: Youth and Crisis* (New York, 1968), 45.

18. Gutman, "Work, Culture, and Society in Industrializing America," 541. For a more elaborate critique of this approach, see James Green, "Behavioralism and Class Analysis," *Labor History* 13 (Winter 1972): 89–106.

19. Thompson, *English Working Class*, 9–11.

20. Ibid., 13.

# Life, Work, and Rebellion
# in the Coal Fields

## CHAPTER I

# "Coal Is Our Existence"

> Ill fares the land, to hastening ills at prey,
> Where wealth accumulates, and men decay.
>
> Oliver Goldsmith
> *The Deserted Village*

Men whom they were never to know saw to it that word came to them of new jobs, new homes, and new opportunities. Faced with hopeless situations where they were and touched by the hopes offered, thousands of them agreed to take up work they knew nothing about—coal mining—in an area they had never heard of—southern West Virginia. They came from southern and eastern Europe, and more came from the South to this new region. All of them knew the culture they were leaving behind with that mixture of grief and expectation that has always characterized migration into and within the United States. None of them was aware of the culture that their arrival in southern West Virginia would help destroy, or of the new one that the capitalists who had enticed them were to provide, or of the one that they themselves were slowly to create from their old traditions and the new realities they would face.

At the end of the 1870s, southern West Virginia was a relatively isolated, underpopulated, agrarian region, occupied by subsistence farmers, hunters, and family clans like the Hatfields and McCoys. The rise of the coal industry in the next decade, however, transformed the area economically, politically, and socially into both an industrialized region and an economic colony. The growth of the coal industry gave the coal operators a dominance in the state government over southern West Virginia until the New Deal. It also broke down the traditional mountain culture, introduced new values, and brought in tens of thousands of southern blacks and Europeans to mix with the native population in the confines of the company town. By 1921 southern West Virginia was a heavily populated, industrial economy dependent upon coal production and linked to national and international markets. A circuit judge told a committee of U.S. senators touring southern West Virginia: "We think and live coal. If you take our coal from us, we shall go back to the days of bobcats and wilderness. Coal is our existence."[1]

People had known of the coal beds of southern West Virginia since colonial times. "In the western country," Thomas Jefferson wrote in his *Notes on Virginia* in 1785, "coal is known to be in so many places, as to have induced an opinion that the whole tract between the Laurel Mountain and Ohio yields coal." For the next century noted geologists surveyed and reported on these rich coal fields.[2]

And for nearly a century these coal lands remained relatively undisturbed. At the time of the Civil War, only 185 mines employing less than 1,600 workers existed in the entire state. Most of these mines were in Kanawha County, where salt manufacturers used coal to boil their brine and blacksmiths used it for smithing.[3] The operators who poured into the area in the 1880s reported that the coal lands were covered with "beautiful fields of waving corn and wheat. Herds of sheep and cattle roamed the hills and were guarded from wolves and bear by the men and boys. They spent their winter months hunting and trapping the then plentiful deer, bear, and other wild animals. From their pelts they made their clothes and feasted upon their meats."[4]

The major impediment to development of these coal fields was transportation for the coal. Although railroads linked the nation by 1870, railroad companies still could not penetrate the Appalachian Mountains. The few commercial mines in Kanawha County shipped their coal down the Kanawha and Ohio rivers in flatboats—a process that was slow, cumbersome, and made precarious by the dangerous conditions of the waterways. During antebellum years, sectional antagonisms had prevented states from allocating the necessary expenditures for internal improvements, and for two decades following the Civil War, the fiscal conservatives who dominated the West Virginia state legislature refused to appropriate money to upgrade the water route.[5]

In the 1880s the money and means became available for the railroads to penetrate the mountains, and the state became a battleground for the railroad robber barons. Realizing potential profits, notables such as J. P. Morgan, John D. Rockefeller, E. H. Harriman, Collis P. Huntington, Henry H. Rodgers, Abram Hewitt, Peter Cooper, and John Camden all competed and connived to build the roads that would carry West Virginia coal.[6] By 1900, southern West Virginia had a massive network of tunnels, bridges, and iron rails[7]—and the coal rush was on.

As the railroads were being built, land and coal speculators poured into the area to grab the land that contained the coal. A local newspaper commented that "capitalists" were "flocking from all parts of the compass to make investments."[8] But these northern capitalists faced an obstacle: the native mountaineers already owned and occupied the land.

An attitude of no respect for the property rights of others allowed the capitalists to take what they wanted. Following the Revolutionary War, the state of Virginia had compensated its soldiers with the "wastelands" of its western counties and sold the rest in bulk to early land speculators. By the 1880s, the deeds to both groups of land had been forfeited to the state because of the failure of the original title holders to pay taxes on the land and to register their ownership. Virginia, and later West Virginia, reclaimed the land and sold it to the people who had settled there. The late nineteenth-century speculators and capitalists now purchased the original deeds and sought help from the courts, claiming that they were the rightful owners.[9]

At least twice the U.S. Supreme Court ruled that the original deeds were invalid and that the local inhabitants were entitled to the land.[10] The northern capitalists then found local federal judges who were noted for their "tender concern for the rights of nonresident landowners," a concern possibly stimulated by bribery, and these judges declared that the original deeds were valid. Facing huge legal fees and years of litigation, neither of which they could afford if they appealed, the natives finally sold their claims, left, or were thrown off their land.[11]

This encounter with the power of wealth left a legacy of distrust and contempt for judges and the judicial system. "Since our lands have become invaluable the slumbering pocket patentees have been awakened," the *Logan Banner* declared in 1890, "and some of them are now taking steps to take from the people of this county . . . lands purchased by them, upon mere technicalities."[12] Twenty-five years later, Willey Sizemore of Maben, Wyoming County, wrote President Woodrow Wilson that he and nearly one hundred other families had been "swindled . . . out of their homes which they have lived on from 10 to 60 years" by the confluences of the South Western Pocahontas Corporation and Judge A. G. Dayton (who later upheld the yellow-dog contract). "By using just such judges, who are friends of corporations, these corporations can do anything they want done in their courts."[13]

The native mountaineers also became aware of the influence of corporate wealth upon politics. The 1894 West Virginia senatorial convention pitted state Senator John Sheppard, a leading defender of the property rights of the occupants, against the attorney for one of the largest land speculators, Henry C. King. A local newspaper depicted the supporters of King. "The men who are seeking to deprive the people of their homes are in the position to furnish the boodle to defeat Senator Sheppard and to elect a Republican, and the Republican Senator in turn is to favor such legislation as will give the homes of the people of Logan, Wyoming and McDowell to the sharks who are setting up their claims to them."[14]

Outside capital proceeded to buy southern West Virginia. A Baltimore banking house, John A. Hubleton and Company, bought 25,000 acres of land at Loup Creek, Fayette County. The Flat Top Land Association purchased 600,000 acres of southern West Virginia. The Norfolk and Western Railroad of Virginia bought 295,000 acres—four-fifths of the Pocahontas coal field.[15] In 1901, a syndicate headed by Morgan started purchasing land in southern West Virginia. By 1919, under Morgan's direction, U.S. Steel had acquired 32,600 acres in Logan and Mingo counties and 50,000 acres in McDowell County.[16] The Guggenheims bought Eccles, Raleigh County. Charles M. Pratt of Brooklyn and a former U.S. senator from Rhode Island, George Wetmore, acquired most of Paint Creek (21,000 acres). The secretary of the treasury under President Warren G. Harding, Andrew Mellon, owned mines in Logan and Mingo counties.[17]

The British banking houses of Vivian, Gray and Company and T. W. Powell of London acquired thousands of acres of the Flat Top Coal Field. An English syndicate headed by Peter Malloy of London purchased part of the New River coal field.[18]

Although individual proprietors, such as W. P. Tams, and at times families, such as the Capeharts, moved into the state and built mines that would be controlled locally, southern West Virginia had generally become the possession of New York, Philadelphia, Boston, Baltimore, and London. By 1900 absentee landowners owned 90 percent of Mingo, Logan, and Wayne counties and 60 percent of Boone and McDowell counties. By 1923 nonresidents of West Virginia owned more than half of the state and controlled four-fifths of its total value.[19]

The coal operators defended this colonization of West Virginia. A West Virginia coal operator residing in Boston asked a U.S. Senate committee, "When did it become improper for us to go down into West Virginia, into a mountainous country that had not a partial [*sic*] of business, and start a mine or mines . . . ? It may be wrong to have alien ownership, but that is a different thing from ownership by inhabitants of one state of property in another state."[20]

Coal in southern West Virginia was as plentiful as Jefferson predicted in 1785 and as rich as the operators had hoped. Its low volatile content and high fixed carbon gave an average British thermal unit of 15,200, which made it the coal with the highest heat unit in the United States and the best steam coal in the country. Its low content of ash and sulphur made it the most fuel-efficient and best coking coal in the country; next to anthracite, it was the best coal for domestic purposes. Geologists soon compared the coal of southern West Virginia to the admiralty coals of Wales as some of the best coal in the world.[21]

Mining coal in West Virginia was also cheaper than in other states. Large, soft seams and exposure on the hillside allowed for easy entry and for use of efficient and economic drift or slope mines. The mines were located in the hills, well above the water level, and pumping water from the mines was unnecessary, as the gentle sloping of the hills provided good drainage. Further, the gentle slopes made hauling the coal out of the mine to the tipple, and from the tipple to the coal market easy. The thick forests of southern West Virginia provided the wood needed for the props in the mines and the construction of company towns—an availability that further reduced the companies' expense and enabled them to undersell their competition in other states.[22]

A coal operator later recalled that at the turn of the twentieth century, coal mining required "relatively small capital investment. All that was required was to build houses for the miners, a store to supply them, and a tipple structure to dump the coal into railway cars." Legend records that two of the largest coal operators in the Winding Gulf coal field began operations "with little more than one mule and a borrowed harness."[23]

Within a couple of decades, the superior and cheaper supercoals of West Virginia had captured the nation's largest coal markets from the older, established midwestern coal fields. Southern West Virginia coal was heating the homes and supplying the energy to run the factories of Indianapolis, Chicago, Detroit, Cleveland, and Dayton. Known for its purity, it was the only coal that could be burned in Washington, D.C., because of the capital's strict ordinances against smoke pollution. Its high heat content, low rate of deterioration in storage, and relative smokelessness made it the primary source of fuel for the U.S. Navy. In 1900 it was rivaling anthracite coal for the lucrative New England market and replacing European coal in South American and Mediterranean countries.[24]

The total coal production of West Virginia increased from 489,000 tons in 1867 to 4,882,000 tons in 1887 and to 89,384,000 tons in 1917. Coal production in Kanawha County increased from 982,000 tons in 1888 to nearly 5,000,000 tons in 1908. In Fayette County coal production soared from less than 900,000 tons in 1889 to over 9,000,000 in 1909. The output of coal in McDowell County skyrocketed from 246,000 tons in 1889 to 3,500,000 tons in 1899 and 12,000,000 tons in 1910. In Mingo County coal production jumped from 95,000 tons in 1895 to nearly 1,500,000 tons ten years later. The state that had relatively no commercial coal mining at the end of the Civil War was second only to Pennsylvania as the nation's leading coal producer in 1920.[25]

The southern West Virginia coal producers exploited the natural advantages of their coal. They conducted extensive advertising campaigns to promote their business and were quick to take advantage of their competitors' problems. "The West Virginia operators might be termed the pirates of the coal trade, standing ready at all times to descend on any fat prize that may appear on the horizon," one observer noted; "let a shortage of fuel appear in the Northwest or in New England and the West Virginia producers will soon be on the job with their high-grade coals gobbling up the main plums."[26]

The first president of Gulf Smokeless Coal Company, W. P. Tams, later reflected upon the aggressiveness of these pioneer operators. "The early operators were a highly individualistic lot, quite different from today's 'organization man.' Many were men of little formal education; only a handful had any training in engineering. However, all successful operators had great drive, high native intelligence, willingness to assume large risks and an unusual capacity for hard and sustained work."[27]

The mining operations that they ruled were a model of efficiency and simplicity. With few officials, usually only a mine foreman, superintendent, and bookkeeper, the coal operator was "often his own salesman, engineer, and trouble shooter." Declared one coal operator in 1911, "I am my own president, superintendent, clerk, office boy, boss mule driver, and carry my office in my hat." Another operator stated, "The old time operator was the company."[28]

The rise of the coal industry and the transformation of southern West Virginia into an industrial society produced swift and vast economic and social changes. These changes, however, had not come as a sudden culture shock to the local people. Although the overwhelming majority of the natives were subsistence farmers and hunters, the production of cash crops such as ginseng had acquainted southern West Virginians with a money economy and had introduced them to the national economy. Furthermore, even before the massive intrusion of the coal industry, approximately 2,500 industries were scattered about the state, mostly in the northern sector. But some, including the mammoth aluminum plants at Glen Ferris in Fayette County, were located in the southern portion.[29]

Logging had been a significant industry in several southern West Virginia counties since the 1830s. With the introduction of the band saw in the 1870s, saw mills sprang up all over the mountainous woody region. Fayette County alone had 150 saw mills. "The entire community was elated at the coming of the mills," Fred Mooney recollected; " 'there will be plenty of work' was the general expression." Lumbering never became a major industry in southern West Virginia, but it did firmly establish a cash income in the region and further connected the economic life of the area with that of the nation.[30]

The railroads had contributed to the economic growth and social transformation of southern West Virginia. Railroad building had brought large numbers of southern blacks and European immigrants into the region.[31] The railroads had also promoted civic development. Before there was a single coal mine in Logan County, the town of Logan had banks, a newspaper, running water, sidewalks, fire hydrants, and a fire department. The area was not New York City, but the people were neither isolated nor wholly ignorant of industrialization and capitalism. No longer was the area, as a journalist covering the Hatfield-McCoy feud a few years earlier reported, "as remote as central Africa."[32]

The development of the coal industry had its greatest effect on the transformation of the traditional way of mountain life. The railroads had driven the wild game back into the hills, thus reducing subsistence hunting. The acquisition of the land by the coal companies caused a stark decline in farming, the mainstay of the native mountaineers for over a century. In the 1880s the typical farm averaged about 187 acres; by 1930 it averaged between 47 and 76 acres. By the 1920s less than 15 percent of Boone and Mingo counties and only about 7.3 percent of McDowell County were given to farming.[33]

Social changes paralleled the economic ones. The growth of the coal industry sparked an increase in education that shook and disintegrated traditional folkways and culture. Inspired and supported by the coal establishment, libraries were built throughout the coal fields, and modern brick schools with modern equipment replaced the little red school house. These educational reforms helped break down clannish thinking and activities. The famous feud fighter and head of the Hatfield clan, "Devil" Anse Hatfield, became a school trustee. Education introduced the native mountaineers to new ways of thinking and to values conducive to an industrial society. A song taught in these coal-field schools went:

> Merrily, merrily work with a will
> Making your fortune with patience and skill
> Plenty of wealth, Life is at best but a rugged ascent,
> Climb it with vigor, you'll never repent.[34]

The impact of coal mining, along with lumbering and railroading, contributed to the decline of the mountain family clan. For nearly a hundred years, the Keeney clan had hunted and farmed the hills of Clay County. In the 1880s the mining and lumber companies claimed the Keeney land, and the clan's farming days were over. One section of the family moved to undisturbed agricultural counties and continued farming. Other members of the family entered the timber industry, while still others moved to Kanawha County, particularly Cabin Creek, and became miners. This part of the family included Frank Keeney, who in 1917 became president of UMWA District 17.[35]

Not only did industrialization cause members of the once tightly knit clans to scatter, it sharply divided them, as family members found themselves on opposite sides of the industrial-capitalistic fence. The Hatfield clan, whose traditional family loyalty had helped spark and sustain the Hatfield-McCoy feud, was split dramatically. Some of its members became merchants in independent towns, some entered medicine, a few went into politics (one became governor and later U.S. senator), several took jobs as railroad detectives, others became coal operators, and others miners.[36]

The character of the population changed dramatically in other ways, too. Because the local mountaineers were not numerous enough to supply a work force, the operators imported a labor force, a move that contributed to the end of traditional mountain culture. Operators sent labor agents to immigration ports and across the Atlantic Ocean to procure European immigrants. They sent more agents to the South to lure blacks to the southern West Virginia coal fields. These efforts proved tremendously successful. In 1880 there was not a single black miner in the state of West Virginia; by 1900 there were nearly 5,000, and by 1910 the number had jumped to almost 12,000. All but a few hundred of them mined coal in the southern part of the state. In 1880 there were 924 European immigrant miners in the state; by 1910 there were 28,000.[37]

The native and imported workers who went to work in southern West Virginia in the 1880s and 1890s found an economic, political, and social organization that was vastly different from that of a decade earlier and different from any other industrial situation in the United States. Company towns were not peculiar to the southern West Virginia coal fields, of course, but several factors made the ones in West Virginia unique. Numerically and proportionately (94 percent), more miners in West Virginia lived in company towns than did miners in any other state. Since the percentage figure includes northern West Virginia, where many miners lived in commercial towns, the proportion for southern West Virginia was probably about 98 percent. Illinois, with 53 percent, was second to West Virginia.[38]

The miner in southern West Virginia found the company town more than the single-city phenomenon it was in Everett, Washington, and Pullman, Illinois. In West Virginia, company towns dominated the entire region. Nor did the miner find big cities or industries close by, as was the case in southern Illinois, Ohio, and Pennsylvania, which offered alternative employment or an opportunity to supplement wages.[39]

The operators defended their building and use of the company town as a necessary instrument for supplying and housing the thousands of workers whom they imported to dig coal. The company town was also economical, and profit was the supreme goal of the coal operators. "We are not running a Christian Endeavor Camp meeting or a Sunday school," an operator wrote his superintendent in 1896; "never lose sight of the fact that the sole purpose of the organization is to make money for their stockholders and matters of conduct . . . that tend to produce a contradictory result should be promptly squelched with a heavy hand."[40] The existence of the company town made it relatively easy to squelch opposition with a heavy hand.

Upon moving into a company town, a miner had to live in a company house and sign a housing contract that the courts of West Virginia subsequently ruled created a condition not of landlord and tenant, but of "master and servant." Consequently, the coal company was allowed to unreasonably search and seize a miner's house without any notice. "If we rent a miner a home, it is incidental to his employment, and if a miner would undertake to keep anyone at that home that was undesirable or against the interests of the company, we would have him leave or have the miner removed."[41]

In evicting these undesirables from their homes, coal operators told the U.S. Coal Commission in 1923 that they had been "considerably more tolerant and considerably slower . . . than the dictates of justice or as humanity requires. . . . In all cases, regard has been paid to the health and comfort of those persons whom it was found necessary to evict. . . . Evictions have universally been carried out in a humane manner."[42] In some cases, this was partially true. Some companies notified striking miners to vacate their houses, although it was not required by law. The miners at Terry, Wyoming County, received a letter that read:

November 26, 1923

Mr. ____________
Terry, W. Va.

This is to notify you that we will not be in need of your services any longer after this date. You are further notified to surrender the possession of the house you now live in on or before January 1st, 1924.

By ____________________,
Supt.
Cook and Carter Coal Co.

Classic was the dismissal notice the superintendent at Widen sent to discharged miners: "I want my house."[43]

Notification was exceptional; the coal companies usually sent mine guards to the miner's house and without warning dumped him, his family, and the furniture onto the company road. While evicting the families on Paint Creek, during the Paint Creek–Cabin Creek strike, the mine guards arrived in the early morning and threw breakfasts out with the furniture. During the process the mine guards destroyed over $40,000 worth of furniture. In the town of Banner, the mine guards came to the house of Tony Seviller, whose wife was pregnant. The head of the squadron shouted, "Get out!" Mrs. Seviller, in bed and in labor when ordered out, responded, "My God! Can't you see I am sick; just let me stay here until my baby is born." The guard leader replied, "I don't give a damn, get out or I'll shoot you out." Mrs. Seviller gave birth to her baby two hours later, in a tent furnished by the UMWA.[44]

The miner who lived in the southern West Virginia company towns worked in the most dangerous coal mines in the United States. Between 1890 and 1912 the mines of West Virginia had the highest death rate among the nation's coal-producing states; its mine-accident death rate was five times higher than that of any European country. Indeed, during World War I the southern West Virginia coal diggers had a higher proportional death rate than the American Expeditionary Force.[45]

If a coal miner survived a month of work in the mines, he was paid not in U.S. currency but in metals and paper (called coal scrip), which was printed by the coal company. Because only the company that printed the coal scrip honored it, or would redeem it, the coal miner had to purchase all his goods—his food, clothing, and tools—from the company store. Hence, the miner paid monopolistic prices for his goods. Journalists and U.S. senatorial investigating committees repeatedly revealed that the region's coal company store prices were substantially higher, sometimes three times higher, than the local trade stores. For the coal company, this difference meant profit. In 1920 a Boone County coal company lost $40,000 in trade, but still made a profit because of the company store. To the miners, it meant, as they later sang, that they "owed their souls to the company store." For some miners, it meant being held in peonage.[46]

The coal miner of southern West Virginia also encountered a political system strikingly different from the American tradition of representative politics. Because the company towns were unincorporated, there were no local political officials, no mayor, no city council, no ward boss to attend to the immediate interests of the miners—there was only the coal operator. A former resident of Widen recalled that the coal operator there "governed completely. He was Mayor, Council, Big Boss, sole trustee of the school, truant officer, president of the bank; in fact he was everything."[47]

A miner's lack of a political voice also involved difficulty in obtaining information to vote for the candidates of his choice. Newspapers critical of the West Virginia coal establishment were banned from company towns. People distributing political information contrary to the political interests of the coal companies were driven out of the company towns, sometimes after being beaten. Because the post office was usually located in the company store with a company official serving as postmaster, a miner's mail might be scrutinized for political or union propaganda. One such censoring postmaster sent a letter addressed to a miner in his town to the attorney general of the United States with a cover letter:

> The writer is also Postmaster at Thayer, West Virginia, as I am also employed as Superintendent for the Ephriam Creek Coal and Coke Company. When the enclosed letter was observed coming thru [*sic*] the Thayer post office we had reason to feel that its contents were suspicious. By holding the enclosed letter before an electric lamp you will observe the contents of the letter can be read very distinctly. As postmaster I took possession of the letter.[48]

A coal miner was not free to attend the political gathering of his choice. The following dialogue between a U.S. senator and C. A. Cabell, a coal operator on Cabin Creek, Kanawha County, speaks for itself:

> Senator: Did they ever try to hold their Socialist meetings there [operator's company town]?
>
> Mr. Cabell: No. If they had, I would not have allowed it.
>
> Senator: That would be a political meeting, would it not?
>
> Mr. Cabell: Well, I reckon it would, from a Socialist point of view; I suppose if properly conducted and all that, for the proper purpose, it would be a political meeting and nobody would object to it; but I think I would be the better judge of that being right amongst the people.[49]

A miner who did obtain political information faced another difficulty —voting for the candidate of his choice. When the miner went to the polls, he found company mine guards serving as pollsters and ready to inspect his ballot. If he voted contrary to company instructions, he was discharged, sometimes violently. Often these mineguard-pollsters simply handed the miner a ballot with his political choices already marked. "We voted 206 men today," a superintendent wrote to his coal operator. Agents for the Department of Justice reported that in southern West Virginia coal company officials "absolutely dominate all of the polling places. Only such voters are allowed to vote as suit the organization. Election results are figured up and given out in advance as to what the county will do." The agents concluded that "the situation . . . is that elections, both primary and general," in the region "are in nowise representative," nor "are the citizens there allowed in any way to express their preference in these elections."[50]

Able to control the political voice of their work forces, the coal companies established powerful political machines. In Logan County, coal operator Walter Thurmond was chairman of the county Democratic committee; the county sheriff, in the pay of the coal operators, was the political boss. In Fayette County, "King Samuel" Dixon, the president of New River Coal Company, reigned. He was the Republican boss of the county and, without question, controlled the voting precincts. "Tales of 'King Samuel' and his political adventures," a coal operator later admitted, "have become part of the folklore of the area." When three coroner juries were established in Fayette County in 1907 to investigate the deaths of miners killed in three separate mine explosions, Dixon selected the members of each jury—all fifteen of them, and all fifteen were coal company officials. In McDowell County the coal operators dominated all local politics. They controlled the county court and the county Republican party. According to one union official, "The political machinery is the willing instrument of the operators." "As long as we have the county officials with us here in McDowell County," a superintendent wrote his operator, "we will have no trouble."[51]

Traditionally, the state political system in West Virginia had been based upon kinship and personal contact, and this type of politics had protected the mountaineer culture. State politics, for example, had blocked the efforts of the state authorities of Kentucky to bring feudfighter Hatfield to justice.[52] This political system began to dissipate as industrialists in northern West Virginia achieved political power by establishing political machines based upon party patronage. At this early stage, according to state historian John Williams, the contests between Republicans and Democrats became a battle of elites, not representative politics.[53]

Retention of political power by farmers in agricultural counties and workers in northern West Virginia prevented the industrial capitalists from totally dominating the state political machinery, but the rise of the coal establishment in southern West Virginia, with the operators' control of local politics, strengthened the capitalists' influence. Consequently, the rise of the coal establishment in southern West Virginia meant that the traditional mountain politics that had protected Hatfield were obsolete. When his son, Cap Hatfield, killed three men on election day in 1896, he was sharply attacked in the press and thrown into the newly built jail in Williamson by the new state political order.[54]

The new political order in southern West Virginia shut the miner off from national and state as well as local politics. From the rise of the southern West Virginia coal establishment in the 1880s until the New Deal of the 1930s, the coal miners in this area found little sympathy and no support from the West Virginia political machinery.[55]

Neither could a miner expect help from his representatives in Congress. The U.S. senators from West Virginia were generally either coal operators or men directly affiliated with the coal establishment. For example, U.S. Senator Clarence Watson (1911–13) was president of Consolidated Coal of Fairmont, the largest coal producer in West Virginia; Senator Howard Sutherland (1917–23) was a coal operator; and Senator William Chilton (1911–17) was the sole owner and manager of the *Charleston Gazette* and a prominent corporation attorney who represented the coal establishment; his firm boasted of representing four-fifths of the corporate interests in the state. Similarly, congressmen representing southern West Virginia usually had connections with the coal industry. Congressman Edward Cooper was a coal operator. He was replaced by Congressman Wells Goodykoontz, the president of the National Bank of Williamson, Mingo County, and a corporation lawyer who had handled many of the land claims of the coal operators.[56]

In 1913, the year of the Paint Creek–Cabin Creek strike, five members of the West Virginia state legislature were arrested for accepting bribes amounting to $30,000 from a coal operator running for the U.S. Senate; his opponent was another coal operator. Part of an editorial in the *New York Press* commented that "the Legislature at Charleston is unable to elect a United States Senator without an open scandal . . . has brought home new realization that the political affairs of that State are dominated by a band of financial adventurers. . . . Men of great capital have bought the oil, coal, and timber lands and rule their domains like barons."[57]

The coal establishment controlled the state's executive branch. In 1888 A. B. Fleming, a corporation lawyer and coal operator, was elected governor. His election inaugurated a series of nine West Virginia governors, in office until 1924, who either were coal company officials or were chosen with the consent of the state's industrialists "after certain understandings had been reached." The economic proclivities of these governors became blatantly apparent, when the union movement developed in southern West Virginia.[58]

The coal establishment's toughest political struggles came in the West Virginia state legislature. Agricultural counties still possessed a significant influence in the state lawmaking body, and workers in northern West Virginia had retained some political rights. Therefore, coal operators and railroad interests often collaborated to control legislation and elections. Southern and northern coal operators were often forced to combine and to influence delegates with gifts like railroad passes and free coal to affect legislation.[59]

If the coal operators were not able to control the legislature totally, they were able to dominate the legislative committees that pertained to the coal industry, particularly the committees of Mines and Mining and Labor and Immigration. In 1915, for example, the chairman of the Committee on Labor and Immigration was the president of the West Virginia Mining Institute.[60] Occupying these strategic positions was of tremendous value. A coal company official appointed to the legislature's steering committee wrote to his coal operator that "in this position, one is in shape to get all the information firsthand. This committee is really the most important of all, in fact, it is the 'meat in the meat' and I have hopes that I will be of service to Mr. Mann [the company superintendent]."[61] Such positions also gave them power. A coal operator, fearing the passage of a certain bill, received comforting words from a delegate: "Colonel Davis, as chairman of the House of Mines and Mining Committee, can take care of it when it comes over to him. He will simply hold it up and let it die with the session."[62]

Organized labor in West Virginia thus had little political power. The near passage of favorable labor legislation did not illustrate the political influence of organized labor, but rather that the labor interests had been duped. As a delegate wrote to coal operator Fleming, "In order to placate the labor interests, certain of their bills were allowed to pass the House. In permitting their passage, we knew full well there would be no time for them to be considered in the Senate." The delegate further explained that although labor had gained "nothing," "the House, and especially some of its officers, received credit for being favorable to labor interests."[63]

The passage of the state's Workmen's Compensation Law in 1913 was a victory for the operators, not the miners. The coal companies were exempted from damage suits by miners injured on the job, which local juries often awarded to the miners. The first state commissioner of Workmen's Compensation, who held the post for fourteen years, was a coal operator, and he did not require the coal companies to pay their assessments. When he left office in 1927, one coal company was $300,000 behind in payments. An injured miner who received a compensation check was required to turn it over to the coal company if he wanted to retain his job and company house. In 1915 a miner wrote the *United Mine Workers Journal (UMWJ)* that "the compensation law in West Virginia . . . is not what the laborers of the state would like for it to be, for it favors the coal companies in most of its reading. Then the companies take advantage of it in every way possible."[64]

If a bill detrimental to the coal industry did pass both houses of the state legislature, the coal establishment had yet another source of power —the state's chief executive. In 1897 the state legislature passed a law that required more demanding qualifications for the state mine inspectors; Governor William A. MacCorkle vetoed it, explaining that it involved "too much risk to our greatest commercial interests. . . .The greatness of West Virginia is founded upon coal. Can the Legislature afford to do anything that would impede, hamper or hinder the progress of this great industry within the borders of our state?"[65]

In 1901 the state legislature passed bills requiring regular mine inspections. Governor George Atkinson vetoed the legislation, acknowledging that "the object of these bills is for the protection of the lives of the men engaged in mining," but claiming that it would duplicate the tasks of the state's Mining Bureau. "I cannot, under my oath of office, approve a bill which is both unconstitutional, conflicting, and inoperative."[66]

Bills that were contrary to the interests of the coal establishment that did become state law were ignored by the coal operators. For example, West Virginia had laws prohibiting the use of coal scrip; yet nearly all the coal companies in southern West Virginia, including the mines owned by U.S. Secretary of the Treasury Mellon, openly used it. After passage of one such law, a superintendent wrote to his coal operator that "none of the operators in the smokeless section are taking this [law] seriously." He himself did "not expect any trouble . . . [because] in the event many of our employees trouble us for redemption in cash, we will make an example of a few of them and soon hear no more it it."[67]

The state also had a law requiring the coal companies to keep a checkweighman on the coal tipple; this law was continuously ignored. Miners who insisted upon this right were quickly discharged and evicted.[68] In 1889 a state law was passed that prohibited any employer from interfering with the peaceful organizing efforts of his employees[69]; for the next forty years, coal operators ruthlessly violated this law. Any miner in southern West Virginia who joined a union was discharged, evicted, blacklisted, usually beaten, and sometimes killed.

West Virginia had child labor laws that mandated a minimum working age of fourteen. These laws, however, contained so many defects and loopholes that the National Child Labor Committee ranked West Virginia thirty-fifth (of the then forty-six states) in child labor restrictions and last among the industrial states. An investigator for this committee branded the child labor laws of West Virginia "absurd." The laws contained no educational requirements for working children; they did not specify any dangerous occupations from which children were prohibited; they did not limit the hours of employment for children. Only one person, the state's Commissioner of Labor, was empowered to enforce the laws.[70]

The major loophole for the coal companies was that the child labor laws of West Virginia did not require documentary proof of age. For a child to obtain work in a coal mine, he simply needed an affidavit from a parent or guardian as proof of age. According to the National Child Labor Committee, if a boy presented himself to a coal company with an affidavit stating that he was beyond the age limit, he was hired "with no questions asked." This provision of the child labor law was used to exempt the companies from responsibility and blame. A National Child Labor Committee investigator observed that the coal operators' "willingness to shift the burden of proof and responsibility upon the parents of boys employed was painfully evident." When a foreman was asked if boys lied about their ages, he replied, "Yes, they do; but if they bring an affidavit, that lets us out!" According to a southern West Virginia coal operator, William Coolidge, "The exploiting of children was not by the manufacturer; the exploiting of children was by the parents."[71]

Generally, the coal companies disregarded the state's child labor laws. Indeed, child labor was so prevalent that an elderly southern West Virginia coal operator later mistakenly recalled that the state had no minimum age laws. And the company policy induced the miners to send their older children into the mines rather than to school. For example, if a boy went to work with his father as a coal loader, even if the child was often too small to be of help, the father was allowed 50 percent more cars; by working overtime, the father could earn a higher wage.[72]

West Virginia mine-safety laws were also deficient. The operators, for example, had to provide fire hoses and to water coal dust only if a mine "generated gas in dangerous quantities." The state allocated less money per ton of coal on mine safety than any other coal-producing state. Governor Atkinson was but one rationalizer of the state's indifference to mine-safety laws, declaring, "It is but the natural course of mining events that men should be injured or killed by accidents." The coal establishment customarily "blamed the victim"; coal company officials claimed that the inexperience and carelessness on the part of the coal miners were responsible for the frequency of mining fatalities. The editors of the *UMWJ* offered another explanation: "West Virginia is noted in the mining world for her insufficiency of proper laws for the insurance and protection of her miners."[73]

The people responsible for enforcing the mine-safety laws that did exist, the mine inspectors, were also unqualified. The state's political historian has pointed out that it was an "open secret among politicians" that the mine inspectors' positions existed merely "to pad the patronage roles and incidentally to promote safety in coal mining." A Logan County miner stated, "There is not an inspector in this state who is not holding his job through the influence of some operator." In 1901 a recently appointed district mine inspector concluded a letter to a coal operator thanking him for the appointment by writing that he would "endeavor by faithful and conscientious attention to duly prove to you and my other good friends . . . my gratitude."[74]

Consequently, the coal companies were free from prosecution for safety violations. Until 1904 there was not a single prosecution in the entire state and none after 1912. When the inspectors did prosecute, they prosecuted the miners, not the coal companies. In 1910 there were 163 prosecutions for safety violations; 159 of these were against miners. When the inspectors attempted to enforce the state's mine-safety laws, the operators went to the local courts, which they controlled, and obtained injunctions blocking the prosecution. "There are coal operators who will endeavor to have a district inspector removed from office," one miner declared, "rather than obey the mining laws, or carry out the recommendations made by the inspector."[75]

The state's chief mine inspector was responsible for supervising the mine inspectors and their work. The governor appointed this office, and hence the inspector's activities often reflected the politics of the governor, not the welfare and protection of the miners. James Paul was chief mine inspector from the early 1890s to 1907. "The administration of his office," declared the editors of the *UMWJ* on the eve of Paul's retirement, "is marked by a long, bloody trail of human slaughter, caused by negligence, by wanton nullification of every mining law in the state." Paul was accused of being a puppet of the coal operators, probably rightly so. In 1901 he wrote to Fleming, requesting the position of chief engineer for the Fairmont Coal Company. Paul assured him that "in the event of my vacating my present position I feel that I would be permitted to suggest a successor."[76]

Neither was there a question about the economic connections and proclivities of John Laing, who was Paul's successor. Laing was president and general manager of Wyatt Coal Company on Cabin Creek and of Morrison Coal Company, Wyoming County, and general manager of MacAlpin Coal Company, Raleigh County. After leaving office, he was elected president of the Kanawha County Coal Operators' Association. Both Governor William Dawson, who appointed him, and Governor William Glasscock, who retained him in office, as Laing testified to a U.S. Senate committee in 1913, knew of his mining connections.[77]

"A powerful and subtle conspiracy between organized capital and the governmental agents of the State [of West Virginia]," wrote the president of the American Federation of Labor, Samuel Gompers, in 1913, "gives indications of the existence of the invisible government that steals from the workers the liberty they . . . are told they have."[78] That same year, a U.S. congressman from Wisconsin toured the coal fields of West Virginia and reported that "one cannot imagine the power of the mining companies. . . . It elects senators and judges. It owns both the Republican and Democratic parties in the state. All laws are made to suit the mine owners. All the judges are elected through their influence, even up to the judges of the Supreme Court."[79]

Seemingly powerless in their new work world and repressed by the controls in the company town, this disparate collection of miners eventually formed a unified social consciousness that slowly evolved into a working-class culture.

## NOTES

1. Quoted in Winthrop Lane, "Black Avalanche," *Survey* 47 (Mar. 25, 1922): 1002.

2. Jefferson is quoted in Phil Conley, *History of the Coal Industry of West Virginia* (Charleston, W.Va., 1960), 95. Geological surveys made during the nineteenth century are cited in John T. Harris, ed., *West Virginia Legislative Handbook and Manual and Official Register, 1924* (Charleston, W.Va., 1924), 398–99; J. T. Peters and H. R. Carden, *History of Fayette County* (Fayetteville, W.Va., 1926), 252–53; Jerry Bruce Thomas, "Coal County: The Rise of the Southern Smokeless Coal Industry" (Ph.D. diss., University of North Carolina, 1971), 50–55; Joseph Lambie, *From Mine to Market* (New York, 1954), 27–28; Lane, "Black Avalanche," 1000–1003.

3. Charles Ambler and Festus Summers, *West Virginia: The Mountain State* (Englewood Cliffs, N.J., 1958), 429–30; Otis Rice, "Coal Mining in the Kanawha Valley to 1861," *Journal of Southern History* 31 (Nov. 1965): 293–315; Jessie Sullivan, "West Virginia's Greatest Industry," *West Virginia Review,* Apr. 1927, 233; D. C. Kennedy, "Kanawha Coal Field," ibid., June 1925, 334; W.Va. State Department of Mines, *Annual Report, 1910,* 36–37.

4. Peters and Carden, *Fayette County*, 561. See also Laurence Leamer, "Twilight of a Baron," *Playboy*, May 1973, 114.

5. Rice, "Coal Mining in the Kanawha Valley," 293–315; Charles Ambler, *Sectionalism in Virginia* (Chicago, 1910), 319; Ambler and Summers, *West Virginia*, 282; Thomas, "Coal County," 36; Adam Shurick, *The Coal Industry* (Boston, 1924), 13.

6. Danny Hubbard, "Rodgers and His Railroad" (M.A. thesis, Marshall University, 1974), chap. 3, and W. P. Tams, *The Smokeless Coal Fields of West Virginia* (Morgantown, W.Va., 1963), 22–25.

7. For the massive construction of railroads in southern West Virginia between 1880 and 1900, see Hubbard, "Rodgers and His Railroad"; Tams, *Smokeless Coal*

*Fields*; Charles Turner, *Chessie's Railroad* (Richmond, Va., 1956); H. Reid, *The Virginia Railway* (Milwaukee, Wis., 1961); Ambler and Summers, *West Virginia*, 429–30; Robert Spence, *The Land of the Guyandot* (Detroit, 1976), 303–5; Lambie, *Mine to Market*, 134; Peters and Cardin, *Fayette County*, 245–46; Edwin Cubby, "Railroad Building and the Rise of the Port of Huntington," *West Virginia History* 33 (Apr. 1972): 234–47; Edwin Cubby, "The Transformation of the Tug and Guyandot Valleys" (Ph.D. diss., Syracuse University, 1962), chap. 8.

8. *Logan* (W.Va.) *Banner,* June 26, 1902.

9. John Alexander Williams, *West Virginia: A Bicentennial History* (New York, 1976), 107–8; Cubby, "Transformation of the Tug and Guyandot Valleys," chap. 7; Otis Rice, *The Allegheny Frontier: West Virginia Beginnings, 1730–1830* (Lexington, Ky., 1970), chap. 6; Thomas, "Coal County," 12–13; Spence, *Land of the Guyandot,* 185–94.

10. Cubby, "Transformation of the Tug and Guyandot Valleys," 196–98. This treatment of the many and complicated court rulings on these land claims is excellent.

11. Williams, *West Virginia*, 107–8, and Willey Sizemore to Woodrow Wilson, Feb. 27, 1915, item 1655095–57, file 50, Record Group 60, General Records of the Department of Justice, National Archives, Washington, D.C. Land companies bought the land and then leased it to coal companies for periods of 30 years. Lambie, *Mine to Market*, 39.

12. *Logan* (W.Va.) *Banner*, June 5, 1890.

13. Sizemore to Wilson, Aug. 23, 1914, item 165095–44, and to Wilson, Feb. 27, 1915, item 165095–57, file 50, Record Group 60, General Records of the Department of Justice.

14. *Logan* (W.Va.) *Banner*, Sept. 13, 1890.

15. Thomas, "Coal County," 75–80; Lambie, *Mine to Market*, 38; Lane, "Black Avalanche," 1004.

16. William Price, "Steel Corporation Mines at Gary," *Colliery Engineer* 34 (Mar. 1914): 464–72; "Garyism in West Virginia," *New Republic* 28 (Sept. 21, 1921): 86–88; Winthrop Lane, *Civil War in West Virginia* (New York, 1921), 119–20; Thomas, "Coal County," 150–52.

17. George Wolfe to Justin Collins, June 5, 1917, Justin Collins Papers, West Virginia University Library, Morgantown, W.Va.; Lawrence Lynch, "The West Virginia Coal Strike," *Political Science Quarterly* 29 (Dec. 1914): 629; Edmund Wilson, *The American Jitters* (Freeport, N.Y., 1968), 167; U.S. Congress, Senate Committee on Education and Labor, *Conditions in the Paint Creek District, West Virginia*, 63rd Cong., 1st sess., 3 vols. (Washington, D.C., 1913), 3:2116 (hereafter cited as *Paint Creek Hearings*).

18. Lambie, *Mine to Market*, 28; *Coal Age* 4 (Nov. 15, 1913): 734; Peters and Carden, *Fayette County*, 491.

19. W.Va. State Board of Agriculture, *Fifth Biennial Report, 1899–1900*, 37; *UMWJ*, Oct. 1, 1923, 2; *Martins Ferry* (W.Va.) *Evening Journal*, Sept. 15, 1923. For listings of absentee holdings other than ones mentioned here, see *Journal of Commerce*, Apr. 23, 1919; U.S. Congress, Senate Committee on Interstate Commerce, *Conditions in the Coal Fields of Pennsylvania, West Virginia, and Ohio*, 70th Cong., 1st sess., 2 vols. (Washington, D.C., 1928), 2:1445 (hereafter cited as *Conditions in the Coal Fields*); U.S. Congress, Senate Committee on Education and Labor, *West Virginia Coal Fields: Hearings . . . to Investigate the Recent Acts of Violence in the Coal Fields of West Virginia*, 67th Cong., 1st sess., 2 vols.

(Washington, D.C., 1921–22), 2:645–56 (hereafter cited as *West Virginia Coal Fields*).

20. Testimony of William Coolidge, *West Virginia Coal Fields*, 2:913. Although the colonizers owned half of the land in the state, and controlled four-fifths of its total value, they paid less than one-third of the state's taxes: see the report of the W.Va. State Tax Commissioner cited in *UMWJ*, Oct. 1, 1923, 2, and *Martins Ferry* (W.Va.) *Evening Star*, Sept. 15, 1923. This condition still plagues the state: see Tom Miller, "Absentees Dominate Land Ownership," in his book *Who Owns West Virginia?* (Huntington, W.Va., 1974), 1–3, and Gary Barkus, "The West Virginia Tax Structure, the People and Coal," *Report*, Appalachian Research and Defense Fund, Nov. 30, 1971.

21. Shurick, *Coal Industry*, 17–18; Floyd Parsons, "Mining Coal on the Virginian Railroad," *Coal Age* 1 (May 18, 1912): 1039–43; *West Virginia Review*, June 1925, 333; S. C. Higgins, "The New River Coal Fields," ibid., Oct. 1927, 26; I. C. White, *The Coal Resources along the Virginian Railway and Its Tributary Regions in West Virginia* (Morgantown, W.Va., 1912), 5; C. F. Carter, "The West Virginia Coal Insurrection," *North American Review* 197 (Oct. 1913): 457; Justin Collins, "My Experiences in the Smokeless Coal Fields of West Virginia," *West Virginia Review*, June 1926, 354–55.

22. Frank Warne, *Union Movement among Coal Mine Workers*, U.S. Bureau of Labor Bulletin no. 51 (Washington, D.C., 1904), 386, 402; L. C. Anderson, letter to the editor, *Outlook* 82 (Apr. 28, 1906): 861; Lambie, *Mine to Market*, 40–42; "Unionism and Coal Production," *Harvard Business Review* 4 (Apr. 1926): 334.

23. Tams, *Smokeless Coal Fields*, 24, and Walter Thurmond, *The Logan Coal Field of West Virginia* (Morgantown, W.Va., 1964), 39.

24. Higgins, "New River Coal Fields," 26; Warne, *Union Movement*, 383–85, 402; Collins, "My Experiences," 354–55; Kennedy, "Kanawha Coal Fields," 334; Harold West, "Civil War in the West Virginia Coal Mines," *Survey* 30 (Apr. 5, 1913): 37–38; Lynch, "Coal Strike," 658–62; Carter, "Coal Insurrection," 458–60; Shurick, *Coal Industry,* 14, 222–23.

25. Compiled from W.Va. State Department of Mines, *Annual Reports, 1900–1920*; Harris, ed., *Legislative Handbook, 1924*, 401–2; Peters and Carden, *Fayette County*, 298; Cubby, "Transformation of the Tug and Guyandot Valleys," 173.

26. W. E. Tissue, "The New River Company," *West Virginia Review*, Oct. 1927, 27; Shurick, *Coal Industry*, 224.

27. Tams, *Smokeless Coal Fields*, 74.

28. Ibid., 28, 72–74, and A. B. Fleming, "A History of the Fairmont Coal Region," in *Proceedings of the West Virginia Coal Mining Institute*, 1911, 255.

29. Ambler and Summers, *West Virginia*, 423–25; Peters and Carden, *Fayette County*, 580–81; Cubby, "Transformation of the Tug and Guyandot Valleys," 129–31; Thomas, "Coal County," 14.

30. Ambler and Summers, *West Virginia*, 437–38; Thurmond, *Logan Coal Field*, 18–20; Spence, *Land of the Guyandot,* 161–84; Peters and Carden, *Fayette County*, 315; Mary Hurst, "Social History of Logan County, West Virginia, 1765–1923" (M.A. thesis, Columbia University, 1924), 24; Fred Mooney, *Struggle in the Coal Fields*, ed. J. W. Hess (Morgantown, W.Va., 1967), 5–6.

31. Cubby, "Transformation of the Tug and Guyandot Valleys," 132–49; Thurmond, *Logan Coal Field*, 20; Williams, *West Virginia*, 105–6.

32. Cubby, "Transformation of the Tug and Guyandot Valleys," 17–24, 161, 239–42, and Charles Simmons, John Rankin, and U. G. Carter, "Negro Coal Miners in West Virginia," *Midwest Journal* 6 (Spring 1954): 60–69; the journalist is quoted in Theron Crawford, *An American Vendetta, a Story of Barbarism in the United States* (New York, 1889), 103.

33. Peters and Carden, *Fayette County*, 617; U.S. Department of Agriculture, *Economic and Social Problems and Conditions of the Southern Appalachians*, Miscellaneous Publications no. 205 (Washington, D.C., 1935), 16; Carter Goodrich, *Migration and Economic Opportunities* (Philadelphia, 1936), 69–70.

34. The rise of education and the coal operators' support of it are discussed more thoroughly in chaps. 3 and 6. For a general discussion of the educational improvements in southern West Virginia during this period, see Charles Ambler, *History of Education in West Virginia* (Huntington, W.Va., 1951), chaps. 7 and 11; Roy C. Woods, "History of the Hatfield-McCoy Feud with Special Attention to the Effects of Education on It," *West Virginia History* 21 (Oct. 1960): 27–33; W.Va. State Superintendent of Free Schools, *Biennial Report, 1899–1900*, 108.

35. Interview with Edgar Keeney, East Bank, W.Va., summer 1975. Keeney is writing a genealogy of his family.

36. Virgil Jones, *The Hatfields and the McCoys* (Chapel Hill, N.C., 1948), 225–33; *West Virginia Coal Fields*, 1:207, 245, 578; Williams, *West Virginia*, 105, 126–28.

37. Compiled from the W.Va. State Department of Mines, *Annual Reports, 1900–1910*, and Thurmond, *Logan Coal Field*, 59.

38. Edward Hunt, F. G. Tyron, and Joseph Willits, eds., *What the Coal Commission Found* (Baltimore, 1925), 136–40.

39. U.S. Women's Bureau, *Home Environment and Employment Opportunities of Women in Coal-Mine Workers' Families*, Women's Bureau Bulletin no. 45 (Washington, D.C., 1925), 12, 51–55.

40. Justin Collins to Jarius Collins, 1896, Collins Papers.

41. For the state supreme court's ruling, see *Paint Creek Hearings*, 2:1348–49. Testimony of M. T. Davis, ibid., 1277; U.S. Coal Commission, *Report of the United States Coal Commission*, Senate Document no. 195, 68th Cong., 2nd sess., 5 vols. (Washington, D.C., 1925), 1:5; Winthrop Lane, *The Denial of Civil Liberties in the Coal Fields* (New York, 1924), chap. 2.

42. Bituminous Operators' Special Committee, *The Company Town*, Report Submitted to the U.S. Coal Commission (n.p., 1923), 36–37.

43. Betty Cantrell, Grace Phillips, and Helen Reed, "Widen: The Town J. G. Bradley Built," *Goldenseal* 3 (Jan.-Mar. 1977): 5. Eviction notices of the towns of Terry, Thurmond, and Hotcoal are in the author's collection.

44. *Coal Age* 4 (July 12, 1913): 66, and *Denver* (Colo.) *Express*, Aug. 22, 1912.

45. Straight Numerical File, item 205194–50–4, Record Group 60, General Records of the Department of Justice.

46. Testimony of D. W. Boone, *Conditions in the Coal Fields*, 2:1739. Also see Tams, *Smokeless Coal Fields*, 25. For reports on peonage in southern West Virginia, see items 156618, 158776, 165722, 165327, 173538, File 50, Record Group 60, General Records of the Department of Justice; Gino Speranze, "Forced Labor in West Virginia," *Outlook* 74 (June 13, 1903): 407–10.

47. Cantrell, Phillips, and Reed, "Widen," 4.

48. John Lyons (postmaster, Sullivan, W.Va.) to Postmaster General, U.S.A., Dec. 2, 1922 (letter in author's collection). See also W. F. Larrison, letter to the

editor, *UMWJ*, Jan. 9, 1906, 7; testimony of Elmer Rumbaugh, *Paint Creek Hearings*, 1:280–90; Syble Keeney Phillips (daughter of Frank Keeney) to author, Aug. 7, 1975.

49. Testimony of C. A. Cabell, *Paint Creek Hearings*, 2:1535.

50. Elliot Northcutt to U.S. Attorney General Harry Daugherty, Dec. 18, 1922, Straight Numerical File, 205194–50–30, Record Group 60, General Records of the Department of Justice; interview with Oliver Singleton, Atlanta, Ga., Apr. 29, 1976; Holley to Justin Collins, June 7, 1916, Collins Papers; Howard Lee, *Bloodletting in Appalachia* (Parsons, W.Va., 1969), 9–11. For further evidence of political corruption in the area, see Straight Numerical File, 182363, Record Group 60, General Records of the Department of Justice.

51. Northcutt to Daugherty, Dec. 18, 1922, Straight Numerical File, 205194–50–30, Record Group 60, General Records of the Department of Justice; Fred Barkey, "The Socialist Party in West Virginia from 1898 to 1920" (Ph.D. diss., University of Pittsburgh, 1971), 123; Tams, *Smokeless Coal Fields*, 84–85, 88; letter to the editor, *UMWJ*, Nov. 5, 1908, 2; William Petry, "Report on the Basic Causes of Industrial Strife in Southern West Virginia," District 17 Correspondence Files, UMWA Archives, UMWA Headquarters, Washington, D.C.; Harris, ed., *Legislative Handbook, 1924*, 155; Wolfe to Justin Collins, Feb. 27, 1913, Collins Papers; testimony of James E. Jones, *Conditions in the Coal Fields*, 2:2122.

52. Williams, *West Virginia*, 123–34; Ronald Eller, "Industrialization and Social Change in Appalachia, 1880–1930," paper presented at the Southern Historical Association Convention, Nov. 12, 1976, Atlanta, Ga.

53. John Williams, "The New Dominion and the Old," *West Virginia History* 33 (July 1972): 318–37, and Williams, *West Virginia*, 117–20. See also Thomas, "Coal County," 89.

54. Williams, *West Virginia*, 123–24; Cubby, "Transformation of the Tug and Guyandot Valleys," 177; *Logan* (W.Va.) *Banner*, Aug. 21, 1897; *Huntington* (W.Va.) *Advertiser*, Nov. 6, 1896.

55. "All the governors back then were coal operators," commented one elderly miner. Interview with John McCoy, Alum Creek, W.Va., summer 1975. Two of the better contemporary analyses of the coal operators' control of the state government are Samuel Gompers, "Russianized West Virginia," *American Federationist* 20 (Oct. 1913): 829, and John W. Brown, *Constitutional Government Overthrown in West Virginia* (Wheeling, W.Va., ca. 1913).

56. Brown, *Constitutional Government Overthrown*, 8–9, and Harris, ed., *Legislative Handbook, 1918*, 410, 412–15.

57. "West Virginia's Bribery Scandal," *Literary Digest*, Apr. 1, 1913, 441–42, and *Appeal to Reason* (Girard, Kans.), Aug. 12, 1913.

58. Williams, *West Virginia*, 122. The economic interests and procoal company stance of most of these governors are discussed in this work.

59. A. B. Fleming to William N. Page, Feb. 20, 1899, A. B. Fleming Papers, West Virginia University Library, Morgantown, W.Va., and Glen Massey, "Legislators, Lobbyists and Loopholes: Coal Mining Legislation in West Virginia, 1875–1901," *West Virginia History* 32 (Apr. 1971): 155, 167–69.

60. J. C. McKinley, *The Coal Crisis*, An Address before the Annual Meeting of the West Virginia Board of Trade (n.p., 1915), 10, and Harris, ed., *Legislative Handbook, 1918*, 436. For the coal establishment's domination of the Committee on Mines and Mining, see Harris, ed., *Legislative Handbook, 1924*, 117–28. Coal

operators held five of the eight positions. For evidence of the coal establishment's domination of southern West Virginia politics, see Harris, ed., *Legislative Handbook, 1918*, 423–42. In that year, 13 of the 16 delegates from southern West Virginia (Kanawha County was excluded) were directly affiliated with the coal industry. Of the three who were not, two of them declared themselves in favor of the coal operator.

61. Wolfe to Justin Collins, Feb. 2, 1913, Collins Papers.

62. William Ohley to Fleming, Jan. 31, 1899, Fleming Papers.

63. For writers stressing the political activism of organized labor in West Virginia, especially its efforts to secure passage of labor legislation, see Barkey, "Socialist Party," and Evelyn Harris and Frank Krebs, *From Humble Beginnings: West Virginia State Federation of Labor 1903–1957* (Charleston, W.Va., 1960). E. P. Monts to Fleming, Mar. 1, 1904, Fleming Papers.

64. Interview with the son of a former company official of Gulf Smokeless Coal Company (name of interviewee withheld), Wyoming County, W.Va., summer 1975. Copies of agreements between the miners and the coal companies in which the miners agreed to turn over their compensation checks are in author's collection. Dow Platt, letter to the editor, *UMWJ*, Dec. 16, 1915, 8; Lee, *Bloodletting in Appalachia*, 9–10.

65. House of Delegates of the State of West Virginia, *Journal, 1897*, 523–24.

66. Senate of the State of West Virginia, *Journal, 1901*, 616.

67. Excerpts of the West Virginia laws prohibiting the use of scrip are in Lane, *Civil War*, 27; also see Massey, "Legislators, Lobbyists and Loopholes," 152–53, 158–61; Lamar Epperly to Justin Collins, July 21, 1925, Collins Papers; Wilson, *American Jitters*, 67.

68. See, for example, Henderson Kelly to Justin Collins, Oct. 27, 1911, Collins Papers.

69. West Virginia State Legislature, *Acts, 1899*, 163.

70. E. N. Clopper, *Child Labor in West Virginia*, National Child Labor Committee pamphlet no. 86 (New York, 1902), 3, 20–24, and E. N. Clopper, *Child Labor in West Virginia in 1910,* National Child Labor Committee pamphlet no. 142 (New York, 1910), 4, 5.

71. Clopper, *Child Labor in West Virginia,* 12–13. For the testimony of Coolidge, see *Conditions in the Coal Fields,* 2:1941.

72. For criticism of the willingness of coal operators to ignore the child labor laws, see W.Va. State Superintendent of Free Schools, *Annual Report, 1914*, 14, and *Biennial Report, 1889–90*, 7. Tams, *Smokeless Coal Fields*, 34, and Owen Lovejoy, "Child Labor in the Soft Coal Mines," *Proceedings of the Third Annual Meeting of the National Child Labor Committee* (New York, 1907), 26–28.

73. William Graebner, *Coal-Mining Safety in the Progressive Period* (Lexington, Ky., 1976), 73, 86–87, chap. 4. The contention that the inexperience of the miners was the cause of the high death rate has been challenged: for example, Keith Dix in *Work Relations in the Coal Industry*, West Virginia Institute for Labor Studies Bulletin no. 78 (Morgantown, W.Va., 1977), chap. 3, shows that the majority of the miners killed in accidents had worked for more than two years in the mines. *UMWJ*, Mar. 15, 1900, 4.

74. Williams, *West Virginia*, 142, and T. E. Thomas to Fleming, Mar. 11, 1901, Fleming Papers.

75. Graebner, *Coal-Mining Safety*, 91, 98–99, and Dix, *Work Relations*, chap. 3. The coal companies were concerned with the number of gas explosions and

supported some efforts to reduce them—for economic, not humanitarian, reasons, however. The day after an explosion at Lick Creek, W.Va., in 1909, a coal operator wrote to his superintendent that "if the West Virginia mining industry is to be a success the people . . . must find the cause of these explosions and, if possible, apply a remedy, or else there will be such drastic legislation, both state and national, that it will be impossible to successfully operate a mine in the state." Justin Collins to Jarius Collins, Jan. 13, 1909, Collins Papers. See also Graebner, *Coal-Mining Safety*, 20, 143.

76. *UMWJ*, Feb. 28, 1907, 4, and Paul to Fleming, Oct. 14, 1901, Fleming Papers. For an example of Paul's pro-operator conduct in office, see Graebner, *Coal-Mining Safety*, 74.

77. *Coal Age* 26 (Nov. 20, 1924): 728; "The Rise of a Call Boy," *West Virginia Review*, Mar. 1929, 206–7; testimony of John Laing, *Paint Creek Hearings*, 1:1650. For Laing's pro-operator conduct in office, see Graebner, *Coal-Mining Safety*, 38. While in office, Laing blamed the miners for the majority of the accidents. See W.Va. State Department of Mines, *Annual Report, 1910*, 16. For union officials' attacks upon Laing's conflict of interests, see *Paint Creek Hearings*, 3:2270. Laing's successor, Earl Henry, was no different. See Graebner, *Coal-Mining Safety*, 92.

78. Gompers, "Russianized West Virginia," 829.

79. Quoted in *Milwaukee Leader*, May 29, 1913.

## CHAPTER II

# "What Kind of Animals"

> Yet what force on earth is weaker than the feeble strength of one.
>
> Ralph Chaplin
> "Solidarity Forever"

"We have often wondered what kind of animals they have digging coal in West Virginia," fumed an article in the *UMWJ*, "and have never been able to successfully solve the problem. Their ignorance must be more dense, their prejudice more bitter and their blindness more intense than that of any other body of miners we have ever heard tell of."[1] This denunciation reflected the UMWA's frustration with the southern West Virginia miners, who for twenty years continuously rejected major efforts to unionize them.

The existence of the nonunion southern West Virginia coal fields was of grave concern to the UMWA. The state's meteoric growth in bituminous coal production and its proximity to the midwestern coal markets made the state a threat to the union's established organized fields. The coal fields represented "a gun pointed at the heart of the industrial government in the bituminous coal industry."[2] UMWA officials, knowing the union could "never rest secure"[3] until West Virginia was organized, sponsored repeated efforts to unionize the state's coal diggers. These attempts met with little success. By 1912 only a small pocket of unionism existed in southern West Virginia, and that foothold proved to be weak and illusive.

Although the UMWA's failure to make any progress among probably the most exploited and oppressed coal diggers in the United States has been neglected by historians, it puzzled interested contemporary observers throughout the United States and prompted various explanations. UMWA officials commonly charged that the apathy of the miners was responsible for the union's ineffectiveness. UMWA Vice-President Frank Hayes declared in 1910 that "West Virginia separated from Virginia during the Civil War period on account of the slavery question. If the miners of West Virginia *were awake* to their interests, both economically and politically there would be another historic separation in the state, and that would be the [miners'] separation . . . from the hands of the coal barons and other organized forces of greed."[4]

Radicals, particularly those in the miners' union, charged that the southern West Virginia coal operators maintained a lobby in the UMWA headquarters in Indianapolis. Others claimed that the problem was in the West Virginia state capital. Samuel Gompers, president of the American Federation of Labor (AFL), stated a widely accepted belief that the coal industry's domination of the state government obstructed the union's success in West Virginia. "The coal operators and the Government [of West Virginia] have been one and the same," Gompers declared; "King Coal and his barons have ruled in and by means of the institutions of society; they own absolutely and control agents and agencies apparently of the people." The most common and widely accepted explanation at the time, and the explanation still accepted by labor historians, is that the southern West Virginia miners were cowed by the presence and activities of the Baldwin-Felts mine guards. A UMWA official expressed this view when he declared that "West Virginia miners will not be organized until they themselves have made up their minds that they are going to be organized in spite of all opposition."[5]

These explanations, while valid, are not complete. In 1912 the miners launched a major union movement in southern West Virginia, and the spirit prompted by that movement increased and broadened over the next two decades, despite the brutal opposition of the mine-guard system and the domination of the state government by the coal establishment.

The failure to study why workers do not organize leaves a void in American labor historiography. A knowledge of why some workers do not unionize is as crucial to the study of labor as a knowledge of why others join unions and radical labor parties. An examination of the southern West Virginia mining force between 1890 and 1912 reveals that the miners remained outside the UMWA, not out of apathy or cowardice, but because the issues and goals of the miners' union did not represent their wants and needs. Furthermore, an exploration of the miners' indifference to the UMWA during these years is a first step in understanding the development of the labor movement in southern West Virginia and the later evolution of its working-class culture.

The people—the native mountaineers and southern blacks and European immigrants—who poured into the southern West Virginia coal fields in the 1890s were overwhelmingly of nonunion and nonmining backgrounds. They were existence-oriented. An elderly coal operator recalled that the "thousands of men, soon tens of thousands . . . who dug that [first] coal . . . were a hard-bitten lot, [they] drank and fought and gambled and whored."[6] They seemed to have cared or planned little for the future—or at least cared little for making the coal fields a permanent home and a decent place in which to live and work.

The native miners were farmers and seemed intent on returning to their farms rather than making a career of mining coal. In fact, many of them kept their farms and returned to them during slack runs or strikes. As a result, they showed little interest in the union. Most of the early southern black miners saw little need for it. Having come to the southern West Virginia coal fields to escape racism and to improve their economic life, they were initially content with conditions in the coal fields, where they found minimal discrimination and, according to one black miner, "made money with ease."[7]

Bent on mining and earning their daily bread, or getting rich and moving on, or returning to their farms, the miners were easily subdued by the coal operators. Labor organizers during the 1897 strike and the organizing drive in 1901 found that the miners were quick to accept both company promises and advice to stay away from the union. This attitude carried over into everyday affairs. Observers pointed out that the miners readily accepted the coal operators' claims that the UMWA was trying to turn the state's coal trade over to the competitive states. The miners, furthermore, refused to protest or even to report to mine inspectors that the companies often forced them to violate state mining laws. A pro-union miner in Oak Hill lamented that the miners in his coal field composed "a class of labor that meets the approval of the operators here . . . they dare not breathe a word without permission from the company."[8]

The immigrant miners in southern West Virginia were mostly from southern and eastern Europe; conspicuously absent were the British miners who had been so instrumental in stimulating union activity in other coal fields and who were the founding fathers of the UMWA. One historian of the coal miners wrote that the British miners "brought with them the political and economic experience that gave [them] . . . the sense of need for solidarity which the American diggers lacked."[9]

In 1900, a decade after the opening of the southern West Virginia coal fields, there were no recorded Welsh or Scottish coal miners in the state. English miners numbered 1,053, or less than 5 percent of the mining force in West Virginia. As few as they were, the English immigrants stressed the importance of unionism to miners of nonunion background. An English immigrant miner in Mingo County in 1901 warned his fellow workers that without the union they would suffer under "a condition of surfdom [*sic*] more appalling, more foreboding to men who honor the name of justice, truth and right than conditions which were forged upon the patriots who fled from the land of English tyranny to seek an asylum in the wilderness of America."[10] The British miners, however, were too few to be effective, and they tired of a hopeless fight.

The next few years witnessed a decrease, both absolutely and proportionally, of British miners in southern West Virginia. They left the state, according to the U.S. Immigration Commission, to escape the invasion of southern and eastern European immigrants, who generally lowered the living and working conditions of coal fields—or at least so the British miners thought. More of them left following the failure of the 1902 strike, when "many of the more active union miners" sought out "areas that would be considered safer grounds." Others left to escape the brutality of the mine guards. Noting the beatings of two UMWA organizers by Baldwin-Felts guards, UMWA District 17 reported that the experienced immigrant miners "have been leaving West Virginia for years and going to other coal fields in other states, in order to avoid coming in contact with such treatment."[11] Finally, the UMWA itself may have been responsible for the decline in British miners. In an effort to cripple the labor supply to the West Virginia coal industry, between 1905 and 1907 the UMWA urged all union miners to stay away from the state.[12]

By 1907, five years before the Paint Creek–Cabin Creek strike, although the number of Welsh miners had increased from zero to 31 (0.06 percent of the mining force) and Scottish miners had increased from zero to 120 (0.21 percent of the mining force), the English miners had dropped from 1,053 to 280 (0.5 percent of the work force). The overall decrease of British miners between 1900 and 1907 was from 1,053 to 431, or from 5 percent to 0.77 percent.[13] Consequently, the union tradition and spirit that developed in southern West Virginia evolved among a work force almost devoid of union experience.

Without the British miners, the early mining force in southern West Virginia lacked the sense of solidarity so vital to a union movement. Prejudice and racism divided the state's miners. There was no sense of mutual responsibility; the miners, for instance, had no qualms about stealing from each other. After spending an entire morning in the hills cutting timber to prop the work place, a miner would return after lunch break to find that another miner had stolen his wood. Upset with such thievery among miners, a coal digger claimed that is was "high time for the miners of West Virginia to wake up and put on robes of honesty and crown themselves with the United Mine Workers of America and stand for justice."[14]

Miners desiring a union found little inspiration from UMWA organizers and officials. On both the national and state levels, UMWA officials left the miners' union for lucrative positions with the southern

West Virginia coal operators. Former UMWA President T. L. Lewis left the union to become secretary of the New River Coal Operators' Association. D. C. Kennedy, after serving as president of UMWA District 17, became commissioner of the Kanawha County Coal Operators' Association. Noting the ease with which union officials became representatives of the southern West Virginia coal industry, Gompers exclaimed in 1912 that the coal operators were making "a direct effort to destroy the miners' organization of . . . [West Virginia] by suborning its officials."[15]

During the 1901 organizing drive, Mary Harris ("Mother") Jones and John Walker discovered that several of the union representatives in southern West Virginia organized in peculiar ways. The men failed to show up for organizing meetings that they had called, and when they did appear, instead of promoting the UMWA, they denounced it. According to Walker, they "berate[d] everybody that is working for the organization." Jones and Walker concluded that the men were "employees of the [West Virginia] coal companies," and demanded their discharge as union organizers.[16]

The case of John Nugent presented the most flagrant example of switching sides. Nugent had been a member of the Knights of Labor. In 1890 it was on Nugent's motion that the Knights of Labor and National Federation of Miners merged to form the UMWA. In 1902 he came to West Virginia as a national organizer and was eventually elected president of UMWA District 17 and later president of the West Virginia State Federation of Labor. In 1907 Nugent resigned his position as District 17 president to take an appointed position as immigration commissioner of West Virginia, an office that was created in 1897 to lure immigrants into West Virginia but that remained vacant for ten years because of the lack of state funds. The governor of West Virginia appointed Nugent to the state office, and the New River Coal Company and later the Consolidated Coal Company paid his salary. Nugent, as commissioner of immigration, in violation of the federal immigration law, journeyed to Europe to encourage miners to emigrate to West Virginia. The governor later admitted that Nugent's expenses on the trip, like his salary, were paid by coal companies. After he returned to the United States, Nugent went to New York to persuade immigrants to go to West Virginia to work in the state's coal mines.[17]

Because of the extraordinarily high accident rate in the West Virginia coal mines, the federal government did not consider the state's mines safe enough to recommend to immigrants. Nugent was not deterred; he handed out statements to the newcomers, marked with the state seal, that read "conditions so far as general safety is concerned are vouched for by the United States Bureau of Mines and the Chief of the Department of Mines of West Virginia."[18] Nugent's actions were detrimental to the union spirit among the state's rank-and-file miners. One wrote to the *UMWJ* that their former district president "has simply degenerated into a transportation agent, hiring men for the 'scab operators' of the state."[19]

The rank-and-file miners became distrustful of UMWA officials and organizers—a fact that the coal operators obviously recognized. UMWA organizer General J. W. St. Clair encountered this problem when he tried to organize miners in MacDonald, Raleigh County. St. Clair delivered an "impassioned harangue on the rights of the miners and how he proposed to fight for those rights." Following the speech, the coal operator, who had been hiding in a laurel thicket, approached St. Clair, congratulated him on the speech, and exclaimed, "We are waiting dinner for you, General, up at my house. Come right along up." The audience of miners muttered "sold out again"; the meeting dissolved, and so did the effort to organize.[20]

Most of the UMWA officials remained loyal to the union and worked for it in southern West Virginia. Yet the miners were no more willing to follow them than the ones who took positions with the coal operators, mostly because the causes that the UMWA promoted did not represent the primary interests and needs of the southern West Virginia miners. Since its founding in 1890, the chief issues of the UMWA were the two traditional issues of American labor unions—higher wages and shorter hours—and these were the issues that were advocated in southern West Virginia.[21] They were not, however, the issues that concerned the coal miners of southern West Virginia.

In the bituminous coal fields, Carter Goodrich pointed out in his classic study, *The Miners' Freedom*, until the 1920s the coal miner was a "disposer of his own time." "The factory hand," Goodrich explained, "works under the boss's eye, his hours are rigidly bounded by the whistle and the time clock," but "the isolated miner sets his own pace and comes and goes pretty much at his own time."[22] Paid on a piece-rate basis, not having to punch a time clock, and working in a drift or slope mine, in contrast to the company-regulated elevators of the shaft mines in many other coal fields, the coal miner in southern West Virginia controlled his own work day. When he had earned enough money for the day, or wanted to leave early to go hunting or fishing, or simply tired of working, the miner walked out the level heading of the drift mine or up the incline of a slope mine and went on his way. The company did not object, for it did not lose any money.

The company was somewhat afraid to restrict this essential aspect of the miners' freedom. A coal operator in McDowell County confessed that the time a miner spends in the mine "depends absolutely on his own volition." This miner, he explained, "begins and quits when he pleases, and the operator does not say a word, and dares not, for the reason that if he did, and the miner should not like it," the miner would pack his bags and "move to another town."[23] This explanation is revealing, for it shows why the miners had little interest in the union's demand for an eight-hour day, and it illustrates that the miners, even before they had a union, possessed a form of protest—geographic mobility. Furthermore, it reveals that southern West Virginia coal operators did not subject their miners to the time-oriented and highly disciplined routine of factory work that troubled other workers when they first encountered industrialization.

The precise earnings of the southern West Virginia miners are difficult to ascertain, as the miners were paid on a piece-rate basis, and each company paid different rates; in addition, a multitude of other variables (e.g., frequency of work) also determined the total income of the miner. In surveying the wage rates of bituminous coal miners in the United States, the Bureau of Labor Statistics was forced to depend upon special records kept by company officials. Hence, these reports are probably no more valid than the coal company propaganda that claimed southern West Virginia miners were the highest paid in the United States or the union propaganda that claimed they were the lowest paid.[24]

While the southern West Virginia miners were not, as one Logan County coal operator claimed, "among the economic aristocrats of labor," neither were they starving to death because of low wages, not, at least, until the 1920s. The wage scales of the southern West Virginia miners generally may have been the lowest in the United States, as UMWA spokesmen, labor historians, and other observers claim, but the softer coal and larger veins allowed the miners to earn more money in less time than miners in states with higher wage scales. The greater frequency of work in southern West Virginia gave miners a larger annual income than miners in other fields who were paid more but worked less. Furthermore, although the company towns were single-industry towns, the miners and their families in southern West Virginia had other ways and means of supplementing the family income.[25]

On the other hand, the income of the southern West Virginia miners was diminished by the high rate of injury and sickness and the large amount of time the miners spent moving from one coal town to another. The lack of railroad and mine cars, always a problem in southern West Virginia, also decreased the miners' total earnings. "How in the world can we support our families and send our children to school," a miner from Meadow Brook asked, when "we get railroad cars only 2 or 3 days a week, so we can work only 2 or 3 days a week?" Reports indicated that miners lost as much as 50 percent of their income because of irregular work.[26]

The greatest drain on the miners' wages was the company store. Coercion, the scrip system, and physical distance often combined to force the miners to deal at the company store, and through the monopolistic control of food and clothing and tools and powder, the coal companies were able to render wage rates and wage increases meaningless to the miners. A miner from Keystone, McDowell County, for example, wrote in 1896 that he had no complaint about his wages, but pointed out that because of the "awfully high prices" of the company store "the miner is compelled to earn sufficient by digging from 3½ to 4 tons daily." Wage advances were always absorbed, "in whole or in part," by price increases at the company store. This was a common complaint among the southern West Virginia miners—wage increases were valueless as long as the company store existed. "The company store is one of the sore spots in the whole scheme of things," a miner complained to a U.S. Senate investigating committee: "When a wage increase is given the miners, the store management invariably increases prices to offset it." He explained that the miners at his mine "won a wage increase at one o'clock, Sunday morning. By Tuesday morning, prices were advanced exactly enough to absorb the wage increases." Another miner wrote the *UMWJ* explaining that "it has always been customary in the coal fields [that] when an increase is given in wages, that the increase is absorbed by a rise in price of the commodities at the company store, and very often superceeded [*sic*] by a rise in food prices before the increase is given."[27]

The company store served other useful purposes. In the company towns, the post office was located in the company store, and the store clerk also functioned as the postmaster. These company officials scrutinized the miners' mail for union and radical literature. While organizing in southern West Virginia, Walker reported that he had to go to the commercial town of Montgomery, often miles from where he was organizing, to pick up his mail because the company store clerk-postmasters read his mail.[28]

It was this overall presence, power, and supervision of the company that really angered the miners and made other issues more important than wage rates and wage increases. The daughter of an early radical in West Virginia, who had mined coal in Logan and McDowell counties, recounted, "They paid him well enough and he didn't mind the work itself, but he often stated that a man couldn't call his soul his own in those communities."[29]

Consequently, wage increases were not a key issue in any of the major strikes in southern West Virginia. During the Paint Creek–Cabin Creek strike of 1912–13, the first major effort on the part of the miners to bring unionism into the area, a reporter noted that an increase in pay "the miners regard as the least vital of all their demands."[30]

The reporter was right. The miners' demands were, in order of importance: (1) recognition of the union; (2) abolition of the mine-guard system; (3) reform in the docking system; (4) a checkweighman representing, and paid by, the miners; (5) trade with any store they pleased; (6) cash wages; and finally, (7) an increase in pay.[31] These demands and their rank are important; they struck at the foundation of the company town system and through that, at the power of the operators; they showed that the miners realized an increase in pay was useless without basic changes in the structure of life and work in southern West Virginia.

Even without a union, the miners had means of coping with their new economic circumstances and ways of life. The coal diggers drew upon the traditions and habits acquired in their earlier, preindustrial days and used them in a meaningful, pragmatic way in the company towns. The miners found their agricultural talents particularly useful. Indeed, gardening and the raising of livestock were widespread among these miners. As late as 1924, the West Virginia Coal Association estimated that over 50 percent of the state's miners planted gardens and kept cows, pigs, and poultry. This estimate seems conservative, as elderly miners recall the times when "all these here hills were covered with crops and pigs and chickens and cows." "Almost all the miners had [a garden]," one miner declared. The Children's Bureau of the U.S. Department of Labor reported that over 70 percent of the miners' families in Raleigh County raised crops and livestock.[32]

Gardening was a practice in which the whole family participated. The father sometimes tended the garden before he went to work in the mornings or on Sundays. But generally the job of caring for the family garden fell upon the miner's sons not yet old enough to work in the mines and his wife and daughters. One miner boasted that his wife and daughter "worked harder in the fields than any man ever did and that's why we grew more stuff in the [company] towns than the farmers on their farms."[33]

Gardening was undoubtedly a continuation of the skills and habits that the miners had acquired before they entered the coal fields. Many of the native state miners had practiced subsistence farming and hunting before entering the coal industry. About half of the southern black migrants who became miners in southern West Virginia had been sharecroppers. And since the majority of the European immigrants came from nonindustrial countries, probably most of them had been farmers of some sort.

The continuation of farming in the southern West Virginia coal fields, however, was not simply a tenacious clinging to a previous way of life. In a region where there were no other local industries in which wives and daughters could supplement the income of the miners, gardening proved an adequate substitute. It was a highly important economic safety valve in an industry plagued with irregular employment and periodic depressions. As one miner recalled, "Our garden saved us during all those depressions, and when I was out of work." "Miners couldn't always depend upon the mine," said a Wyoming County miner; "therefore we would have to raise a garden to make sure we always ate."[34]

Gardens meant the miners did not have to spend as much money on food. "We needed a garden," Clyde Scarberry explained; "we didn't always have enough [money] for food even when I was working." This situation was especially true for the miners with large families. J. H. Vernatter, a miner in Logan County who had ten children, said that "the only way we could feed all of them on my pay was to raise a garden." A miner with six children simply stated: "If we didn't raise hogs, corn and potatoes, we didn't eat." The Children's Bureau, while appalled by the generally destitute conditions in mining areas, noted that the miners in the Winding Gulf coal field always had fresh vegetables with their meals. In fact, the bureau pointed out that the poorly paid miners, many of them out of work, ate better than workers and many other higher paid people in the cities. The miners also grew these crops, especially vegetables, to decrease their dependency and hence indebtedness to the company store. Bill Mullins recalled, "We grew everything so we had to buy very little from it." "My family was never in debt to the store," a miner's wife stated; "we were careful with our money and were sure to plant crops every year."[35]

The raising of gardens and livestock in the company towns helps to explain why the miners were not as interested in the UMWA's promise of higher wages as were miners in other communities. P. M. McBride, on a tour of southern West Virginia in 1896, wrote about the trouble in organizing miners in Kanawha County. "There does not exist the hunger and suffering here that is found in . . . [other coal fields]. Every available spot of ground seems to have received attention from the plow or spade, the houses resemble the homes of the market gardener. . . . This explains their comparatively comfortable position. They raise all the vegetables they require and this assures them that the wolf shall be kept from the door."[36]

The miners' retention of their agricultural skills proved valuable when they finally decided to fight for the union. Gardens and livestock served as powerful sustaining forces during strikes. In the 1919 strike, for example, a miner from Kayford, Kanawha County, wrote to UMWA President John L. Lewis that "we're not worrying about strike benefits . . . because we are killing hogs and gathering corn and other crops and squirrel hunting."[37] (Even today, when miners begin planting extra gardens, it means that contract time is near.)

One part of life in southern West Virginia company towns, which has been considered a preindustrial carry-over among other industrial workers, was the making and consumption of alcohol. The southern West Virginia miners did drink. They drank to celebrate the birth of a child, christenings, weddings, Christmas, New Year's Day, and saints' days (which they called "Big Sundays"). These special celebrations often continued into the work week. "Work was down the past week," a superintendent wrote to his coal operator; "it was Easter and we had more drunks than usual." The miners also drank to celebrate payday, which "was announced every two weeks by the playing of cards and the drinking of whiskey." Payday was a "big bash," a Raleigh County coal operator recalled; alcohol was "poured into a large galvanized tub, probably used during the week for bathing and washing clothes, and then the night would be spend drinking and singing songs." This practice was also carried over into the work week. A company official explained to his coal operator that very little was mined on a particular day because it was "Pay Monday and most of our workers are not here." "Monday, Tuesday, and Wednesday after paydays," another superintendent declared, "are the drunks' holidays."[38]

The miners did not need any special day or excuse to drink, however. On Saturday nights they gathered on the steps of the company store and sang songs and told jokes while they drank their homemade liquor. At the end of a work day, they went to the local pub and drank. Saloons were numerous in both the commercial and company towns in the coal fields. A visitor to Fayette County noted that "there were saloons up every hollow." The town of Mt. Hope, Raleigh County, with a population of only a few hundred had eight to ten taverns "where a man could get whiskey for a dime a drink, and a dollar a quart." The saloons in southern West Virginia had a very profitable business because of the miners. The saloon in Decota, on Cabin Creek, Kanawha County, averaged $300 a day from the miners. One particular payday, the tavern had a staggering net profit of $2,160.[39]

The miners made their own alcohol, too, using almost anything, including the crops they grew in their gardens. Most of them specialized in homemade beer called "homebrew." With ten pounds of meal, ten pounds of sugar, three cakes of yeast, water, and something for flavor—

usually evaporated peaches—they got "the worst kick you could ever have." Different ethnic groups made their own favorites. Italian miners, for example, were fond of making a potion they called "pickhandle" from raisins, yeast, and water; it contained 10 to 11 percent alcohol, and, according to a local reporter, had a "wicked kick."[40]

Most of the miners made their liquor in the hills, but stills could be found throughout the company town, in miners' homes, in barns, and even in the coal mines, where the miners often made liquor while they worked or during lunch break. The latter practice was so common that the chief of the West Virginia Department of Mines, concerned with the safety hazards that could result, issued a circular that ordered mine inspectors to check for stills in the mines and to report them to the police.[41]

Company officials were also concerned with the miners' drinking habits. The coal operators of southern West Virginia were early proponents of prohibition, claiming that liquor lowered the moral standards of their workers, increased accidents, and decreased production. "Coal mining seems to be about the only industry in the country," a superintendent declared, "where a fellow is allowed to lose money for his employer two or three days every two weeks [because of drunkenness] and then go back to work as if nothing had happened." The president of a West Virginia coal company explained, "I am not a prohibitionist, but saloons . . . hurt coal production. The coal states, at least, should be dry. I believe the operators are unanimous on this question."[42]

This concern with the miners' drinking habits represented a clash of cultural values—not of preindustrial and industrial ways of life, but a clash between upper-class and working-class cultures. Early into the formation of the southern West Virginia work force, the making, drinking, and selling of alcohol became a useful and beneficial practice for the miners.

Saloons in the coal fields, for example, served a social purpose. George Korson explained, rather romantically, that saloons in the bituminous coal regions were an "oasis in a desert of wretchedness and oppression, an escape from a harsh environment." Calling attention to the polyglot nature of the early mining work forces, Korson continued, "A boisterous welcome was extended to every adult mine worker regardless of race, creed, personal appearance, or previous condition of servitude. There was no qualification for membership and the only dues were a nickle for a high-collared beer or a dime for a glass of the hard stuff."[43]

Alcohol met the physical and psychological needs of the miners. After studying the drinking behavior of Welsh coal miners, a psychologist gave an analysis that could well apply to the miners of southern West Virginia. Following a day of toil in the mines, a miner was tired and "thirsty from losing more fluid by perspiration than he has drunk to replace [because of the poor water system in the mines]; his clothes [are] cold and damp with sweat. His tired body calls for a sedative. . . . Hence, an alcoholic beverage is the ideal to meet his present need." The industrial secretary of the YMCA, Ira Shaw, in arguing for better recreational facilities for the company towns, told a gathering of West Virginia coal operators that "the man who labors at hard tasks, such as a miner performs, craves something of a stimulating nature when his daily toil is done. The dangerous character of his work adds to this feeling of need for stimulation, as he works under nerve as well as muscle strain, owing to the dangerous character of his work." Because most mining communities in southern West Virginia before 1913 did not make provisions for amusement and relaxation, Shaw explained, "It's no wonder that [the miner] takes the most easily procured substitute, alcoholic stimulation."[44]

The making and drinking of alcohol also served economic purposes: like gardens, liquor supplemented the miners' income. In replacing calories, liquor was for the miners, as it was for Oscar Handlin's uprooted, cheaper than bread. But the miners also sold it; in fact, miners in southern West Virginia claim they probably sold more of the liquor they made than they consumed. According to one miner, moonshine could be made for about fifty cents a gallon and sold for fourteen to fifteen dollars a gallon. Another miner claimed that he received twenty dollars a gallon. It is of interest that the miners sold most of their moonshine to company officials, who drank some of it and took the rest to Huntington and Charleston to resell. "We didn't make nothing from mining," a miner explained; "we had to make it someplace." Another miner recalled, "Making moonshine was the only way we had of making money during slack runs and strikes"—a statement that illustrates that the making and selling of liquor became another powerful sustaining force during strikes.[45]

The miners made and sold moonshine to supplement their strike rations, which may have been why the state police seemed to have enforced the state's prohibition laws more strictly during strikes. During the 1919 strike, which lasted for less than a month, state police made several "still raids" in the Harts Creek area of Mingo County, which had always been "a notorious moonshine area," confiscated a sixty-gallon still and six other smaller stills in Raleigh County, three stills on Cabin Creek, and several dozen more in Logan and Boone counties. When federal soldiers were sent to Mingo County during one strike, one of their first activities was to conduct a "still hunt."[46]

Gardening and moonshining, then, represented an adjustment, not a resistance, to industrialization. The miners' retention of these past habits was not a tenacious clinging to previous "subcultures of work and leisure" in a society that demanded a "new human nature."[47] They were the preservation of traditions and skills that eased the transition to an industrialized society; they were realistic ways of supplementing low wages, maintaining decent health, and beating the company store.

The nature of the work in the southern West Virginia coal fields did not produce the tensions and clash of cultural work values—such as those that accompanied preindustrial workers into the factory system—that might have produced an early collectivist protest leading to a nascent union effort. Coal mining did not "entail a severe restructuring of working habits—new disciplines, new incentives, and a new human nature upon which these incentives could bite effectively"[48]—as the early coal miners in southern West Virginia were not part of an industrial world regulated by time clocks, supervisors, and demands for speed-ups; they retained their work tools and controlled their work hours and work place.

In an era when industrial capitalists were moving toward a more "rational system of management," as David Montgomery argues, in an effort to circumvent the control that the autonomous craft worker exercised over the production process, coal mining remained a "cottage industry (only the cottage is a room in the mines instead of a home)," and the coal miner retained control over his work place. He had not been separated from his tools. Indeed, company policy required each coal digger to purchase his own tools. The failure of the coal industry to automate as early as other industries retarded the specialization and division of labor that occurred in other industries. Most important, the absence of scientific management in the coal fields allowed the miners to retain an "individualistic independence," which was the most important component of the miners' freedom.[49]

Because of the physical structure of the coal mine—a honeycombed tunnel extending in all directions for miles underneath the earth—the coal miner was "an isolated piece-worker" who saw "his boss less often than once a day." Consequently, the miner was relatively free from management supervision and regulation. And he resented any effort to supervise or regulate his work, a fact recognized by observers in other industries as well as by the miners and operators. The editor of a management magazine, for instance, warned other industries about hiring former coal miners. Noting the miners' lack of supervision on the job, the editor explained that "the coal miner is trained to do as he pleases." The "ex-miner resents all suggestions as to his working methods, resents all efforts to compel continuous application, and assumes in general a hostile attitude toward supervision." Wrote Goodrich, "The miners' freedom is unique in American industrial life and the discipline of the mines is so utterly unlike the order and regimentation of a plant like Ford's that it is hard to believe that the two are continuing to exist side by side."[50]

Unlike other industrial workers, coal miners relied on their own skill and judgment. A coal operator later denigrated the work of an early miner by explaining that his was "not a highly skilled occupation"; it required "little more than a strong back and average intelligence to become a good miner." From a miner's perspective, however, the job demanded ability and involved highly technical decisions. "The miner was his own boss," John Brophy recollected; "his judgment was at work as well as his muscles." The miner "made his own decisions—how deeply to undercut the face, how much powder to use, how to pace himself in loading the car, and many other things." Analyzing a time study that the U.S. Coal and Coke Company had made on the miners at Gary, West Virginia, Goodrich explained that it "gives a fair impression of the variety of the loader's work and a suggestion of the great number of decisions for which the day's work calls."[51] And upon such decisions depended not only his own life, but the lives of his fellow workers.

The miner's feelings of ease at his job were impossible for other industrial workers to share. In hot, dry mines, miners often stripped to their underwear while doing their day's work. In dry mines without water or mud, the miner urinated on the dirt he used to pack his powder. He found an intriguing mysteriousness about the coal, unlike the routine of an assembly line, which bored factory workers. "Your mine is a strange mistress," a Mingo County miner stated. "You think you get to know her, and then, come to find out you don't." The ever-present dangers of gas explosions and slatefalls gave the coal digger a "certain fascination" with his work, which the miners admitted was "closely akin to the lure of a sailor's life." Hence, one of the miners' favorite songs was:

Miner's life is like a sailor's
  'Board a ship to cross the wave;
Every day his life's in danger,
  Still he ventures being brave.
Watch the rocks, they're falling daily,
  Careless miners always fail;
Keep your hand upon the dollar
  And your eyes upon the scales.

The early coal miner was, as one observer noted, "a male in a masculine industry where others fear to tread."[52]

The coal miner was not alienated from his work or product. He took pride in his career—once a miner, always a miner. He possessed a "proud sense of occupational identity" that, according to psychiatrist Edmund Ross, helped him to define himself and gave him an identity that seemed to be lacking among other industrial workers. He understood his materials, the work he did, the strategy of extracting coal, "how his job fit into the overall pattern." His daily travel underground gave him an overview of the entire operation, past and present. In an industrial nation dependent upon the energy resource he produced, the coal miner knew he was essential to the nation and that he was necessary for the common good.[53]

Further inhibiting both an early collective mentality and protest was geogaphic mobility, the miners' form of individual protest. American workers have historically been geographically mobile people; Stephan Thernstrom has written that workers on the move have been "not a frontier phenomenon, or a big-city phenomenon, but a national phenomenon." The southern West Virginia miners were extremely mobile. While quantitative evidence (e.g., census and company records) is not available to document their mobility precisely, it seems the mobility rates of the southern West Virginia miners far exceeded that of other American workers. "We moved all the time," recalled George Scales, a native of South Carolina, who drew "fifteen paychecks at fifteen different mines. . . . It was easy, all you needed was a pair of gloves, overalls and experience and you could get hired anywhere." A miner's wife in Raleigh County told the Children's Bureau that miners moved "every time the moon changed."[54]

The rates of labor turnover in the southern West Virginia coal fields reflected the miners' mobility. On November 16, 1904, fifty-eight men worked at the Acme Mine of the Stevens Coal Company on Paint Creek. Ten months later, on September 30, 1905, although the number of miners had jumped to sixty-five, only twenty of the original fifty-eight men were still employed at the mine. By March 15, 1906, the number of original miners had dwindled to twelve.[55]

Investigators were astounded by how much the miners moved around. After a study of mining families in the Winding Gulf coal field, the Children's Bureau reported that more than one-third of the miners moved every two years. In five of the company towns they visited, the bureau discovered that more than one-half of the families had lived in the town for less than a year. The U.S. Immigration Commission referred to the southern West Virginia mining force as a "floating population."[56]

This mobility was not unique to southern West Virginia. Brophy wrote that the miners in western Pennsylvania led "a gypsy life." His father moved about, Brophy explained, not because of restlessness, "he was too good and steady a man to jeopardize his family's welfare for a whim," but in search of regular employment. The mobility rates of the southern West Virginia miners, however, far exceeded those of other coal fields—at least as reflected in the turnover rates. A U.S. Coal Commission survey revealed that of the thirty-three major bituminous coal districts in the United States, the five coal fields of southern West Virginia ranked first, second, fourth, fifth, and sixth—or five of the top six—in coal fields with the greatest percentage of labor turnover. For coal fields with stable labor forces, excluding the Winding Gulf coal field, the southern West Virginia coal fields occupied the lowest five positions.[57]

These mobility rates concerned state and county officials. The superintendent of the Fayette County schools, for example, claimed that the miners' constant moving was detrimental to the educational system. "Owing to the ever shifting population [of the miners], we are continually required to instruct those who have just come among us, and will not remain with us long enough to show the fruits of our labor." Under these conditions, the county superinentdent explained, teachers were not inspired to do their best.[58]

Company officials were also concerned. They blamed the constant movement on the "shiftless methods of living" of the native mountaineers who were "not accustomed . . . to a continuous and sustained labor." But two-thirds of the work force in southern West Virginia had come from outside the state, and they, too, were constantly moving. One observer noted that 50 percent of black miners, most of whom were from the South, changed residence "once a year" and that the European immigrant miners moved "so frequently as to make a conclusion a mere guess."[59]

Although some miners, usually the more experienced ones, did leave southern West Virginia for other coal fields or other forms of work, the great majority remained in the area. Their explanations for staying are significant. Foremost was "the lure of the mines," the tenacious idea of once a miner, always a miner. Ascencion Aranjo, a native Mexican who spent forty years digging coal in southern West Virginia, said, "I enjoyed working in the mines, I had no desire to leave it." Other elderly miners explained that they did not expect conditions to be any better in the cities, that they "feared" the cities, and that they were repulsed by the idea of working on an assembly line. Hence, the miners preferred the miners' freedom to factory discipline. A black miner in southern West Virginia told Goodrich that he, like many other miners, had returned to the mines after a trial at factory work because "in the mines, the supervisors, they don't bother you none."[60] Although the miners did not bind themselves to any particular company or company town, they were binding themselves to a region and to mining as a way of life and work.

Geographic mobility in southern West Virginia was primarily a phenomenon of movement from company town to company town. Like other American workers, the miners moved for economic improvement; they sold their bodies and skills to the highest bidder. Indeed, the coal companies, always in need of good miners, often procured workers from other companies by offering them bonuses and individual contracts. Such practices were detrimental to the formation of a union mentality among the miners; later the UMWA was careful to write into its contracts provisions condemning the "voluntary paying of more than the contract price, either by bonuses or otherwise, which is done ordinarily for the purpose of enticing employees from other mines."[61]

The miners moved for other reasons. Sydney Box, for example, living in Glen White, moved to Glen Rodgers to increase his wages. But his wife was unhappy in Glen Rodgers because the town was dirtier and the mines were more dangerous. So the family returned to Glen White, although it meant a return to lower wages. Immigrant miners often sought company towns with the best schools rather than highest wages so that their children could learn English. Immigrant miners also commonly moved about to escape a particular coal company's prejudice against the foreign-born miner.[62]

The miners moved not only for economic improvement, but also because of discontent with a company's racial policies or educational facilities and because of discrimination. Hence, the early miners, even before the union, did have a form of protest—they packed their bags and left. The Children's Bureau observed that where there was no union, the miners had "no redress from conditions which may be intolerable, except to move to another camp"[63]—and they moved and moved again.

While effective as a form of protest on an individual basis, geographic mobility was detrimental to unionism, inhibiting a common perspective and a need for a unified response. As a UMWA District 17 contract acknowledged, it "create[d] discord and disorder" in the southern West Virginia coal fields.[64]

Paradoxically, the miners' mobility would ultimately produce a strong, collective mentality as it made a single, gigantic community out of the five coal fields in southern West Virginia and the hundreds of isolated company towns scattered across them. "In each community each individual knew everyone else," Fred Mooney recalled. "From 1917 to 1925 one could go to Cabin Creek Junction and ask anyone who happened to be present, 'Where does George Carr live now?' 'Why he still lives in Kayford,' would be the answer. . . . This knowledge of one's neighbors and their whereabouts over hundreds of square miles was prevalent in all communities throughout southern West Virginia."[65] It is surely no coincidence that it was during these years that massive, violent resistance broke out throughout southern West Virginia.

Before that time, when a common idea existed—for example, the belief that the miners had the right to regulate their own working day—the coal companies had to bend or lose a major portion of their work force. As a McDowell County coal operator confessed, the miners regulated their own work hours and the company did not object for fear the miners would pack their bags and move to other towns. For the most part, however, until 1912, the miners were concerned primarily with individual issues, complaints, and aspirations; consequently, their protests were individualistic.

The miners' failure to participate in the UMWA strikes and their refusal to join the miners' union illustrate this individualism and the evolution of a working class that possessed wants, needs, and goals that diverged dramatically from the UMWA and its policies. In 1894 the UMWA called a nationwide coal strike in an effort to halt a series of wage reductions in the nation's coal fields, which had been caused by the business depression of the 1890s and an overproduction of coal. The nonunion coal diggers of southern West Virginia refused to suspend work. As a result, trainloads of nonunion West Virginia coal poured into the Midwest and broke the national strike, and the UMWA was left destitute and demoralized.[66]

After that disaster, the UMWA became very aware of the need to organize southern West Virginia. Not only could the state's nonunion workers break a national coal strike, but also they threatened the jobs and wages of mine workers in the organized coal fields. The major coal consumers throughout the country, largely because of the strike, were turning to the superior, cheaper, nonunion West Virginia coal. Consequently, the UMWA regarded West Virginia as the "keystone" of its next national strike.[67]

The continuing business depression of the 1890s and the increased competition of West Virginia coal had forced coal operators around the country, especially in the Midwest, to continue wage reductions. On July 2, 1897, the UMWA issued another call for a nationwide strike in an attempt "to prevent any further reduction [in wages] from taking place." This call met with unprecedented success in nearly all the coal fields in the United States, as tens of thousands of coal miners, unorganized as well as organized, laid down their picks and shovels—except in southern West Virginia, which quickly became the "crucial point" in the struggle.

Gompers declared, "The success of this strike depends on one factor . . . that the miners of West Virginia will suspend operations."[68]

Labor leaders from all over the United States poured into the mountain state to organize the miners and to get them to support the 1897 strike. Gompers himself spent several weeks in the Kanawha and New River districts. He was joined by, among others, James Sovereign, grand master workman of the Knights of Labor; Henry Lloyd, the president of the United Brotherhood of Carpenters; James O'Connell, president of the International Machinists Union; and Eugene V. Debs, founder of the American Railway Union. The UMWA sent its president, Michael Ratchford, and that fiery angel of the miners, Mother Jones, in addition to union organizers Richard L. Davis and Chris Evans. A participant later asserted, "Love for the principles of trade unionism was never better illustrated than was shown in the miners' late conflict" in West Virginia.[69]

The coal establishment in West Virginia was prepared for organized labor's invasion of the state. Organizers were driven out of the company towns by company police and arrested, jailed, and fined by the local authorities for criminal behavior such as violating the Sabbath. The coal operators obtained injunctions from the courts that banned organization efforts.[70]

Ratchford called for a general meeting of prominent trade union officials from all over the country in Wheeling, West Virginia, hoping that such a "war council" would place national pressure on the state government and courts that were denying the miners and labor organizers their constitutional rights.[71]

The proposal was endorsed by Gompers, and the labor leaders met on July 27. Represented were the executive councils of the AFL and the Knights of Labor, the UMWA, and twenty-eight other national and international unions. The war council sent letters and then a committee to the governor of West Virginia to protest the violation of civil liberties and to demand the rights of free speech and free public assembly for the union organizer. It issued an appeal to all trade unions and the American people to hold mass meetings to rally support for the strike and to pressure the courts to withdraw their injunctions.[72]

The American workers and people responded more favorably than the West Virginia government to these requests. Labor unions sent thousands of dollars and more organizers to West Virginia. "Indignation mass meetings" were held throughout the country to present the miners' case to the public and to condemn the courts' interference with the constitutional rights of the organizers and miners in West Virginia. Exuberant over the popular support for the strike, Ratchford optimistically wrote to Gompers, "The downtrodden miners of West Virginia will take heart, and, as if by magic, they will stand erect and assert their rights as free men."[73] The magic, however, never appeared, and the miners never took heart, much less stood erect. The strike found more support in distant places than among the coal diggers in West Virginia.

The only success that the UMWA's strike call had in West Virginia came in McDowell County, where about 1,000 miners, many of whom had belonged to the Knights of Labor, suspended work. These men were quickly discharged and evicted from their company houses. Some violence occurred, and a few of the operators requested state troops. However, the sheriff of McDowell County decided to handle the strikers himself, probably because there were more deputy sheriffs than strikers in the county. Davis, whom the UMWA had sent to organize the black miners in the county, wrote that trying to organize miners in the face of the county sheriff, deputies, and company police "was taking one's life in his hands. . . . While we never had any injunctions issued against us, we had men and winchesters against us which were in most cases just as effective." Company guards chased and shot at then UMWA Vice-President (later President) John Mitchell, who finally escaped certain death by swimming an "icy mountain stream."[74]

Throughout the rest of West Virginia, the miners remained at work, uninterested in the union's appeals. By August 2, 1897, AFL organizer Frank Weber was already writing to Gompers that the strike in southern West Virginia was doomed: "The movement is somewhat of a see-saw, so long as the organizers are there the men quit . . . the next day the operators or their henchmen visit the homes of the miners and make large promises to the wives and hold out a rosy view of the future for them to secure the influence on their husbands."[75] The miners remained at work and outside the union.

The success of the 1897 strike in the Midwest, however, filled the UMWA's depleted treasury, increased the union's membership tenfold, and, most important, led to the development of an agreement system. The operators of Ohio, Illinois, Indiana, and western Pennsylvania agreed to meet with UMWA officials and to determine wage rates and work conditions by joint conference. The resulting Central Competitive Field served as the basis of all future UMWA contracts. In return, the operators obtained a promise from the UMWA officials to organize the southern West Virginia coal fields. Because these miners had refused to strike, the southern West Virginia operators refused to participate in the Central Competitive Field, and, consequently, the West Virginia coal operators, already underselling other coal fields, were in an even better competitive position than before, as they were not subject to union contracts, union wage scales, union-called strikes, and other union restrictions. The UMWA and the operators in the Central Competitive Field knew that "without West Virginia the joint conferences would either remain on the defensive or collapse."[76]

Unable to interest the southern West Virginia miners in the UMWA, the union officials turned to external pressure to force the state's coal operators into signing a union contract. In January 1898 Ratchford declared a national boycott against West Virginia coal, requesting organized labor to withhold its patronage and "to refuse to handle such coal for shipment." The ambitious effort met with some success, as various trade unions throughout the country, as well as UMWA local unions, endorsed the boycott. The Building Trades Council of Cincinnati, following an address by a UMWA official on the "degrading conditions that exist among the miners of West Virginia," agreed to "patronize coal" only from areas "where miners live in a manner befitting an American citizen." The boycott also found support outside of union circles. For example, all of the forty-six coal dealers in Dayton, Ohio, who sold West Virginia coal, agreed to the ban. The state of Michigan agreed to buy only union coal instead of the nonunion coal from West Virginia that it had been using. The UMWA also attempted, without success, to dissuade the U.S. Navy from buying West Virginia coal.[77]

The boycott failed. Moral persuasion and appeals could not overcome the advantages of the superior, cheaper West Virginia coal. Within two years, West Virginia coal was again entering the midwestern markets in bulk.

With the failure of the boycott, the UMWA then returned to the problem of trying to persuade the southern West Virginia miners to join the union. In 1901 the union launched a major organizing drive in the region, spending thousands more dollars and sending an army of organizers into the southern portion of the state. To the surprise of the organizers, the coal companies offered little resistance. Walker, an organizer from Illinois, wrote Mitchell, "They [the coal companies] have always opposed organization but have always had everything their own way so they feel secure."[78]

The operators of the coal companies knew their miners. The organizers often had to hold their meetings in saloons because the miners would not leave them to attend meetings. When the miners did attend, their lack of interest was even more apparent. Evans reported that on several occasions he had persuaded the miners to join the union, but then the company superintendent appeared and simply talked the men into "returning to work without the union, and I was left to mourn." Walker explained that he managed to enlist several miners, but once the men discovered that they had to pay initiation fees, they quickly lost interest.[79]

Unionism without either financial or physical cost seemed the only grounds on which the miners were willing to join the UMWA. An organizer in McDowell County explained that the miners in his area were "too timid" to show up for the meetings. Another organizer related that he had established a local at Simmering, McDowell County, but it dissolved because none of the miners was willing "to take the local offices" for fear of being discharged, evicted, and blacklisted.[80]

The miners knew their operators. When local unions were formed and miners took offices in the locals, the coal companies reacted swiftly and harshly. "The truth is," a more successful organizer wrote Mitchell, "that as fast as men organize, the companies are discharging them and turning them out of their houses."[81]

The coal companies also reacted against the organizers when they became successful. Most of the organizers were threatened, some of them were beaten, and nearly all of them had their efforts halted by injunctions or were jailed, often for holding organizing meetings on Sunday and thereby violating the Sabbath. Furious over the injustice and outrages committed against the miners and organizers, Walker attempted to press charges against the coal companies and local law officers. The miners quickly told him that such efforts would be futile because "the court is practically controlled by the operators . . . and surely the supreme court is in the same fix."[82]

Officials at the UMWA convention in 1902 reported that the drive in 1901 succeeded in establishing eighty local unions and in enrolling 5,000 miners.[83] Both of these figures seem exaggerated, however, especially when one reviews the correspondence between Jones and Mitchell during the final days of the organizing campaign. During the first days of the convention, a UMWA official assailed the southern West Virginia miners for their failure to take part in the 1897 strike and for their refusal to join the union in subsequent years. The quick-tempered Jones exploded, "We are sent out to raise our craftsmen in the public mind, and not lower them." The official who made such public pronouncements was "as ignorant as the West Virginia miners." Jones demanded that any further announcements about their organizing efforts in West Virginia be favorable toward the miners. Mitchell's agreement to comply with this directive makes the convention's figures somewhat suspect.[84]

Furthermore, the UMWA organizers' gloom during the drive suggests that they were not meeting with the success that had been claimed at the convention. After six months in the area, Evans wrote Mitchell, "I am under the impression that our work here will be largely a repetition of former years." Toward the end of the campaign, Jones wrote to Mitchell, "I am forced to the conclusion that the best interests of the West Virginia miners will be served by teaching them that they must, to some extent, depend upon themselves." In April 1902, she wrote Mitchell that "these people must be made to know this is their fight and that we have no nursing bottle for them." A sympathetic Mitchell responded, "We cannot expect to do all the work ourselves, or see it done even in our lives, and I hope that you will not destroy your health and usefulness in the labor movement by overexerting yourself." The support for the miners' union was still so minimal that the organizers warned Mitchell not to call a strike in the state then because not enough miners would support it "to seriously cripple" the coal operators.[85]

The organizers' reports might have been the reason that Mitchell opposed carrying the 1902 anthracite (hard coal) strike into West Virginia—a position that earned him the wrath of radicals then and of labor historians today. Contrary to the desires of their president, the delegates to the 1902 convention favored a general sympathetic strike and wanted it to include West Virginia. If the West Virginia miners were not called out, the delegates feared that West Virginia coal would be sent east, breaking the hard coal strike and allowing the West Virginia operators to capture the anthracite markets. Consequently, the delegates overruled Mitchell and ordered the West Virginia miners to cease work. Thus, West Virginia entered the 1902 coal strike, a strike for which it was ill prepared and one that the state's miners later admitted was a "mistake."[86]

Several thousand West Virginia miners, mostly in the New River field, responded to the strike call. Again, the coal operators responded by discharging, evicting, and blacklisting. Again, the operators went to the courts and obtained sweeping injunctions against the union, union organizers, and striking miners. Again, union officials and striking miners were arrested, jailed, and fined. Warrants were issued for the arrest of UMWA Secretary-Treasurer William B. Wilson (later a U.S. secretary of labor) before he entered the state. The operators also resorted to violence. Jones later recalls, "Men were shot. They were beaten. Numbers disappeared and no trace of them was found."[87]

Although several thousand miners had struck, too many southern West Virginia miners remained at work to "seriously cripple" the coal industry, just as Jones and Walker had predicted. Except in Kanawha County, where the local miners gained a contract, the coal operators made no concessions, and the UMWA lost all the local unions that it had scattered about the state. The UMWA's failure in West Virginia in 1902 balanced its success in other coal fields. Within two years, the booming business of nonunion West Virginia coal forced the UMWA to accept general wage reductions in the organized coal fields in order to save its agreement systems.[88]

The foothold that the UMWA gained in Kanawha County proved fragile and illusive. Although the miners had a contract, the union continuously lost ground during the next years as the miners with a union seemed no more interested in unionism than they had before. Mooney, who had joined the UMWA at that time in Kanawha County, remarked that he "did not know what for nor why but everyone was 'jinin' the union, so why not I? Several years elapsed before an inkling of what was meant by unionism began to draw upon me." Until that time, "I plugged along, paying my dues grudgingly and occasionally attending a meeting."[89]

In 1904 the refusal of coal operators to grant the demands of District 17 officials for a closed shop and checkoff for the union miners on Cabin Creek in Kanawha County led to a strike. In less than ten days the miners returned to work, on a nonunion basis. During the financial panic of 1907, union miners in parts of Kanawha County managed to fight a wage cut, although the UMWA's weakness was illustrated when many miners quit the union and returned to work early in the contest when the coal companies began evicting strikers from company houses.[90]

In an effort to maintain and rebuild the UMWA in Kanawha County between 1907 and 1910, union officials from the district and national headquarters held numerous organizing meetings and sponsored picnics and other social activities throughout the county to arouse the miners' interest. These efforts were fruitless. District 17 officials blamed the continuing business depression for the loss of membership and local unions. They claimed that the poor paying, irregular work in the area prevented miners from being able to pay initiation fees and monthly dues. One local official wrote that the coal trade has been "so dull and run so slack the miners at best live a hand to mouth existence . . . men cannot fight the money power on empty stomachs." Consequently, District 17 officials lowered their initiation fees and monthly dues. This move also failed; the problem was not only the miners' lack of money—they simply did not care to support the union. A union miner in Kanawha County complained to the *UMWJ* that "a large per cent of the men working under the union contract refuse to pay dues . . . .because they know that with the great number of non-union miners surrounding this district, that the offices are unable under the contract to force them into the union." In July 1909 the pro-union miners in Kanawha County capitulated, passing a resolution that asked for the Brotherhood of Railway Trainmen to use their "moral influence" to organize the Cabin Creek coal field. Their efforts had failed.[91]

Despite their oppression and exploitation, the miners had no desire to join the union that the UMWA offered. Between 1897 and 1910 the UMWA had spent nearly a million dollars on organizing southern West Virginia, sent in scores of organizers, and conducted various types of boycotts; District 17 officials sponsored picnics and meetings and had lowered initiation fees and monthly dues. Yet most miners remained outside of the union, and those who did belong to the UMWA were "grudgingly paying their dues" and only "occasionally attending a meeting."

Although the southern West Virginia miners had shown little interest in the appeals of UMWA organizers, the coal establishment was alarmed by the threat of unionization. Three times the operators had watched the UMWA invade their territories and attempt to gain the allegiance of the state's miners. Twice they had witnessed this union call for national boycotts in an effort to cripple their business. In 1902 they saw this force attempt to persuade the miners to strike in the interests of another coal field to keep West Virginia coal from capturing the eastern markets. Unionization threatened not only their feudalistic controls over their workers, but also wanted to place their work forces in the hands of a hostile power.

The coal operators were now determined to prevent the UMWA from again invading the state. Believing, as did Justin Collins, that "if the coal operator undertakes to fight the union, he will have to fight force with force," more and more operators formed police barricades against these invaders by the hiring of Baldwin-Felts detectives. By 1910 Baldwin-Felts guards could be found in nearly every company town in southern West Virginia. The principal work of these guards, according to the head of the agency, Tom Felts, was "to protect the operators against organized labor."[92]

In protecting the coal establishment from union agitators, the Baldwin-Felts guards were effective and brutal. Whereas in earlier years, during nonunion periods, national organizers moved relatively unmolested around the state, by 1907 they were surrounded and harrassed by Baldwin-Felts guards from the time they boarded a train in Cincinnati headed for West Virginia. The arrival of UMWA organizers often meant headlines for local labor newspapers. The February 10, 1910, issue of *The Labor Argus* read:

> HORRIBLE BUTCHERY OF REPRESENTATIVES OF UNITED MINE WORKERS BY BRUTAL AND BLOODTHIRSTY "GUARDS" IN THE NEW RIVER FIELD.

The September 8, 1910, edition of the paper headlined:

GUARDS BRUTALLY ASSAULT UNION MEN; TWO MEMBERS OF THE INTERNATIONAL EXECUTIVE BOARD OF UNITED MINE WORKERS OF AMERICA BEATEN UP BY COAL COMPANY THUGS.

The *UMWJ* stigmatized West Virginia as the "Black Belt"—"we do not mean that they employ nothing but colored men, but that it is a region where it is not safe for an organizer to . . . let his business be known." By 1912 UMWA presidents openly admitted that they no longer encouraged organizers to enter the state because of the Baldwin-Felts guards.[93]

The vigilance of the Baldwin-Felts guards against union agitators carried over into the everyday, internal affairs of the company town. If a miner showed too much independence or became inquisitive or complained, he was suspected of being a union agitator and might be beaten and driven out of the town. The guards forcibly prohibited the assembly of more than three miners at night, a practice they called "bunching," for fear that one of the miners might be an organizer preaching unionism. Describing the guards' activities, a miner recollected, "They rode horses and wouldn't let anyone they didn't like walk on the railroad tracks. Worse though, you couldn't go into the office and tell about mistakes or complain about the store. They would just kick you off the porch."[94]

These guards gained a national reputation for their beating, maiming, and murdering of union miners and organizers. In fact, Edward Levinson in his book "*I Break Strikes*," a general study of industrial spy and strikebreaking agencies throughout the United States, awarded the Baldwin-Felts Agency the "blue ribbon for wanton, cold-blooded murders." The miners claimed that the guards had kicked the fetus out of a miner's pregnant wife and had cut the breasts off of another woman.[95]

Miners throughout the country reported that they were appalled by the activities of the Baldwin-Felts guards in West Virginia, declaring that guards in Pennsylvania, Illinois, and other coal-producing states were "nothing like those in West Virginia." Indeed, in the speech that precipitated the massacre, by miners, of nineteen mine guards at Herrin, Illinois, a miner declared, "We must show the world this ain't West Virginia." Because of the general use of Baldwin-Felts mine guards in Colorado and West Virginia, miners referred to these two states as "Hell and Repeat."[96]

The southern West Virginia miners hated the Baldwin-Felts guards more than did any other group of miners. During the Paint Creek–Cabin Creek strike, when the miners unleashed their wrath against these guards, a miner told a journalist, "I never had to kill a man and hope never to be compelled to kill one, but I would kill a dozen of these guards as I would kill a dozen rats if they should attempt to lord it over us as they have been accustomed to do." When Mooney informed the people in District 17 headquarters that seven Baldwin-Felts guards had been killed in a gunfight with the pro-union forces in Matewan, "almost all seemed pleased." During the Paint Creek–Cabin Creek strike, a newspaper reporter correctly analyzed that "no class of men on earth are more cordially hated by the [southern West Virginian] miners than these same guards who are engaged to protect them from outsiders."[97]

The irony of the situation was that the very effectiveness of the Baldwin-Felts guards helped to create the class consciousness that the miners needed to make a commitment to unionism. Instead of holding down the union movement, as contemporaries and historians claim, the activities of these guards made it relevant. These men were hired to guard against agitation, and they did so effectively and brutally. Indeed, so effective and so brutal were their actions that they stimulated class hostility among the coal diggers and convinced them of the need for collective security and collective action. And it was the Baldwin-Felts guards who daily exposed and represented the injustices of the economic setup that by 1912 the miners no longer were willing to tolerate.

The presence and activities of the Baldwin-Felts guards also caused the miners to question the political system that not only tolerated them but also endowed the guards with authority from the state. Referring to the brutal beatings of two UMWA organizers in the area, a UMWA District 17 resolution declared that "the Governor of this state, the various county officials and the lawmakers of this state permit men to be beaten into insensibility, crippled and maimed for life without cause except . . . that they are members of the United Mine Workers."[98]

In his message to the West Virginia State Legislature in 1907, Governor William Dawson had blasted the Baldwin-Felts mine-guard system. The miners were not fooled by his rhetoric; although they approved of the speech, they pointed out that "Mr. Dawson continued them in the office which he had appointed them and cloaked them with the protection of the state."[99]

Furthermore, by sealing off the area to UMWA organizers, the Baldwin-Felts guards insured that any uprising of the southern West Virginia coal diggers would begin with the miners themselves and guaranteed that the union spirit that evolved in southern West Virginia—and that the ideas and ideals the miners formulated about unionism and about what unionism should represent—would reflect the character and values of the rank-and-file miners, not the international union.

## NOTES

1. This journal article is reprinted in *National Labor Tribune*, Apr. 13, 1899.
2. Selig Perlman and Phillip Taft, *History of Labor in the United States*, 4

(New York, 1966), 326. Labor unions had been active among the few thousand coal miners in southern West Virginia prior to the 1890s. The National Protective Union, Miners' and Laborers' Benevolent Association, the National Federation of Miners, and especially the Knights of Labor had made numerous attempts to organize the miners and had met with "indifferent success." In 1880 a brief, local strike in Raleigh County even brought the Molly McGuires into southern West Virginia. For the next two decades, a number of miners, particularly in McDowell and Kanawha counties, remained in those unions and, against increasing obstacles, fought tenaciously for unionism in southern West Virginia. But with the transformation of southern West Virginia into a major coal-producing region, the influx of tens of thousands of new miners of nonmining and nonunion backgrounds, and the rise of the coal establishment, nearly all traces of unionism disappeared. By 1890 neither the West Virginia miners nor the coal operators were participating in interstate joint conferences. In the year that saw the birth of the UMWA, southern West Virginia, designated UMWA District 17, was unorganized. Ibid., 326–27; Kenneth Bailey, "Hawks Nest Coal Company Strike, January, 1880," *West Virginia History* 30 (July 1969): 625–34; Charles Ambler and Festus Summers, *West Virginia: The Mountain State* (Englewood Cliffs, N.J., 1958), 445.

3. Elsie Gluck, *John Mitchell* (New York, 1927), 243–44.

4. Letter to the editor, *UMWJ*, Sept. 22, 1910, 2.

5. McAlister Coleman, *Men and Coal* (New York, 1969), 78; Samuel Gompers, "Russianized West Virginia," *American Federationist* 20 (Oct. 1913): 826; *UMWJ*, July 15, 1909, 1, 2, 4.

6. Laurence Leamer, "Twilight of a Baron," *Playboy*, May 1973, 120.

7. The geographic and economic background of the black miners and their reasons for migrating to West Virginia are discussed in chap. 3.

8. Andrew Roy, *A History of the Coal Miners of the United States* (Columbus, Ohio, 1907), 328; Arthur Suffern, *Conciliation and Arbitration in the Coal Industry of America* (Boston, 1915), 68; W. F. Larrison, letter to the editor, *UMWJ*, Jan. 9, 1908, 1. See also C. A. McNeil, letter to the editor, ibid., Mar. 2, 1905, 2.

9. The countries of origin of the immigrant miners are compiled from W.Va. State Department of Mines, *Annual Reports, 1900–7*. Coleman, *Men and Coal*, chap. 4, "Founding Fathers." For the role and importance of British miners in the formation of miners' unions, see Roy, *History of Coal Miners*, 60–65; John Brophy, *A Miner's Life* (Madison, Wis., 1964), 73; Frank Warne, *Union Movement among Coal Mine Workers*, U.S. Bureau of Labor Bulletin no. 51 (Washington, D.C., 1904), 380.

10. S. S. Morrison to John Mitchell, July 10, 1901, John Mitchell Papers, Catholic University of America, Washington, D.C.

11. *UMWJ*, Mar. 10, 1910, 3; U.S. Immigration Commission, *Report of Immigrants in Industry*, 1 (Washington, D.C., 1911), 535; Fred Barkey, "Socialist Party of West Virginia from 1898 to 1920" (Ph.D. diss., University of Pittsburgh, 1971), 41.

12. "Boycott of a State," *Engineering and Mining Journal*, Oct. 12, 1907, 698, and Evelyn Harris and Frank Krebs, *From Humble Beginnings: West Virginia State Federation of Labor, 1903–1957* (Charleston, W.Va., 1960), 19.

13. Compiled from W.Va. State Department of Mines, *Annual Report, 1907*.

14. Henry Stephenson, letter to the editor, *UMWJ*, Aug. 11, 1898, 5, and A. E. Harris, letter to the editor, ibid., Oct. 15, 1908, 1. See also Charles Syden-

shicker, letter to the editor, ibid., Mar. 2, 1905, 4.

15. Anna Rochester, *Labor and Coal* (New York, 1931), 183; Winthrop Lane, *Civil War in West Virginia* (New York, 1921), 102; Roy, *History of Coal Miners*, 352–53. For other union officials who made the switch, see *UMWJ*, Sept. 1, 1924, 11; testimony of John Laing, U.S. Congress, Senate Committee on Education and Labor, *Conditions in the Paint Creek District, West Virginia*, 63rd Cong., 1st sess., 3 vols. (Washington, D.C., 1913), 2:1661 (hereafter cited as *Paint Creek Hearings*). Gompers, "Russianized West Virginia," 282.

16. Walker to John Mitchell, Dec. 3 and 7, 1910, and Jones to Mitchell, Dec. 2, 1901, and May 6, 1902, Mitchell Papers.

17. Roy, *History of Coal Miners*, 279; Harris and Krebs, *From Humble Beginnings*, 17–20; Coleman, *Men and Coal*, 53; testimony of John Nugent, *Paint Creek Hearings*, 2:2076; Gompers, "Russianized West Virginia," 827–28.

18. Harris and Krebs, *From Humble Beginnings*, 19–20; testimony of John Nugent, *Paint Creek Hearings*, 2:2076–78; Gompers, "Russianized West Virginia," 827–28; Suffern, *Conciliation and Arbitration*, 74; File 018, item 22201, Record Group 70, Records of the U.S. Bureau of Mines, National Archives, Washington, D.C.; John Mitchell to Commissioner of Immigration D. J. Keefe, May 24, 1917, Mitchell Papers.

19. W. F. Larrison, letter to the editor, *UMWJ*, Jan. 9, 1908, 1.

20. W. P. Tams, *The Smokeless Coal Fields of West Virginia* (Morgantown, W.Va., 1963), 46–47.

21. David McDonald and Edward Lynch, *Coal and Unionism* (Silver Spring, Md., 1939), 30, and Roy, *History of Coal Miners*, 333.

22. Carter Goodrich, *The Miners' Freedom* (Boston, 1925), 41–43, 103.

23. L. C. Anderson, letter to the editor, *Outlook* 82 (Apr. 28, 1906): 861–62.

24. U.S. Department of Labor, *Hours and Earnings in Anthracite and Bituminous Coal Mining*, Bureau of Labor Statistics Bulletin no. 316 (Washington, D.C., 1922), 1; U.S. Department of Labor, *Wages and Hours of Labor in Bituminous-Coal Mining: 1933*, Bureau of Labor Statistics Bulletin no. 601 (Washington, D.C., 1934), 3; Clarence Bonnett, *History of Employers' Associations in the United States* (New York, 1956), 463; testimony of William Warner, *Paint Creek Hearings*, 2: 2062–66; Bituminous Operators' Special Committee, *The United Mine Workers in West Virginia*, Statement Submitted to the U.S. Coal Commission, Aug. 1923 (n.p., n.d.), 5; William Coolidge, *Brief in Behalf of Island Creek Coal Company* (Boston, 1921), 10–11; Roy, *History of Coal Miners*, 333; Perlman and Taft, *History of Labor*, 20; Barkey, "Socialist Party," 36; Suffern, *Conciliation and Arbitration*, 72–73.

25. Walter Thurmond, *The Logan Coal Fields of West Virginia* (Morgantown, W.Va., 1964), 65; interview with Marion Preece, Delbarton, W.Va., summer 1975; Suffern, *Conciliation and Arbitration*, 72; Roy, *History of Coal Miners*, 333; testimony of William Warner, *Paint Creek Hearings*, 2:2062–66. See also chap. 1.

26. *Report of the West Virginia Mining Investigation Commission, appointed by Governor Glasscock on the 28th Day of August 1912* (Charleston, W.Va., 1912); U.S. Bituminous Coal Commission, *Majority and Minority Reports, 1920* (Washington, D.C., 1920), 44–45; Sam Boykin, letter to the editor, *UMWJ*, Feb. 15, 1920, 8. For the problem of irregular work in the bituminous coal fields, see Louis Bloch, *The Coal Miners' Insecurity* (New York, 1922); F. G. Tyron, "The Irregular Operation of the Bituminous Coal Industry," *American Economic Re-*

*view* 11 (Mar. 1921): 57–73; F. G. Tyron and W. F. McKenny, "The Broken Year of the Bituminous Coal Miner," *Survey Graphic*, Mar. 25, 1922, 1012.

27. A. B. Smoot, letter to the editor, *UMWJ*, June 4, 1896, 5; Lane, *Civil War*, 28–30; *UMWJ*, Nov. 29, 1917, 17. See also Sam Boykin, letter to the editor, ibid., Feb. 15, 1920, 8.

28. William Warner to Attorney General A. Mitchell Palmer, Nov. 25, 1919, Straight Numerical File, item 205194–50–33, Record Group 60, General Records of the Department of Justice, National Archives; Walker to Mitchell, July 9, 1901, Mitchell Papers. Also see testimony of Elmer Rumbaugh, *Paint Creek Hearings*, 1:199–203.

29. Interview, Fred Barkey with Elizabeth Franklin Matheny, Fairmont, W.Va., quoted in Barkey, "Socialist Party," 49.

30. Quoted in Harold West, "Civil War in the West Virginia Coal Mines," *Survey* 30 (Apr. 5, 1913): 46.

31. Harris and Krebs, *From Humble Beginnings*, 73.

32. West Virginia Coal Association, "Gardens in Mining Towns," *West Virginia Review*, Oct. 1924, inside cover; interview with Walter Hale, Eckman, W.Va., summer 1975; Nettie McGill, *Welfare of Children in the Bituminous Coal Communities in West Virginia*, Children's Bureau, U.S. Department of Labor (Washington, D.C., 1923), 52–54.

33. Interviews with Elsie Turner, Sophia, W.Va., and Joe Pauver, Gary, W.Va., both in summer 1975. The miners without gardens appear to have been, for the most part, single miners who did not have a family to take care of it. Interviews with David Cross, Alpaca, W.Va., and Mary Johnson, Wilcoe, W.Va., both in summer 1975.

34. Interviews with Bill Main, New Richmond, W.Va., and Clyde Scarberry, Keystone, W.Va., both in summer 1975. Some of the gardens in the company towns were valued at $150; see *Coal Age* 3 (Aug. 22, 1914): 312.

35. Interviews with Clyde Scarberry, Keystone, W.Va., J. H. Vernatter, Robinette, W.Va., Willie Richardson, Sabine, W.Va., Bill Mullins, New Richmond, W.Va., Mrs. Clyde Tolliver, Sabine, W.Va., and John Stowall, Sabine, W.Va., all in summer 1975. See also McGill, *Welfare of Children*, 52–54.

36. Letter to the editor, *UMWJ*, May 28, 1896, 1.

37. W. A. Cleve to Lewis, Nov. 4, 1919, District 17 Correspondence Files, UMWA Archives, UMWA Headquarters, Washington, D.C. For the dependence on gardens during work stoppages, see P. H. Kelly to Justin Collins, Mar. 11, 1910, Justin Collins Papers, West Virginia University Library, Morgantown, W.Va.

38. Herbert Gutman discusses the "preindustrial carry-over" of the making and consumption of alcohol. "Work, Culture, and Society in Industrializing America," *American Historical Review* 58 (June 1973): 558–60. Tams, *Smokeless Coal Fields*, 61; George Korson, *Coal Dust on the Fiddle* (Philadelphia, 1943), 60; *Coal Age* 4 (Nov. 15, 1913): 741–42; Jarius Collins to Justin Collins, Feb. 28, 1901, and a mine superintendent to Justin Collins, Mar. 27, 1911, both in Collins Papers; Margaret Jordan, "A Plea for the West Virginia Miner," *Coal Age* 6 (Dec. 5, 1914): 914–15; West Virginia Mine Superintendent, "Payday Drinking," ibid. 4 (Oct. 4, 1913): 478–79.

39. Robert Holiday, *Politics in Fayette County* (Montgomery, W.Va., 1958), 160; Leamer, "Twilight," 120; *Coal Age* 4 (Nov. 15, 1913): 741–42.

40. Interviews with Oscar England, Mullins, W.Va., John Schofield, Alpoca,

W.Va., and Herbert Lee McKinney, Herndon, W.Va., summer 1975. See also *Charleston Gazette*, Nov. 3, 1919.

41. Interview with Tom Canady, Anawalt, W.Va., summer 1975, and W.Va. State Department of Mines, *Circular to District Mine Inspectors*, Feb. 22, 1922.

42. *Coal Age* 8 (Aug. 29, 1918): 408; "Payday Drinking," 478–79. See also *Charleston Daily Mail*, Apr. 8, 1921, and *Coal Age* 8 (Feb. 16, 1918): 557, 8 (Mar. 23, 1918): 556, and 8 (Nov. 19, 1918): 347.

43. Korson, *Coal Dust*, 61–62.

44. Edgar Collis, "The Coal Miner: His Health, Diseases and General Welfare," *Journal of Industrial Hygiene* 7 (May 1925): 235–36, and Ira Shaw, "Welfare Work among Miners," *Coal Age* 1 (July 5, 1913): 21.

45. Interviews with five individuals, all of whom wish their names withheld. The interviews were conducted in the summer of 1975 at five towns in West Virginia: Anawalt, New Richmond, Beckley, Mullins, and Matewan. See also William Warner to Governor Cornwell, Sept. 22, 1917, John J. Cornwell Papers, State Department of Archives and History, Charleston, W.Va. Moonshining is a powerful leveling force. The Bureau of Agricultural Economics of the U.S. Department of Agriculture pointed out that moonshining carried no stigma in communities where it offered one of the few sources of money. Bureau of Agricultural Economics, Bureau of Home Economics, *Economic and Social Problems and Conditions of the Southern Appalachians*, U.S. Department of Agriculture Bulletin no. 205, Jan. 1935 (Washington, D.C., 1935), 135. For other examples of miners making and selling moonshine to supplement wages, see Reginald Millner, "Conversations with the 'Ole Man,'" *Goldenseal* 5 (Jan.–Mar. 1979): 58–64.

46. *Charleston Gazette*, Aug. 3, 12, 20, 23; Nov. 12, 25; Dec. 21, 1919; Howard Lee, *Bloodletting in Appalachia* (Parsons, W.Va., 1969), 73; U.S. Congress, Senate Committee on Education and Labor, *West Virginia Coal Fields: Hearing . . . to Investigate the Recent Acts of Violence in the Coal Fields of West Virginia*, 67th Cong., 1st sess., 2 vols. (Washington, D.C., 1921–22), 2:660; William Cornwell to Governor Cornwell, Sept. 22, 1917, Cornwell Papers.

47. E. P. Thompson, "Time, Work-Discipline, and Industrial Capitalism," *Past and Present* 38 (Dec. 1969): 57; Gutman, "Work, Culture, and Society in Industrializing America," 558–60. These two and other recent social and labor historians have argued that the workers' retention of preindustrial traditions illustrated a resistance to industrialization. The struggle to maintain these traditions, they claimed, showed that workers were not molded easily or swiftly by industrialism or industrialists; workers attempted to maintain traditions and habits that gave meaning and integrity in a new, harsh, and demanding work environment. This may have been true for the nineteenth-century British workers and the European immigrants of peasant backgrounds who entered factory systems in New York, Chicago, and other urban industrial areas. But in the southern West Virginia coal fields, the southern blacks, European immigrants, and native miners, overwhelmingly of preindustrial agrarian backgrounds, were forming new styles of life and work. The traditions and habits they preserved were traditions and habits that eased, smoothed, and aided their adjustment to an industrialized way of life and work.

48. Thompson, "Time, Work-Discipline, and Industrial Capitalism," 50–57.

49. David Montgomery, "Workers' Control of Machine Production in the Nineteenth Century," *Labor History* 17 (Fall 1976): 485–508; R. Dawson Hall, "Have Mining Engineers Accepted All the Developments in Machinery for Han-

dling Coal Imply?" *Coal Age* 12 (July 7, 1921): 13; Carter Goodrich, "Nothing but a Coal Factory," *New Republic* 44 (Sept. 16, 1925): 91–93. See also Goodrich, *Miners' Freedom*.

50. For the miners' resentment of supervision, in addition to Goodrich, *Miners' Freedom*, 14, and "Nothing but a Coal Factory," 91–94, see Coleman, *Men and Coal*, 9–10; William Graebner, *Coal-Mining Safety in the Progressive Period* (Lexington, Ky., 1976), 116; H. A. Haring, "Three Classes of Labor to Avoid," *Industrial Management*, Dec. 1921, 370.

51. Tam, *Smokeless Coal Fields*, 34; Brophy, *Miner's Life*, 42; Goodrich, *Miners' Freedom*, 30.

52. Interviews with Marion Preece, Delbarton, W.Va., and Frank Blizzard, Dry Branch, W.Va., both in summer 1975. Winthrop Lane, "Black Avalanche," *Survey* 47 (Mar. 25, 1922): 1044; E. N. Clopper, *Child Labor in West Virginia*, National Child Labor Committee pamphlet no. 86 (New York, 1908), 6; M. H. Ross, "Life Style of the Coal Miner," *Appalachian Medicine*, Mar. 1971, 6, 7; Korson, *Coal Dust*, 413.

53. Ross, "Life Style," 6–7.

54. Stephan Thernstrom, *The Other Bostonians* (Cambridge, Mass., 1976), 227; interview with George Scales, Keystone, W.Va., summer 1975; McGill, *Welfare of Children*, 7–8.

55. Stevens Coal Company, Acme Mine Payroll Books, Nov. 1, 1904, to Mar. 15, 1906 (original in author's collection).

56. McGill, *Welfare of Children*, 7–8, and *Immigrants in Industries*, 7:230.

57. Brophy, *Miner's Life*, "On the Move" (chap. 5), and Boris Emmet, *Labor Relations in the Fairmont, West Virginia Bituminous Coal Fields*, U.S. Bureau of Labor Statistics Bulletin no. 361 (Washington, D.C., 1924), 62–65.

58. A. R. Dickerson, "Report of the County Superintendent of Fayette County," in W.Va. Superintendent of Free Schools, *Annual Reports, 1901–2*, 1.

59. George Fowler, "Social and Industrial Conditions in the Pocahontas Coal Fields," *Mining Magazine* 27 (June 1904): 383, and G. T. Surface, "Negro Mine Laborer: Central Appalachian Coal Field," *Annals of the American Academy of Political Science* 33 (Mar. 1908): 116–17.

60. *UMWJ*, Mar. 10, 1910, 3; A. B. Smoot, letter to the editor, *UMWJ*, June 4, 1896, 5; testimony of miner no. 2, U.S. Congress, Senate Committee of Interstate Commerce, *Conditions in the Coal Fields of Pennsylvania, West Virginia and Ohio*, 67th Cong., 1st sess., 2 vols. (Washington, D.C., 1928) 1:57; interviews with Ascencion Aranjo, Eskdale, W.Va., Marion Preece, Delbarton, W.Va., and Willie Richardson, Sabine, W.Va., all in summer 1975; Goodrich, *Miners' Freedom*, 15–16.

61. Isaac Hourwich, *Immigrants and Labor* (New York, 1972), 421; Helen Harvey, "From Frontier to Mining Town in Logan County" (M.A. thesis, University of Kentucky, 1942), 60; *UMWJ*, May 1, 1920, 9.

62. Interviews with Sydney Box, Glen White, W.Va., and Ascencion Aranjo, Eskdale, W.Va., both in summer 1975, and Ellen Murray, "Why Foreign Miners Are Restless," *Coal Age* 7 (Oct. 13, 1917): 620–21.

63. McGill, *Welfare of Children*, 7.

64. *UMWJ*, May 1, 1920, 9.

65. Fred Mooney, *Struggle in the Coal Fields*, ed. Fred Hess (Morgantown, W.Va., 1967), 142–44.

66. Suffern, *Conciliation and Arbitration*, 66–67; Warne, *Union Movement*,

389–90; Bonnett, *Employers' Associations*, 419; Perlman and Taft, *History of Labor*, 20–22; *New York Times*, June 5, 1894; Roy, *History of Coal Miners*, 85, 307; Kenneth Bailey, "Tell the Boys to Fall in Line," *West Virginia History* 32 (July 1971): 330–37.

67. Warne, *Union Movement*, 389–90; Roy, *History of Coal Miners*, 305–7; Perlman and Taft, *History of Labor*, 22.

68. Perlman and Taft, *History of Labor*, 22–25; Bonnett, *Employers' Associations*, 418. The midwestern coal operators pleaded with UMWA to postpone the strike: "The only people who are likely to benefit are the West Virginia operators who will very probably do as they did at the time of the great strike three years ago, when they secured a great extension of their trade." *Engineering and Mining Journal*, July 13, 1897, 1. Samuel Gompers to W. B. Prescott, July 19, 1897, Letterbooks of Samuel Gompers, Manuscript Division, Library of Congress, Washington, D.C.

69. Chris Evans, "The Miners' Strike," *American Federationist* 4 (Dec. 1897): 241; Roy, *History of Coal Miners*, 230–31; A. B. Smoot, letter to the editor, *UMWJ*, June 4, 1896, 5; Ratchford to Gompers, July 23, 1897, reprinted in *American Federationist* 4 (Aug. 1897): 120.

70. Gompers to Ratchford, July 19, 1897, Letterbooks of Gompers; Perlman and Taft, *History of Labor*, 23; Gompers, "Russianized West Virginia," 826.

71. Evans, "Miners' Strike," 241; Perlman and Taft, *History of Labor*, 23; *American Federationist* 4 (Aug. 1897): 120–21.

72. Gompers to James Wood, Aug. 6, 1897, to Governor Atkinson, Aug. 2, 1897, to Ratchford, Aug. 6, 1897, Letterbooks of Gompers; Perlman and Taft, *History of Labor*, 24–25; "Conference with Governor Atkinson," *American Federationist* 4 (Aug. 1897): 122.

73. "Appeal for the Miner," *American Federationist* 4 (Aug. 1897): 119; Gompers to Ratchford, July 12, July 30, 1897, to Robert Askew, July 12, 1897, to W. B. Prescott, July 19, 1897, to Ernest Crosby, July 13, 1897, to John O'Sullivan, July 13, 1897, Letterbooks of Gompers; Ratchford to Gompers, July 23, 1897, reprinted in *American Federationist* 4 (Aug. 1897): 120.

74. A. B. Smoot, letter to the editor, *UMWJ*, June 4, 1896, 6; Bonnett, *Employers' Associations*, 419–20; Floyd Parsons, "Coal Mining in Southern West Virginia," *Engineering and Mining Journal*, Nov. 9, 1907, 881–94; Jerry Bruce Thomas, "Coal County: The Rise of the Southern Smokeless Coal Industry and Its Effects on Area Development, 1872–1910" (Ph.D. diss., University of North Carolina, 1971), 241–44; Richard Davis, letter to the editor, *UMWJ*, Sept. 9, 1897, 1; Gluck, *Mitchell*, 40.

75. Weber to Gompers, Aug. 2, 1897, and Gompers to Ratchford, Aug. 2, 1897, Letterbooks of Gompers; Roy, *History of Coal Miners*, 327.

76. Perlman and Taft, *History of Labor*, 25; Suffern, *Conciliation and Arbitration*, 3; Edward Coxe, "The Competition of West Virginia with Ohio Coal," *Engineering and Mining Journal*, Oct. 8, 1898, 424; Robert Wiebe, "The Anthracite Strike of 1902: A Record of Confusion," *Mississippi Valley Historical Review* 48 (Sept. 1961): 231.

77. Roy, *History of Coal Miners*, 333–34; Warne, *Union Movement*, 393; *UMWJ*, June 23, 1898, 7, July 14, 1898, 7, Aug. 4, 1898, 7, Aug. 25, 1898, 14, and Sept. 29, 1898, 7; File 8426–Z, Record Group 80, Records of the Secretary of the Navy, National Archives; Wiebe, "Anthracite Strike," 239.

78. Evans to Mitchell, July 15, July 23, 1901, Feb. 7, 1902, Mitchell to Walker,

Dec. 8, 1901, S. S. Morrison to Mitchell, July 10, 1901, Walker to Mitchell, July 9, 1901, Mitchell Papers.

79. Mother Jones to Mitchell, Feb. 7, Mar. 14, 1902, Evans to Mitchell, July 23, Oct. 18, 1901, Walker to Mitchell, July 9, 1901, Mitchell Papers.

80. Ed Cahill to Mitchell, July 22, 1901, W. H. Crawford to Mitchell, July 15, July 20, Sept. 14, 1901, Mitchell Papers.

81. W. H. Crawford to Mitchell, July 15, July 21, Sept. 14, 1901, Mitchell Papers.

82. I. W. Coleman to Mitchell, July 22, 1901, Crawford to Mitchell, July 20, 1901, Walker to Mitchell, July 9, 1901, Mitchell Papers.

83. Thirteenth Annual Convention, UMWA *Minutes* (Indianapolis, Ind., 1902), 41. These figures have been blindly accepted by historians. See Perlman and Taft, *History of Labor*, 327, and Roy, *History of Coal Miners*, 364–65.

89. Jones to Mitchell, Feb. 7 and Mar. 6, 1902, Mitchell Papers.

85. Evans to Mitchell, July 23 and Oct. 21, 1901, Jones to Mitchell, Mar. 6 and Apr. 4, 1902, Mitchell to Jones, May 10, 1902, Walker to Mitchell, July 9, 1901, Mitchell Papers.

86. Wiebe, "Anthracite Strike," 232; Lawrence Lynch, "The West Virginia Coal Insurrection," *Political Science Quarterly* 29 (Dec. 1914): 628; Thirteenth Annual Convention, UMWA *Minutes*, 144, 55; Harold West, "Civil War in the West Virginia Coal Mines," *Survey* 30 (Apr. 5, 1913), 39.

87. Sheldon Harris, "Letters from West Virginia: Management Version of the 1902 Coal Strike," *Labor History* 10 (Spring 1969): 228–31; *Charleston Gazette*, June 7–13, 25, July 4–12, 25–30, 1902; Perlman and Taft, *History of Labor*, 328; *New York Times*, July 7, 1902; Jones to Mitchell, July 25, 1902, Mitchell Papers; Mary "Mother" Jones, *The Autobiography of Mother Jones*, ed. Mary Parton (Chicago, 1972), chap. 9; interview, William Blizzard with William Austin Billups, *UMWJ*, Mar. 1–15, 1975, 13.

88. Wiebe, "Anthracite Strike," 239–40; Harris, "Letters from West Virginia," 230–31; Perlman and Taft, *History of Labor*, 329; Charles P. Anson, "A History of the Labor Movement in West Virginia" (Ph.D. diss., University of North Carolina, 1940), 216.

89. Mooney, *Struggle*, ed. Hess, 8, 9, 15.

90. C. F. Carter, "The West Virginia Coal Insurrection," *North American Review* 198 (Oct. 1913): 463; Lynch, "Coal Insurrection," 630; Coleman, *Men and Coal*, 81; *UMWJ*, Dec. 31, 1908, 5; Roy, *History of Coal Miners*, 364–65.

91. *UMWJ*, Sept. 24, 1908, 3, Oct. 15, 1908, 3, July 15, 1909, 5, May 26, 1910, 1; G. A. McNeil, letter to the editor, ibid., Mar. 2, 1905, 5; W. J. Campbell, letter to the editor, ibid., Sept. 9, 1909; Tom Cairns to Mitchell, May 28, 1908, Mitchell Papers; Tom Felts to Justin Collins, Mar. 22, 1908, Collins Papers.

92. Justin Collins to F. M. Jackson, Aug. 25, 1908, and Felts to Justin Collins, Aug. 1, 1911, and Jan. 18, 1910, Collins Papers.

93. *Labor Argus* headlines are reprinted in Harris and Krebs, *From Humble Beginnings*, 58. Gluck, *Mitchell*, 206, and *UMWJ*, Nov. 21, 1909, 12. For T. L. Lewis, see "Report" of the President, *UMWJ*, Jan. 19, 1911, 4; for John White, see *New York Call*, Mar. 9, 1913.

94. West, "Civil War," 43; interview, Joe Mino with Fred Barkey, Leewood, W.Va., May 28, 1968, in Barkey, "Socialist Party," 49; Arthur Gleason, "Company-Owned Americans," *Nation* 110 (June 12, 1920): 794–95.

95. Edward Levinson, "*I Break Strikes*" (New York, 1935, reprinted in

*American Labor from Conspiracy to Collective Bargaining*, ed. Leon Stein and Philip Taft, Series 1 [New York, 1969]), 151; *UMWJ*, Feb. 17, 1910, 4, Feb. 24, 1910, 4; *Paint Creek Hearings*, 3:2282.

96. G. Stewart, letter to the editor, *UMWJ*, Oct. 24, 1912, 12; Coleman, *Men and Coal*, 120; Korson, *Coal Dust*, 295. In Herrin, Ill., during the 1922 coal strike, miners captured, disarmed, and murdered nineteen strikebreakers and company guards of the Southern Illinois Coal Company after it had tried to operate its mines on a nonunion basis and its guards had attacked and killed four strikers. Perlman and Taft, *History of Labor*, 483.

97. West, "Civil War," 43–45, and Mooney, *Struggle*, ed. Hess, 76. See chap. 4 for an account of the Paint Creek–Cabin Creek strike.

98. *UMWJ*, Mar. 17, 1910, 13. See also *UMWJ*, Aug. 5, 1909, 3, and Mar. 10, 1910, 3.

99. "The Guard Question," *Proceedings of the West Virginia State Federation of Labor, 1911 State Convention* (Charleston, W.Va., 1911), 44–45.

CHAPTER III

# Class over Caste: Interracial Solidarity in the Company Town

> You live in a company house
> You go to a company school
> You work for this company,
> according to company rules.
>
> You all drink company water
> and all use company lights,
> The company preacher teaches us
> what the company thinks is right.
>
> Carl Sandburg
> "The Company Town"

Although the UMWA made little headway in the southern West Virginia coal fields between 1890 and 1911, a powerful social force was already producing a collective mentality among the miners. By its contrived and rigid structure, the company town, while giving the coal operators extraordinary forms of power over the miners, precluded the development of a social and political hierarchy based on color ethnicity, that is, a caste system, within the working-class community. Its standardized living and working conditions prohibited socioeconomic competition and mobility, and its highly rigid capitalistic structure established distinct class lines, based not on an ethnicity or race, but on occupation. The company town quickly and ruthlessly destroyed old cultures and stimulated the development of a new one. The nature of the company town focused the workers' discontent, not on each other nor on a racial or ethnic group, but upon the employer—the coal operator—enabling the miners to develop that sense of group oppression necessary for class feeling and behavior. As Jeremy Brecher explains, "Class consciousness involves more than an individual sense of oppression. It requires the sense that one's oppression is a function of one's being part of an oppressed group, whose position can be dealt with by the action of the entire group."[1]

The influence of the company town in producing interracial harmony was evident in the earliest stages of the creation of the southern West Virginia work force. Under the impact of the migration of blacks, northern cities exploded in racial conflict, as thousands of blacks became economic and social competitors with whites. Between 1890 and 1900 alone, northern white workers conducted over thirty strikes protesting the hiring of black workers, and hundreds of northern labor unions declared themselves "lily-white" by prohibiting blacks from membership. Later thousands of northern white workers joined the Ku Klux Klan, and race riots between black and white workers flared in East St. Louis, Illinois, Newark, New Jersey, Philadelphia and Chester, Pennsylvania, Washington, D.C., Omaha, Nebraska, and Chicago, Illinois.[2]

In contrast, racial troubles remained conspicuously absent in southern West Virginia. Recognizing the migration of thousands of blacks into southern West Virginia, the U.S. Department of Labor reported that "the tradition of harmony and reciprocal good will remains, and intense and bitter race feelings have not developed in West Virginia."[3]

Part of the explanation lies in the type of migrant that the coal fields attracted. Black migration to the north was essentially a rural to urban movement, of farmers and sharecroppers accustomed to the southern agrarian way of life moving into industrial cities.[4] The migration into southern West Virginia, the only state below the Mason-Dixon line to have an increase in black population between 1890–1910, was a movement into a rural area, apparently equally from two sources.[5] Southern industrial black workers, as well as agrarian blacks, flooded the state's coal fields, seeking escape from southern racism. An elderly black miner recalled that he moved from Alabama to West Virginia in 1902 because "the Negro counted for nothing in the south, the south wasn't like these [West Virginia] coal camps." Similarly, a migrant black miner in Omar, West Virginia, wrote a friend in Alabama that a "collered [*sic*] man stands just as good as a white man here." A black miner in southern West Virginia wrote the *UMWJ* that he was glad he had left Alabama because "there is none so hampered as the colored toilers in the Alabama coal fields. The colored man is so badly treated and browbeaten he hardly knows which way to turn for real friendship." Southern industrialists recognized this cause for defection. The president of Bessemer Coal, Iron and Land Company, angered over the loss of his black employees, told the Chamber of Commerce in Birmingham, Alabama, "I don't blame the Negroes for leaving Birmingham. The treatment that these unfortunate Negroes are receiving from the police is enough to make them depart for Kentucky or West Virginia."[6]

The blacks who moved from the southern farms to the southern West Virginia coal fields migrated generally for economic purposes. A black miner in McDowell County, who had been a sharecropper in Virginia, stated that he moved to West Virginia because he "could see it [money] better. The miner got paid once or twice a month. On the farm you had to wait till the fall of the year when you gathered your crops, and then the other fellows, the landowner, merchants, etc. got it all and that way we didn't make nothing." The migrants were even more impressed with the "high pay." "On the farm I was no[t] making anything," explained Bill Deering, "[but] in West Virginia I made a dollar on my first day and I thought I was rich!" Another miner recalled that he made two dollars on his first day and declared, "I'll never go back to farmin'." A black migrant miner in Holden, Logan County, wrote his former minister in Alabama that he now "made money with ease. I am saving my money while spending part of it." Rumors spread throughout southern farmlands, "You can get rich in the West Virginia coal mines."[7]

That the black migrants from southern industrial areas had been accustomed to more oppressive conditions and that the migrants from the farms were content with economic conditions in their new homes may account for the white miners' initial hostility to them during the early formation of the southern West Virginia work force. A white miner in Oak Hill, Fayette County, angrily declared that he did not care for a recently hired group of blacks because "it is a class of labor that meets the approval of the operators here. . . . They are used to low wages and therefore grateful for the increase in pay and they dare not breathe a word without permission from the company."[8]

The controlled conditions of the company town, however, contained the hostility of the whites to the blacks. When four white miners in a company town in McDowell County, for example, struck in protest of the hiring of a black motorman, the superintendent immediately fired and blacklisted them. In a letter to the coal operator, the superintendent explained, "I had to show them who was boss." An elderly black miner supported this observation: "If the company felt a black man could do a job better, cause they wanted profits, they put the black man on the job and nothing was said. It was up to the company. . . . I don't know if the whites resented it, if they did, it didn't matter cause they couldn't do anything about it."[9]

The controlled and standardized economics of the company town prevented the economic competition between the races that produced so much racial strife in the northern cities. Because of hostility of northern whites, blacks were segregated and kept away from the better jobs. The great majority of them were forced to work at menial, dirty, and poorly paid positions that the native American or Americanized immigrant did not want. Racial conflicts erupted when black workers attempted to rise above these jobs or follow their trade, thus entering into competition with white workers and appearing to depress the job market.[10]

In the company towns blacks did not compete economically with the white workers, nor did they depress the job market. The black migration coincided with the meteoric rise of West Virginia as a coal-producing state. Although thousands of blacks entered the coal fields, the demand for miners, at least till 1920, was greater than the supply. In 1910 the number of miners in the state was triple that of the previous decade (62,189 versus 20,287 in 1900). Yet the companies were still 15,000 miners short of their needs.

The companies' standardized economic policies insured future economic racial parity. The companies' system of payment, for instance, did not allow discriminatory practices. Over 90 percent of the miners in southern West Virginia were paid on the "piece-rate basis"; less than 50 percent of the rest of the nation's miners were paid this way. West Virginia miners were paid according to the amount of coal they mined, not according to the color of their skin.[11]

Furthermore, in the company town the miners worked side by side inside the coal mine. Writing about West Virginia, one labor historian has noted that "segregation either by occupation or by place of work, unlike most factory industries in the South, is conspicuous by its absence in the coal mines." Such economic togetherness undoubtedly contributed to interracial unity. A team of psychiatrists explained that working inside a coal mine produces a peculiar social cohesion among miners. Inside the mine, they have suggested that

> the miner finds himself in a world of tension . . . [he is] potentially expendable, close to danger . . . under protracted tension because he may be the next man to get it. . . . Thus while on the job, both potential and actual threats are all around the miner, and this has to be dealt with in such a way as to minimize "wear and tear." Miners, like other groups, deal with such stress by strong social cohesion on the job. An example of such unity under stress may be seen in army and civilian groups during war-time. When such groups face dangers, prejudice, selfishness, pettiness, hatred and other frailties of personality melt in a "common defense." There is a "social security" among men who team up, not only to keep working, but to keep alive.[12]

The migration of southern blacks into the southern West Virginia coal fields differed from the northern migration in its familial patterns and its effects upon the migrant's family. The move to the industrial cities of the north, according to E. Franklin Frazier, caused the migrants to suffer a deep sense of loss in migration because "customary familial attachments were broken." Between 70 and 80 percent of the black migrants in the north, one observer estimated, were without family ties. Black families were furthered weakened after settlement. Because of the ethnic pecking order of the American job market and other discriminatory, economic policies (e.g., low wages), black women were forced to become breadwinners while their husbands cared for the children and house, thus producing unstable family arrangements.[13]

Unlike the migration to the North, the migration to southern West Virginia was primarily a family movement. The migrant black miners brought their families with them to the company town. In the Elkhorn District of McDowell County, family structure (husband and wife) characterized 68 percent (431 of 660) of the black households. Of the remaining 32 percent, almost one-half were married men who had left their families in their native state until they could earn enough money to bring them to West Virginia. It took Walter Hale thirteen years of mining during the winter and returning home during the summer to earn enough money to bring his family to West Virginia, but he was eventually successful.[14]

The company town stabilized the already family-oriented black migrants—as the company town sustained the patriarchal family—because the pecking order of the job market in the coal fields was sexual, not racial. Both company policy and superstition prevented women from working in the mines; therefore, other than doing laundry for the single miners or company officials or taking in boarders, the miner's wife was housewife and mother. Unlike the situation in the northern industrial areas, the black male in the southern West Virginia company town was always the breadwinner.[15]

The coal companies pursued policies that further secured the family unit. Managers preferred married men because they worked harder and longer under the piece-rate system to feed their families, and they were also less mobile than single men. The U.S. Women's Bureau reported that "only the presence of the family can keep the mine worker in the area. . . . The bunk house and the lodging house long ago proved themselves inadequate to attract a requisite number of workers to maintain a stable labor supply . . . [because] the mine worker cannot at will substitute the restaurant for the family table as can the wage earner in other industries." Because of their preference for married men, the coal companies generally laid off workers during slack periods according to marital status rather than seniority or color. Herbert Spencer, a single black miner in Gary, McDowell County, recalled being laid off with a number of single white miners, while his father and other married miners, both black and white, retained their jobs.[16]

Stable family structures became extremely important as the black migrants remained in southern West Virginia. Not only did married men make the best miners, they later made good union men. Single miners often left struck coal fields for more peaceful areas where they could make more money. The married miner viewed the coal fields as more of a permanent home. They found it more difficult to move and recognized that they had a vested interest in improving their working conditions. Consequently, it was the married miner who stuck out the strikes and fought the hardest for the union, and their families helped sustain them.[17]

Company housing promoted a class over caste perspective. When the southern blacks flooded the northern cities, the housing situation was one of the gravest problems they encountered. Throughout the North, the black newcomers were jammed together in the worst houses, forced to pay excessively high rent, and suffered from an inequality of public services.[18] In the southern West Virginia company towns, the migration of miners was neither haphazard nor excessive. The companies regulated the influx according to their needs of a work force and the availability of houses; hence, overcrowding never became a problem.[19]

Most of the company towns were integrated, although some of them, depending on company policy, were segregated. Even when segregated, the town was too small, its population too familiar, and social interaction too great to allow racial stereotyping and social distance and, hence, a culture of discrimination to flourish. The population of a company town usually ranged from 200 to 500 people. When the company town was segregated, blacks never lived more than a few hundred yards "up the hollow" from the whites, and they still worked together in the integrated mines and went on picnics or played baseball together on Sunday afternoon in the company ball park. On Saturday night black and white miners went to integrated whorehouses or drank homebrew together in front of the company store. Assuredly their children played together in the company playground or in the company streets.[20]

Segregation, where it existed, was not an ironclad rule, as coal companies often violated their own segregation policies when it was convenient or necessary. For example, when a coal operator in the Winding Gulf coal field obtained the services of white cokemen, who were difficult to obtain, he placed them in the empty houses in the black section of the town because the houses for whites were filled. The extreme geographic mobility of the coal miners and the ease of employment because of the labor shortage allowed the black miner, if bothered by residential segregation, to move to a company town with integrated housing. Rogers Mitchell lived in eleven different company towns during his forty years as a coal miner in southern West Virginia. Only one of the towns, Ed Wight, was segregated. In that town, Mitchell explained, "all the coloreds lived on the hill, the whites lived on the bottom. I felt different so I didn't stay. . . . I drew one paycheck then left."[21] Likewise, if a white person objected to living beside blacks in an integrated town, he had two choices, to tolerate the situation or leave. There was no other section of town where he could move.

Whether voluntary or coerced, integration worked remarkably well among the various ethnic, racial, and religious groups in the company towns, where the miners and their families developed a strong community spirit and neighborly comradeship. An aged black miner related that his family had lived beside an Italian family, and "we were great friends." The Italian miner's wife "would cook Italian dishes and bring them over to our family to eat. And my wife would bake corn bread for both families and took part over to them. . . . Our young people had great times together. They would play and dance together. We helped each other out like good neighbors."[22]

The coal companies' standardized housing practices insured social and racial parity, for housing conditions in the company town did not consider race, religion, creed, or previous condition of servitude. The coal companies built all their houses alike, usually an A-frame, Jenny Lind–type, with the intention of housing the most people possible at the lowest cost. The result was some rather shabby, poorly built houses. John Schofield, a black miner, recalled the time when, during a heavy rainstorm, the roof of his house collapsed, hitting his wife on her head and knocking her unconscious for several minutes. Sydney Box, a white migrant from England, remembered that on his arrival in southern West Virginia he thought the miners' shacks were chicken houses, because they were so small and poorly built. The black and white miners' houses were equally bad. Further, the black migrants were not hampered by excessively high discriminatory rent, as the coal companies charged a standardized rent based on the number of rooms to a house, and not on the color of the occupant.[23]

Black and white miners suffered equally from the same insecurity with regard to keeping their housing once they had it. The housing contract, which created a relationship of "master to servant," was, as the U.S. Coal Commission claimed, "obnoxious and inconsistent with the spirit of free local communities," but it served to create friendliness and solidarity among the miners.[24] It limited a miners' friends to his fellow workers because of the companies' suspicion of strangers in the company towns, and it created a common anger against the employer.

Lack of home ownership had other important effects. Neither white nor black miners cared to invest, financially or physically, in the repair or improvement of their homes; therefore, housing competition did not exist between neighbors. Because both races lived in the same style of house and neither bothered to upgrade the houses, investigators reported that even in the segregated towns both black and white sections of the town consisted of "rows of frame houses monotonously similar" and that the houses of blacks were "as good as those of the white people."[25] Because there was no home ownership, there was no fear of lowering of property values when blacks moved into a company town.

In the company town there was no discrepancy in the quality or quantity of public services between the black and white miners. In Iaeger, McDowell County, for example, black families lived about one-third of a mile "up the road" from the whites. But both races received the same public services—or suffered from the lack of them. In 1922 Iaeger had no telephones, sidewalks, paved roads, electricity, or garbage collection. (Garbage was "thrown to the pigs.") Outdoor water service was available to all miners free—only company officials had inside water. Outdoor privies in both sections were cleaned once a year without charge—only company officials had indoor toilets. The company provided free coal for fuel to all employees, but charged fifty cents per delivery. Both races supplemented their income by raising gardens and poultry on the vacant land that the company provided for those purposes. The only store from which the miners could purchase their needs was the company store—the closest commercial store was fifteen miles away and off-limits to the miners.[26]

The company town quickly and ruthlessly dissolved the traditional cultures and the time-honored social institutions of the migrant miners (whether native white, immigrant white, or southern black) that might have encouraged the continuation of ethnic or racial, instead of class, perspectives in the newly created work force. This aspect of life in the company town is most vividly illustrated by the fact that the company town broke down one of the most important and enduring institutions for blacks in American society—the black church and the black preacher.

Even during slavery blacks had constantly looked to their ministers for secular as well as spiritual guidance. When the southern blacks moved to the North, they brought their church and its leadership patterns with them. Denied access to most of the intellectual, political, and commercial channels of expression in the northern communities, black migrants, George Haynes explained, "found an untrammeled outlet in their churches. They built up its form, made its rules and traditions, handled its finances and picked its leaders unhindered by the surrounding [white] world."[27]

In southern West Virginia the black preacher and the black church suffered a "fall from grace." In the company town, the coal company, not the blacks, built the church, "made its rules and traditions, handled its finances and picked its leaders." The minister in the company town church retained the pulpit at the sufferance of the company, which meant the preacher dared not address himself to the immediate material needs and wants (for example, unionism) of his congregation, or he would lose his job, if not his life. After a black minister, the Reverend Alfred Eubanks, made a pro-union talk in a Saturday night sermon in McDowell County, he was attacked by mine guards on his way to Sunday service the next morning, "severely beaten over the head with a pistol," and then arrested on the charge of resisting an officer. In Logan County, after a black minister advised members of his flock to join the UMWA, he was attacked by the county sheriff, punched in the ribs, and warned, "be the _____ _____ sure you don't do anything here but preach." In the majority of the cases, the coal company simply fired the pro-union minister or prohibited him from preaching on company property.[28]

If the black preacher wanted to remain in the coal fields, he not only had to preach what the company desired but also had to serve as a company "lick." During week days or slack season, the black preacher served as a labor agent who went south to recruit miners for his company. At times the preacher spied for the company and, during strikes, served as a mine guard. A black miner in southern West Virginia complained:

> We have some negro preachers in the district who are nothing more than stool pigeons for the coal operators, and instead of preaching of the Gospel of the Son of God, they preach the doctrine of union hatred and prejudice, but I thank God the tide is fast changing and we are beginning to see the light for ourselves and realize the fact that they are only selling us out to the bosses for a mere mess of the porage.[29]

The black preacher's affiliation with the company earned him the disrespect, if not the contempt, of his congregations. The minister often had to maintain order during services "at the point of a pistol." Frequently, gambling and liquor selling took place in church, even during services. Probably the most revealing incident occurred when a group of black miners got drunk on a Saturday night and went to church. Halfway through the service, the inebriated churchgoers stood up and began shooting at the minister, who ran across the room and jumped out the window eight feet above the ground, breaking a leg and an arm. Rather than rising to the defense of their minister, the remainder of the congregation broke out in laughter and went home. On another occasion, a black minister literally "died of fright" when he saw his flock arm themselves during a strike. The mine superintendent in the town lamented the minister's death because, as he wrote to the coal operator, "he was one of the best leaders we [the coal company] had." Consequently, in the company towns the black miners as well as immigrants and native whites developed new social institutions, new patterns of leadership and prestige, and a new religion.[30]

One of the most conspicuous, integrative institutions in the company towns was, paradoxically, the segregated school system. This contention is difficult to argue because, as Raymond Mack points out, "most Americans now alive have grown up with the assumption that citizens opposed to desegregation are anti-Negro."[31] In effect, the Supreme Court's reasoning in *Brown v. Board of Education* ("To separate them from others of similar age and qualifications solely because of their race generates a feeling of inferiority as to their status in the community that may affect their hearts and minds in a way unlikely ever to be undone") has become generally accepted throughout the land.

In 1872 West Virginia enacted a provision to its constitution that called for racially segregated schools. For the next twenty years, the state's schools remained separate and unequal. In the 1890s, however, as the coal companies began to dominate both the people of southern West Virginia and the state government, they ushered in what elderly black teachers today call the "Golden Age of Negro education in West Virginia." Concerned with the need for a more literate work force to reduce accidents, to increase productivity, and to stabilize a very mobile work force, the coal establishment secured the passage of progressive educational legislation. Largely because of the efforts of the coal industry, teachers' salaries became based on qualifications, not color. Because black teachers, especially those in the coal fields, were better qualified, the salaries for black teachers were, on the average, higher than for whites. Further, West Virginia was one of the few states in the nation that paid more per black pupil ($111.47) than for white ($100.63) for education.[32]

Black spokesmen in West Virginia were not reticent in pointing out the state's separate but equal educational policies and the advantages that they offered to black students. The director of the Bureau of Negro Welfare and Statistics for West Virginia remarked, "For thirty years West Virginia has been busily engaged in educating Negroes whose education had been sadly neglected by their native states." The black West Virginian and professor at West Virginia State College, Thomas Posey, while recognizing that the state constitution "provided for separate schools," pointed out that "the people of the state have fought for equal facilities under the law. The success they have achieved is evident by the fine educational and eleemosynary institutions in the state." Posey was no Uncle Tom; he had earned a Ph.D. in economics from the University of Wisconsin, where he had studied under Selig Perlman, and was a member of the Socialist party of America.[33]

The coal companies exercised their greatest influence on educational policy at the local level. Taking advantage of the decentralized school system, the coal companies funneled tremendous amounts of aid into the local schools. They subsidized the building of schools and provided them with equipment. Public education in McDowell County in 1885 consisted of nine log-cabin schools, worth less than $100 each, and the school term was only three months a year. In 1904, largely as a result of the efforts of the coal companies, the county had seventy-eight schools, worth between $300 and $600 each, and the school term was eight months. The coal companies in Logan County spent nearly $100,000 on education within a four-year period.[34]

The coal companies also supplemented the teachers' incomes with a monthly bonus, to induce the better teachers to remain in the coal fields. Coal companies in McDowell County subsidized the teachers' salaries by almost $20 a month. In Logan County the coal companies gave the teachers monthly bonuses that amounted to $6,000 over a four-year period. Combined with the state's equal-pay policies for teachers, the subsidization of salaries made the pay for teachers high. Consequently, well-qualified black teachers throughout the eastern United States (including the son of a U.S. Senator B. K. Bruce of Mississippi) went to the southern West Virginia coal fields to teach.[35]

The contribution of the coal companies to public education in southern West Virginia was recognized throughout the country. Winthrop Lane, a New York journalist and no friend of the coal operators, conceded that "one respect in which the mining companies . . . are trying to improve conditions is in regard to schools." The black leader and scholar Carter G. Woodson praised the black schools in the McDowell County company towns as "well-equipped" and having as "well-qualified" teachers as any black schools throughout the country. The state supervisor of Negro Schools in West Virginia annually praised "the willingness of companies operating mines to provide adequate school facilities for children living in their camps."[36]

The children of black miners took advantage of the educational opportunities that the coal companies offered. In 1910 nearly 80 percent (1,639 of 2,067) of the black children between the ages of six and fourteen in McDowell County attended school. Native white children of the same age in the county had an attendance rate of 75 percent, and the children of white immigrants had an attendance rate of 78.6 percent. The editor of the *McDowell Times*, a black newspaper, claimed that the blacks' higher rate of attendance was an indication of the aspirations of black boys and girls when they were given "the advantage of the very best common school education."[37]

The separate but equal educational facilities were important in the lives of the black miners and their children. Although the educational level of the coal-field school remained relatively low as compared to northern education (high schools, for example, were conspicuously absent in the coal fields), the schools provided blacks with the sense of dignity and pride that comes with the knowledge of one's past and helped to make it possible for the blacks to live, work, and cooperate with whites on an equal basis. As long as the black teachers did not discuss labor affairs, they had complete control over their schools on the administrative level as well as in the classroom. As a result, reported a former teacher in Wyoming County, "The segregated school offered the opportunity to teach black history and culture and to emphasize the positive contributions of blacks to American society and to instill racial pride." Able to select their own texts, teachers used books such as Booker T. Washington's *Up from Slavery*. The black teachers in the southern West Virginia coal fields were in constant communication with Woodson and used his ideas, materials, and books in their classes. One of these ideas was the staging of annual "Negro History Week," for which teachers had their students make posters that illustrated the history of blacks in American society and place the posters throughout the town, in the mine offices, company stores, and the school. "Negro History Week" culminated with a large parade with floats, and students dressed as famous black Americans.[38]

This was the "proper education of the Negro race" that W. E. B. Du Bois claimed could not be obtained in the integrated, but white-dominated, northern schools. He pointed out that he had graduated from the great white institution of Harvard "without the slightest idea that Negroes ever had any history." A graduate of the coal-field education tells of going to Temple University to work for his master's degree and being astonished at the northern blacks' lack of knowledge of their own history: "Why they had no idea who Crispus Attucks and John Henry were, people we learned about in grade school!"[39]

The presence of schools in the coal fields offered another important benefit to the black miners. Ever cognizant of the problems and needs of members of their race, the black teachers, without pay, offered free night courses to the illiterate blacks working in the mines. Columbus Avery, born in 1871 in Alabama, had never attended school and was illiterate when he came to the West Virginia coal field in the 1890s. In Mingo County he took night classes and worked in the mines during the day. Avery learned to read and write, and he eventually held office in the union.[40]

Black teachers had the respect of both the whites and blacks in the town, but like black preachers, they could never serve as leaders because they also could not address themselves to the miners' economic needs. If black (or white) school teachers discussed labor affairs, they lost their monthly bonus and jobs, because the coal operators, who often controlled the local school boards, had the right to fire as well as hire teachers. Where the coal companies did not control the school board, they controlled the school. The coal companies, after building the school on company property, leased or deeded the land to the county under the condition that "if used for any other purpose than educational or religious services, the land shall revert to the owner [the coal company]."[41]

As differences in housing, religion, and education existed between the North and southern West Virginia, so were there differences in politics. After settling in the North, the black migrants found a source of protection, prestige, and mobility in politics. Residential segregation in the northern industrial cities produced voting blocs of blacks, which enabled the blacks to elect members of their race into office who, to a limited extent, were able to respond to the wants and needs of the black community. Ambitious blacks became party bosses because of their ability to exploit the voting blocs for their personal advancement.[42]

Partly because of the Appalachian tradition—"mountaineers are always free"—partly because of the number and proportion of blacks in West Virginia at the time, but most important because their support gave the Republicans a majority in the legislature, blacks in West Virginia also obtained some political power. Several commentators and historians have emphasized the political rights and power of blacks in this state. Posey, for instance, acknowledged that "there is no state in the Union where the Negro has a larger share in the party councils or enjoys the political prestige of our own colored citizens." Because of their political power, blacks were elected to state offices and won considerable concessions from the Republicans, including state colleges for blacks, several state orphanages, and the formation of the Bureau of Negro Welfare and Statistics. Blacks were appointed to national as well as state offices. And blacks became political bosses in several commercial towns.[43]

It may be too easy to exaggerate the influence of black miners on the Republican party and erroneous to write of their political consciousness. If blacks enjoyed such political acuteness and power as other writers have claimed, surely the Republican party would have passed progressive mining laws, especially safety legislation, and would have abolished the hated mine-guard system. It is much more likely that neither the black nor white miners possessed enough political power to threaten, even remotely, the power of the coal companies. Officers and managers of these companies still rendered politics meaningless to the miner. Denied access to political information (other than what the companies supplied) and ordered how to vote or given a ballot already marked, the miners did not have a political voice in southern West Virginia.[44] Consequently, politics did not provide opportunities for mobility and prestige, nor did it produce any black leaders in the company town.

The role of middle-class blacks in state politics and in the Republican party did have some beneficial consequences for the black miners that should not be ignored. Their influence made the Republican party a bulwark against "Jim Crow" laws in West Virginia, led to the passage of progressive legislation, helped to ban *Birth of a Nation* and other movies that would "arouse racial feeling," and aided the establishment of important black institutions.[45]

Little is known about the role of social clubs and fraternities in the lives of the black miners in the company towns, but it must have been significant. In 1915 over 10,000 blacks belonged to at least one fraternal organization, and by 1922 the total membership reached 33,000, or one-half of the black population in West Virginia. One observer noted that it was "the exceptional Negro miner who did not belong to at least one of the various lodges."[46]

Many blacks joined the social organizations for practical reasons. The director of the state's Bureau of Negro Welfare and Statistics explained that large numbers of blacks joined because three-fourths of the blacks in West Virginia were "engaged in the hazardous occupation of coal mining" and therefore unable to obtain life insurance. "The only provision which a large number of them can make for sickness and death and care for dependents . . . is to become a member of one or more fraternal and benevolent societies." In the year from July 1, 1924, to June 30, 1925, these organizations paid out $32,640 in death benefits to black miners' families. Many black miners joined simply because "it was the thing to do in the coal fields."[47]

The coal companies encouraged fraternities and often allowed them to use space on company ground for their meetings; a few companies even provided buildings for the social clubs. These fraternities relieved the monotony of life in the company town and provided "moral uplift" (immorality, adultery, or drunkenness, for example, was reason for expulsion). The lodges were usually headed by the more established, middle-class, anti-union conservative blacks from commercial towns; the coal operators undoubtedly hoped that their black workers would emulate the officers of the lodge. A black state official and an ardent supporter of the coal companies explained that the lodges were valuable for black miners because "they stand for a high degree of morality, for peace, law and order and progress. They teach industry and thrift. . . . More than any other agency working among Negroes, they have taught them the value of unity."[48] Herein lay another inherent contradiction in the operators' efforts to control their miners.

In the social clubs the blacks, especially those from the southern farms, learned and experienced the values of cooperation and the principles of brotherhood. A. L. Booker, a charter member of the local union at Helen, Raleigh County, claims that belonging to the Masons helped him and probably other miners to "become good union men because both organizations taught many of the same values; that a person should not hurt a fellow miner, he should not discriminate nor take another man's job. The Masons taught us to aid your brother in the time of need." George Hairston joined the Masons at War Eagle, Mingo County, a few years before the violent labor war in that county. He recalls that belonging to the Masons helped make him a "better union man" and helped prepare both black and white miners for the conflict because "it drawed us closer together; it brought us together in brotherhood and taught us not to mistrust or mistreat your brother." At a massive Thanksgiving service of the Colored Odd Fellows Lodge at Eckman, McDowell County, the principal speaker, a middle-class black from Keystone, instructed his audience of miners on the values of coal capitalism by quoting from Psalm 133: "Behold how good and how pleasant it is for brethren to dwell together in unity."[49]

Within the constraints of the company town, socioeconomic class lines became distinct and solidified. Florence Reese's classic labor song, "Which Side Are You On," vividly describes the class arrangements in the company town as a man, either black or white, was either a miner or a company "lick." Several observers and scholars have commented on this aspect of the company town, and the National Coal Association probably analyzed it best when it urged the abolition of the company town, believing that the "influence of small-town life will be most beneficial to the miners and their families. . . . It will be the best thing in the world for the miners' children to mingle with the children of the carpenter and the storekeeper and the doctor and also for the miners to participate and take an interest in civic affairs." The result would be to "create a civic rather than a class consciousness."[50]

The class structure in a company town prevented the development of a middle-class black leadership with conservative, racially conscious tendencies. Middle-class professional blacks (e.g., dentists, doctors, and ministers) who lived in company towns were not looked upon as symbols of individual or racial progress, but as company "licks" and spies, neither to be trusted nor emulated. Moreover, professional blacks did not live among the miners, even in the segregated towns; rather they lived on "silk-stocking row" in the company town or moved to commercial towns and traveled daily to work. There was no inspiration for social mobility that has so often splintered the American working class and prevented solidarity.[51]

The black miners developed antagonisms toward the middle-class blacks who lived in the commercial towns. Black organizations, such as the National Association for the Advancement of Colored People, were held in contempt because they represented "upper colored groups and not the common miner." The black miners also despised the anti-union attitudes of middle-class blacks. The black newspaper the *McDowell Times* was decidedly in favor of the coal operator and proclaimed that the black miners in West Virginia possessed the "best working conditions" and "most comfortable homes" and received higher pay than any other black workers in the country. It urged the black miners to eschew the union and not to strike because the operators were "their best friends." The superintendent of the Winding Gulf Coal Company described the editor of the *West Virginia Clarion*, another black newspaper, as a "real friend of the coal industry . . . who does more to keep us posted as to conditions than any agent [labor spy] covering the field."[52]

With the decline in influence and prestige of the black preacher and cut off from politics and distrustful of the middle-class, professional blacks, the black (and white) miners in the southern West Virginia coal company towns developed new sources of prestige and leadership patterns. One source proved to be on the job, that is, down in the coal mine, where the miners who displayed an "ability to meet emergencies" and "efficiency in performance" were most admired.[53] By the 1910s the miners accorded their highest prestige to the men who held office in the only institution that the miners regarded as their own and the one that promised to end the exploitation and oppression of the company town—the UMWA. When the union came to southern West Virginia, it would reflect the interracial harmony and egalitarian brotherhood that the miners had developed in the company towns. Blacks would hold office throughout the District 17 organization; they would be represented on the District's Executive Board, would work as District organizers, and would commonly hold office in the local union. In one instance, the president, vice-president, and secretary-treasurer of a local would all be blacks.[54]

More significant was the spirit of those locals. A miner in Mount Claire wrote the *UMWJ* that "we have 150 miners in our local and among them are white and Negro Americans, Horvats, Hungarians, Slavs, Croations, Italians. We get along well in our local for having so many nationalities and races represented in our membership." Fred Ball, the president of the UMWA local at Silush, a white man, wrote the journal telling about the members' cooperation and success during a strike. He finished the letter by exclaiming, "So brothers you can call [us] . . . Negroes, or whites or mixed. I call it a darn solid mass of different colors and tribes blended together, woven, bound, interlocked, tongued and grooved and glued together in one body."[55]

That cohesive union spirit promised to abolish the system that held all miners, regardless of race, nationality, or religion, in industrial slavery. The southern blacks who migrated to the West Virginia coal fields may have been initially content with their higher wages and the lack of discriminatory practices, but after several years of living and working in the company town, their attitudes and values changed dramatically. The exploitation and oppression that they encountered, as a class and not a race, in the company town, and the social, political, and economic power that the coal operators exercised over them reminded the black miners of the darkest pages in black history in the United States—slavery—but this time the slaves were whites as well as blacks.

This comparison of conditions in the company town to those in slavery may be exaggerated, but it was real to the miners who lived and worked in the southern West Virginia coal fields; they constantly compared their plight to slavery. One black miner called the company town a "damnable, slave-driving system" and asked for the return of "Grant's Army" to emancipate the miners from the "greatest octupus of serfdom encompassing any set of men on the American continent." A miner from Raleigh County wrote:

> The boss said stand up boys
> And drive away your fears,
> You are sentenced to slavery
> For many long years.
>
> So pick up your shovels
> And do your work well.
> You are right here where we want you
> And we'll work you like hell.[56]

The most revealing comparisons came from black miners who had been slaves or who were children of slaves. George Echols, a black miner in Mingo County and a former slave, told a U.S. Senate committee:

> I know the time when I was a slave and I feel just like *we* feel now. . . . My master and my mistress called me and I had to answer. We claim that we are citizens of the United States of America according to the amendment in the Constitution. You know that that guarantees us free and equal rights and that is all we ask.

The chairman of the committee asked, "Just what rights do you feel are getting away from you?" Echols responded:

> If we get together and are talking, we are ordered to scatter and move out. If we go out for a walk, two by two, we are ordered to scatter and move out. If we go to town . . . we have to tell our wives good-bye for we know we might not come back. . . . [It is like it was] before the Constitution prohibited slavery. Let us be free men. Let us stand equal and I do not mean that there is any prejudice against colored people here.[57]

Columbus Avery stated that "miners possibly had it worse than the colored folk under slavery, if I understood my parents [who] had been slaves right." Walter Hale recalled discussions with his parents, who were former slaves, who had told him he had it worse as a miner than they did under slavery.[58]

The black miners realized, however, that there was a major difference between plantation slavery and company-town industrial slavery. Plantation slavery was based on color, company-town industrial slavery was based on class. A black miner in southern West Virginia, who described himself as a union miner "from my head to my feet" wrote,

> The American white miner is in a worse state today than the Negro was in slavery. He has to feed himself and family. The Negro did not have to do all that. When he wanted to go for a long distance he went to his master and asked for a mule to ride and it was all right. He has a full stomach to sleep on. No man working for a coal operator ever has a full stomach. Abe Lincoln signed a proclamation to set the colored slave free. Now the white miner is in slavery. Who will set him free?[59]

The black miners also recognized that because this time they were aligned with working whites, they had the opportunity to fight for their freedom. A black miner in southern West Virginia explained, "There was a time when we colored people could not fight for our freedom, but I thank God that the day is come that every man can fight for his own rights. My dear colored friends let us all stand firm and show the world that we have grown to be men of America, and that we are not afraid to fight for our rights."[60]

And fight for those rights they did, on Paint Creek and Cabin Creek. They fought with, not against, their fellow white workers, on a working-class basis, for the one institution that offered them power, prestige, leadership, and escape from bondage. The company town was not the only force in southern West Virginia conducive to racial harmony and working-class solidarity. However, it prevented the black migration of the 1890s from creating in southern West Virginia many of the racial antagonisms that affected other coal fields and the northern cities. Nor would the UMWA's egalitarian union policies have been as successful without these necessary preconditions.

## NOTES

1. Jeremy Brecher, *Strike!* (San Francisco, 1972), 247. As shown in chap. 1, both numerically and proportionally, more miners in West Virginia lived in company towns than in any other state in the nation—94 percent. The number of blacks employed in the southern West Virginia coal industry was substantially higher than in any other state. In 1920, for example, 18,371 West Virginia coal miners were black. Second to West Virginia was Alabama, with 11,723 black miners, and third was Pennsylvania, with 2,288. Indeed, in 1920 43 percent of all black miners employed in the bituminous coal industry worked in West Virginia. Sterling Spero and Abram Harris, *The Black Worker, 1915–1930* (New York, 1974), 208; Edward Hunt, F. G. Tryon, and Joseph Willits, eds., *What the Coal Commission Found* (Baltimore, 1925), 136–40; Darold Barnum, *The Negro in the Bituminous Coal Mining Industry* (Philadelphia, 1970), 1–10; W.Va. Bureau of Negro Welfare and Statistics, *Annual Reports, 1925–29*, 22–24; James Laing, "The Negro Miner in West Virginia," *Social Forces* 36 (Mar. 1936): 418, and "Negro Migration to the Mining Fields of West Virginia," *Proceedings of the West Virginia Academy of Science* 10 (1936): 171–81; U.S. Department of Labor, Women's Bureau, *Home Environment and Employment Opportunities of Women in Coal-Mine Workers' Families*, Women's Bureau Bulletin no. 45 (Washington, D.C., 1925), 12, 51–55.

2. William Tuttle, *Race Riot, Chicago in the Red Summer of 1919* (New York, 1974), 108–55; Allan Spear, *Black Chicago, the Making of a Negro Ghetto* (Chicago, 1974), 201–22; Spero and Harris, *The Black Worker*, 57–62, 162; Kenneth Jackson, *The Ku Klux Klan in the City, 1915–1930* (New York, 1967); Henderson Donald, "The Negro Migration of 1916–1918," *Journal of Negro History* 6 (Oct. 1921): 383–498. Sociologists claim that it is a "socio-cultural law" that the migration of a "visibly different group" into a given area increases prejudice, discrimination, and the "likelihood of conflict." Gordon Allport, *The Nature of Prejudice* (Reading, Mass., 1954), 227–29; Robin Williams, *The Reduction of Inter-Group Tensions* (New York, n.d.), 57–58; Gerhart Soenger, *The Social Psychology of Prejudice* (New York, 1953), 99.

3. "Industrial Relations and Labor Conditions: Economic Condition of the Negro in West Virginia," *Monthly Labor Review* 16 (Apr. 1923): 713. The coal fields generally experienced less racial strife than other industrial areas. Some labor historians have claimed that the egalitarian, industrial union policies of the UMWA were the determining factor. See, for example, Herbert Northrup, *Organized Labor and the Negro* (New York, 1944); Herbert Gutman, "The Negro and the United Mine Workers," in *The Negro and the American Labor Movement*, ed. Julius Jacobson (Garden City, N.Y., 1968), 49–127; Spero and Harris, *The Black Worker*; Charles Simmons, John Rankin, and U. G. Carter, "Negro Coal Miners in West Virginia," *Midwest Journal* 54 (Spring 1954), 60–69; Barnum, *The Negro in the Bituminous Coal Mining Industry*, 1. This explanation, while useful, is not definitive as racial turmoil hindered the success of the UMWA in many coal fields. In northern West Virginia racial conflict broke out in the early 1920s. One mine was wrecked because of the employment of blacks, while another mine was named "the Ku Kluxer's mine" because the superintendent and white mine workers allegedly had an agreement against employing blacks. A district official in northern West Virginia declared that racial intolerance among the native whites constituted "a serious menace to the United Mine Workers"; a rank-and-file miner from Clarksburg in northern West Virginia lamented that "in our miner's union there is not enough of that spirit of oneness. The union must lay aside the color of the skin and count men as men who honestly struggle to play the part of a man." Boris Emmet, *Labor Relations in the Fairmont, West Virginia Bituminous Coal Field*, U.S. Department of Labor, Bureau of Labor Statistics Bulletin no. 361 (Washington, D.C., 1924), 24–26; T. H. Seals, letter to the editor, *UMWJ*, Sept. 15, 1924, 7. For racial strife among Illinois miners, see Agnes Wieck, "Ku Kluxing in the Miners' Country," *New Republic* 38 (Mar. 26, 1924): 122–24, and Carter Goodrich, *The Miners' Freedom* (Boston, 1925), 113. For Alabama, see Robert David Ward and William Rodgers, *Labor Revolt in Alabama* (Tuscaloosa, Ala., 1965), 21, and *UMWJ*, Jan. 15, 1919, 5, for a description of how the Alabama operators played black and white miners against each other. For the problems in Pennsylvania, see William Donovan, letter to the editor, ibid., Dec. 15, 1932, 11, and Aug. 9, 1917, 4. For Tennessee and Kentucky, see Ronald Lewis, "Race Relations and the United Mine Workers' Union in Tennessee," *Tennessee Historical Quarterly* 36 (Winter 1977): 524–36.

4. P. O. Davis, "Negro Exodus and Southern Agriculture," *American Review of Reviews* 68 (Oct. 1923): 401–7. Also see George Haynes, "The Movement of Negroes from Country to City," *Southern Workman* 42 (Apr. 1913): 230–36; Donald, "Negro Migration," 407; Abram Harris, "Negro Migration to the North," *Current History* 20 (Sept. 1924): 921–25. For a discussion of the difficulty that agrarian blacks had in adjusting to northern industrial norms, see Spero and Harris, *Black Worker*, 163–65. For a general discussion of this problem, see Herbert Gutman, "Work, Culture, and Society in Industrializing America, 1815–1919," *American Historical Review* 78 (June 1973): 531–88.

5. This view is purely impressionistic and is based on interviews with elderly black miners and other research. Because the company towns were unincorporated, they were not labeled by census enumerators, and therefore an accurate statistical study of the black migration into southern West Virginia is impossible. Because the 1900 census shows the black migrants were from Virginia, North Carolina, and Alabama, it is easy to assume that they had been sharecroppers. See Jerry Bruce Thomas, "Coal County: The Rise of the Smokeless Coal Industry

and Its Effect on Area Development, 1872-1910" (Ph.D. diss., University of North Carolina, 1971), 76. For a similar view about the patterns of black migration as given in this chapter, also based on interviews, see Peter Gottlieb, "Making Their Own Way: Southern Black Migration to Pittsburgh" (Ph.D. diss., University of Pittsburgh, 1977).

6. Interview with Rogers Mitchell, Institute, W.Va., May 17, 1975; W. L. McMillian to R. L. Thorton, Nov. 2, 1916, Straight Numerical File 182363-231, Record Group 60, General Records of the Department of Justice, National Archives, Washington, D.C.; T. H. Seals, letter to the editor, *UMWJ*, Sept. 15, 1924, 17. The president of Bessemer Coal, Iron and Land was Henry L. Budland, and his speech of Sept. 22, 1916, is in Migration Study, Negro Migrants Letters, FRB, 1916-18, Papers of the National Urban League, Manuscript Room, Library of Congress, Washington, D.C. (hereafter cited as Negro Migrants Letters, National Urban League Papers).

7. Interviews with Thomas Cannady, Anawalt, W.Va., Abe Helms, Eckman, W.Va., and Bill Deering, Williamson, W.Va., summer 1975; letter, no name or date, to Dr. [no name], Negro Migrants Letters, National Urban League Papers. Also see the letters and testimonies in Straight Numerical File 182363, Record Group 60, General Records of the Department of Justice.

8. W. F. Larrison, letter to the editor, *UMWJ*, Jan. 9, 1908. For other accounts of early racial antagonisms, see John R. Williams to William Thomas Brynawel, Nov. 10, 1895, in Alan Conway, ed., *The Welsh in America: Letters from Immigrants* (Minneapolis, Minn., 1961), 204-10, and Gutman, "Negro and the United Mine Workers," in *Negro and Labor Movement*, ed. Jacobson, 49-127. In this early period there was still much cooperation between black and white miners. See, for example, *UMWJ*, May 17, 1894, 6, and Kenneth Bailey, "Tell the Boys to Fall in Line," *West Virginia History* 32 (July 1971): 224-37.

9. W. J. Elgin to Justin Collins, May 9, 1912, Justin Collins Papers, West Virginia University Library, Morgantown, W.Va.; interviews with John Drew, Williamson, W.Va., Curt Smith, Mullins, W.Va., and John Schofield, Alpoca, W.Va., summer 1975. Other elderly miners, both black and white, made similar statements.

10. Emil Frankel, "Occupational Classes among Negroes in Cities," *American Journal of Sociology* 35 (Mar. 1930): 718-38; Spero and Harris, *Black Worker*, 155-57; Guy Johnson, "Negro Migration and Its Consequences," *Social Forces* 11 (Mar. 1924): 404-8.

11. Laing, "Negro Miner," 418-20. For discussion on types of payment in the bituminous coal fields, see Hunt, Tyron, and Willits, *What the Coal Commission Found*, 164-229.

12. Northrup, *Organized Labor*, 159; Lewis Field, Reed Ewing, and David Wayne, "Observations on Relations of Psychosocial Factors to Psychiatric Illness among Coal Miners," *International Journal of Social Psychiatry* 3 (Autumn 1957): 133-45.

13. Gilbert Osofsky, *Harlem: The Making of a Ghetto* (New York, 1966), 147-48; E. Franklin Frazier, *The Negro Family* (Chicago, 1969), chap. 13; James Blackwell, *The Black Community: Diversity and Unity* (New York, 1975), chaps. 2 and 3; Charles A. Valentine, *Culture and Poverty: Critique and Counter-Proposals* (Chicago, 1968); Donald, "Negro Migration," 407-8.

14. U.S. Department of Commerce and Labor, Bureau of the Census, Population Census, McDowell County, Elkhorn District, 1900 (manuscript), National

Archives. All the towns in this census district were coal company towns. Interview with Walter Hale, Eckman, W.Va., summer 1975. It appears that most of the single black migrants into West Virginia went to the coal fields in the northern part of the state; see W.Va. Bureau of Negro Welfare and Statistics, *Annual Reports, 1923-24*, 8-9. For the role of the single black miner in southern West Virginia, see Randall G. Lawrence, "Here Today, Gone Tomorrow: Coal Miners in Appalachia, 1880-1940," paper presented at the meeting of the Organization of American Historians, Apr. 1-4, 1981, Detroit, Mich.

15. In 1924 West Virginia had the highest percentage of black males employed in the United States, and the lowest percentage of black females. W.Va. Bureau of Negro Welfare and Statistics, *Annual Report, 1924*, 25; Women's Bureau, *Home Environment*, 37-55.

16. Women's Bureau, *Home Environment*, 16-19; Nettie McGill, *The Welfare of Children in Bituminous Coal Mining Communities in West Virginia*, Children's Bureau, U.S. Department of Labor publication no. 117 (Washington, D.C., 1923), 8-10; interview with Herbert Spencer, Keystone, W.Va., summer 1975. For the company preference of hiring by families, see George Wolfe to Collins, May 18, 1916, and A. H. Herndon to Collins, Dec. 15, 1909, Collins Papers; Josiah Keeley, "The Cabin Creek Consolidated Coal Company," *West Virginia Review*, June 1926, 349. Similarly, see the letters and testimonies in Straight Numerical File 182363, Record Group 60, General Records of the Department of Justice, especially the testimony of George Bowman, Oct. 23, 1916.

17. Interview with Willie Anderson, Holden, W.Va., summer 1975.

18. Spero and Harris, *Black Worker*, 162; George Haynes, "Negro Migration," *Opportunity*, Oct. 1924, 303-6; Donald, "Negro Migration," 436-38; Eli Ginzberg and Dale Hiestand, *Mobility in the Negro Community*, U.S. Commission on Civil Rights, Clearinghouse Publications no. 11 (New York, 1968), 21.

19. This was only a southern West Virginia company town experience, for the black migrants who moved into the commercial cities (e.g., Charleston) and northern West Virginia encountered the same housing problems as did the migrants who moved into the northern industrial centers. See W.Va., Bureau of Negro Welfare and Statistics, *Negro Housing Survey of Charleston, Keystone, Kimball, Wheeling, and Williamson* (Charleston, 1938).

20. Jack Rodgers, "I Remember that Mining Town," *West Virginia Review*, Apr. 1938, 203-5; interviews with Willie Betz, Keystone, W.Va., Tom Cannady, Anawalt, W.Va., and John Schofield, Alpoca, W.Va., summer 1975; Mary Simmons, interview with Pauline Haga, *Beckley* (W.Va.) *Post Herald* (Diamond Jubilee Edition), Aug. 2, 1975; Louise Coffee interviewed in *Charleston Gazette*, July 28, 1975. Sociologist Raymond Mack has pointed out that the most damaging aspect of segregation is that it "minimizes one's exposure to members of a minority who do not conform to the stereotype of what they should be like," thus producing a "culture of discrimination" and race prejudice. "Riot, Revolt, or Responsible Revolution: Of Reference Groups and Racism," *Sociological Quarterly* 10 (Spring 1969): 147-56. In the same vein, Sidney Mintz has explained that it is not race nor segregation but the "perception of race differences" that accounts for the absolute as well as relative well-being of a minority within a given society. "Toward an Afro-American History," *Journal of World History* 13 (1971): 317-33, and his foreword to *Afro-American Anthropology*, ed. Norman Whitten and John Szwed (New York, 1970).

21. W. J. Elgin to Collins, May 9, 1912, Collins Papers, and interviews with John Drew, Williamson, W.Va., and Rogers Mitchell, Institute, W.Va., summer 1975.

22. Ralph Minard, "Race Relations in the Pocahontas Coal Fields," *Journal of Social Issues* 8 (1952): 37.

23. W. P. Tams, *The Smokeless Coal Fields of West Virginia* (Morgantown, W.Va., 1963), 51; interviews with John Schofield, Alpoca, W.Va., and Sidney Box, Glen White, W.Va., summer 1975; R. H. Hamil, "Design of Buildings in Mining Towns," *Coal Age* 11 (June 1917): 1045–48.

24. Committee on Coal and Civil Liberties, "A Report to the U.S. Coal Commission," Aug. 11, 1923. U.S. Department of Labor Library, Washington, D.C. The housing contract is discussed in chap. 1.

25. Women's Bureau, *Home Environment*, 9, 29; Tams, *Smokeless Coal Fields*, 52; Thomas J. Morris, "The Coal Camp: A Pattern of Limited Community Life" (M.A. thesis, West Virginia University, 1950), 9–11, 18, 19.

26. U.S. Coal Commission, 1923, Records of the Division of Investigation of Labor Facts, Living Conditions Section, Mining Communities Schedules, A, Schedule for Iager, Box 30, Record Group 68, Records of the United States Coal Commission, National Archives.

27. George Haynes, "Negro Migration," *Opportunity*, Sept. 1924, 273, and Oct. 1924, 304, 305. See also Charles Hamilton, *The Black Preacher* (New York, 1972), especially 12–13, and W. E. B. Du Bois, *Souls of Black Folk* (New York, 1961), 141, who noted that "the black preacher is the most unique personality developed by the Negro on American soil. A leader, a politician, an orator, a 'boss,' an intriguer, an idealist—all these he is, and ever too, the center of a group of men, now twenty, now a thousand in number."

28. *UMWJ*, Jan. 1, 1925, 3, and Jerome Davis, "Human Rights and Coal," *Journal of Social Forces* 3 (Nov. 1924): 103–6.

29. Spero and Harris, *Black Worker*, 378; W.Va. Bureau of Negro Welfare and Statistics, *Report, 1922*, 77; Laing, "Social Status," 563; "Negro Miner," letter to the editor, *UMWJ*, June 1, 1916, 9. Also see the resolution passed by the UMWA local at Morrisvale, W.Va., to protest the companies' ousting of a nearby pro-union minister, ibid., July 15, 1922, 7.

30. W.Va. Bureau of Negro Welfare and Statistics, *Report, 1921–22*, 77–78; interview with Columbus Avery, Matewan, W.Va., summer 1975; Jarius Collins to Justin Collins, June 8, 1916, Collins Papers; Laing, "Social Status," 563–66. For the disruption of the traditional institutions of native whites and European immigrants, see chaps. 1 and 6, respectively.

31. Mack, "Riot, Revolt," 155.

32. Interview with Mr. and Mrs. Carl Hazzard, Mullins, W.Va., summer 1975. Charles Ambler, *History of Education in West Virginia* (Huntington, W.Va., 1951), chap. 7; "Biennial Survey of Education in the United States, 1944–1946," in Mark Rich, *Some Churches of Coal Mining Communities of West Virginia* (New York, 1951), 12; W. W. Sanders, "Report of the Division of Negro Schools," W.Va. State Superintendent of Free Schools, *Annual Report, 1930*, 84–86; Wolfe to Justin Collins, July 14, 1913, Collins Papers; U.S. Congress, Senate Committee on Education and Labor, *West Virginia Coal Fields: Hearings . . . to Investigate the Recent Acts of Violence in the Coal Fields of West Virginia*, 67th Cong., 1st sess., 2 vols. (Washington, D.C., 1921–22), 1:27 (hereafter cited as *West Virginia Coal Fields*); Edwin Cubby, "The Transforma-

tion of the Tug and Guyandot Valleys" (Ph.D. diss., Syracuse University, 1962), 296–97; John Peters and F. Carden, *A History of Fayette County* (Fayetteville, W.Va., 1926), 326–28.

33. W.Va. Bureau of Negro Welfare and Statistics, *Annual Report, 1926*, 6, and *Annual Report, 1932*, 17.

34. Walter Thurmond, *The Logan Coal Field of West Virginia* (Morgantown, W.Va., 1964), 81–82; Ambler, *History of Education*, chap. 7; Superintendent, McDowell County, "Report, in W.Va. State Supervisor of Free Schools, *Annual Report, 1906*, 17, 217; W.Va. State Supervisor of Free Schools, *Annual Report, 1914*, 75, 217, and *Annual Report, 1930*, 47.

35. Thurmond, *Logan Coal Field*, 81–82; Ambler, *History of Education*, chap. 7; McDowell County "Report," in W.Va. State Supervisor of Free Schools, *Annual Report, 1906*, 17, 217; W.Va. State Supervisor of Free Schools, *Annual Report, 1914*, 75, 217, and *Annual Report, 1930*, 47; interview with Mr. and Mrs. Carl Hazzard, Mullins, W.Va., summer 1975.

36. Winthrop Lane, *Civil War in West Virginia* (New York, 1921), 34; W. W. Sanders, Supervisor of Negro Schools, "Report," in W.Va. Bureau of Negro Welfare and Statistics, *Annual Report, 1926*, 140. Also see Raymond Murphy, "A Southern West Virginia Mining Community," *Economic Geography* 9 (Jan. 1933): 48–50.

37. *McDowell* (W.Va.) *Times*, May 9, 1913.

38. Interviews with Mr. and Mrs. Carl Hazzard, Mullins, W.Va., Mrs. A. L. Booker, Mullins, W.Va., and John Drew, Williamson, W.Va., summer 1975. For a discussion on the importance of "Negro History Week," see Carter G. Woodson, "The Celebration of Negro History Week, 1927," *Journal of Negro History* 12 (Apr. 1927): 103–9.

39. W. E. B. DuBois, "Does the Negro Need Separate Schools?" *Journal of Negro History* 10 (July 1925): 328–35, and interview with John Drew, Williamson, W.Va., summer 1975. In the 1960s, progressive black leaders and radicals were making similar arguments for segregated schools. Roy Innis, assistant director of the Congress of Racial Equality, acknowledged that "people today talk about control of their community schools. Integration is contrary to the mood of the black people." The Black Panthers argued that black students "should be taught black economics and black history in black schools, with black teachers." Quoted in Mack, "Riot, Revolt," 154–55.

40. Interview with Columbus Avery, Williamson, W.Va., summer 1975. For the importance of these night schools in decreasing black illiteracy, see Director of Extension Work among Colored Schools, "Report," in W.Va. State Superintendent of Free Schools, *Annual Report, 1916*, 89–91, and Ambler, *History of Education*, 213.

41. Testimony of M. T. Davis in U.S. Congress, Senate Committee on Education and Labor, *Conditions in the Paint Creek District, West Virginia*, 63rd Cong., 1st sess., 3 vols. (Washington, D.C., 1913), 2:1280; Wolfe to Justin Collins, July 14, 1913, Collins Papers; Homer Morris, *The Plight of the Bituminous Coal Miner* (Philadelphia, 1934), 94. For an example of a public school teacher losing his job for supporting a strike, see *West Virginia Coal Fields*, 1:27, 33. For the miners' early resentment of the company-controlled school, see Henry Stephenson, letter to the editor, *UMWJ*, July 21, 1898, 17. As late as 1933, when the UMWA called an organizing meeting at a Mingo County school, the superintendent of U.S. Coal and Coke at Gary, W.Va., informed the county board of

education that holding the meeting in the school "will automatically and immediately cancel the lease on the lot on which this school house is built. . . . It is not a question as to whether or not the Board of Education grants . . . permission for the holding of a labor meeting in this school house, or on these school grounds, but the crucial fact is whether or not such a meeting is held." William Stratton to Mingo County Board of Education, July 10, 1933, District 17 Correspondence Files, UMWA Archives, UMWA Headquarters, Washington, D.C. (copy also in author's files).

42. John Hope Franklin, *From Slavery to Freedom* (New York, Random House, 1967), 524–26.

43. Spero and Harris, *Black Worker*, 370–74; *McDowell* (W.Va.) *Times*, May 6 and 16, 1913; Thomas Posey, "Political Activity of West Virginia Negroes," W.Va. Bureau of Negro Welfare and Statistics, *Annual Report, 1929*–32, 40–45; Kenneth Bailey, "A Judicious Mixture," *West Virginia History* 34 (Jan. 1972): 152–54; Fred Barkey, "Socialist Party of West Virginia from 1898–1920" (Ph.D. diss. University of Pittsburgh, 1971), 168.

44. For a discussion on the backwardness of mine-safety legislation in West Virginia, see William Graebner, *Coal-Mining Safety in the Progressive Era* (Lexington, Ky., 1973), chap. 1. Political coercion is discussed in chap. 1. For individual examples of political corruption, see the interview with Oscar Roebuck, Cabin Creek, W.Va., summer 1975; Wolfe to Justin Collins, Feb. 27, 1913, Collins Papers; *UMWJ*, Jan. 1, 1925, 3.

45. W.Va. Bureau of Negro Welfare and Statistics, *Annual Report, 1922*, 72–73; Spero and Harris, *Black Worker*, 370–74; Bailey, "Judicious Mixture," 155.

46. *McDowell* (W.Va.) *Times*, Aug. 20, 1915; W.Va. Bureau of Negro Welfare and Statistics, *Annual Report, 1924*; Laing, "Social Status," 566.

47. W.Va. Bureau of Negro Welfare and Statistics, *Annual Report, 1922*, 59, 83; *Annual Report, 1924*, 81; letter, no name or date, to Dr. [no name], Negro Migrants Letters, National Urban League Papers.

48. C. A. Cabell, "Building a Mining Community," *West Virginia Review*, Apr. 1927, 210; interviews with Frank Hunt, former vice-president of the Winding Gulf Coal Operators' Association, Beckley, W.Va., and Oscar Pennel, Mead, W.Va., summer 1975; W.Va. Bureau of Negro Welfare and Statistics, *Annual Report, 1922*, 5.

49. Interviews with A. L. Booker, Mullins, W.Va., and George Hairston, Mead, W.Va., summer 1975; *McDowell* (W.Va.) *Times*, May 16, 1913. Similarly, Barkey discovered many socialist miners who belonged to social lodges. "Socialist Party," 104.

50. National Coal Association, Bulletin no. 725, Aug. 13, 1927, copy in author's files. Also see Goodrich, *Miners' Freedom*, 57, 58; Archie Green, *Only a Miner* (Urbana, Ill., 1974); George Korson, *Coal Dust on the Fiddle* (Philadelphia, 1943), 20, 21; Winthrop Lane, "The Black Avalanche," *Survey* 47 (Mar. 25, 1922): 1002–6. For similar interpretations about company towns in England, see A. L. Lloyd, *Come All Ye Bold Miners* (London, 1952), and G. D. H. Cole, *Labour in the Coal Mining Industry* (London, 1923), 6–7.

51. Ralph Minard, "Race Relations," 29–44; Spero and Harris, *Black Worker*, 304–6; Seymour Lipset and Reinhart Bendix, *Social Mobility in Industrial Society* (Berkeley, Calif., 1963), 106.

52. M. L .Shrumm, letter to the editor, *McDowell* (W.Va.) *Times*, May 30,

July 4, Aug. 28, 1913; Laing, "Social Status," 567.

53. Laing, "Negro Miner," 418–20; Minard, "Race Relations," 31, 32, 36; Morris, *Plight*, 28.

54. Testimony of Frank Ingham, *West Virginia Coal Fields*, 1:26–30; interviews with Rogers Mitchell, Institute, W.Va., George Hairston, Mead, W.Va., Columbus Avery, Williamson, W.Va., and Carl Hazzard, Mullins,W.Va., summer 1975; Helen Norton, "Feudalism in West Virginia," *Nation* 123 (Mar. 1931): 154–55; Fred Mooney, *Struggle in the Coal Fields*, ed. Fred Hess (Morgantown, W.Va., 1967), 26–29.

55. Ball, letter to the editor, *UMWJ*, Sept. 1, 1921, 15, and A. E. Grimes, letter to the editor, ibid., Mar. 15, 1921, 17. See also Anthony Gray, letter to the editor, ibid., Oct. 15, 1913, 15.

56. George Edmunds, Paint Creek, W.Va., letter to the editor, ibid., June 20, 1912, 5. The ballad cited in text was written by Fred Niswander of Freeman, W.Va.; it was found in his letter to John L. Lewis, Nov. 12, 1934, District 17 Correspondence Files, UMWA Archives. For another example in which the southern West Virginia miners compared their plight to the slaves, see Artie Surber, letter to the editor, *UMWJ*, Jan. 15, 1922, 7. For other ballads, see "Miner's Lifeguard," and Orville Jenks, "In the State of McDowell," in Korson, *Coal Dust*, 304–5, 414–15. During the "armed march on Logan," a reporter asked one of the miners, "What do you boys really think you can do?" The marcher-miner replied, "Well, John Brown started something at Harper's Ferry, didn't he?"; Heber Blankenhorn, "Marchin' through West Virginia," *Nation* 113 (Sept. 14, 1921): 288.

57. Testimony of George Echols, *West Virginia Coal Fields*, 1:469–71.

58. Interviews with Columbus Avery, Williamson, W.Va., and Walter Hale, Eckman, W.Va., summer 1975.

59. Anthony Gray, letter to the editor, *UMWJ*, Apr. 1, 1931, 9.

60. Emmett Lemons, letter to the editor, ibid., May 15, 1921, 15.

CHAPTER IV

# "Solidarity Forever"

When the union's inspiration through the
workers' blood shall run,
There can be no power greater anywhere
beneath the sun.

Ralph Chaplin
"Solidarity Forever"

In April 1912, the union spirit that had been maturing for the past two decades emerged in the Kanawha and New River coal fields. A contract dispute led to a walkout of thousands of nonunion as well as union miners in the Paint Creek district. They were soon joined by 7,500 nonunion miners on Cabin Creek, Kanawha and Fayette counties, who quit work, stating that they "could not stand the oppression of the coal barons any longer." In the process that Jeremy Brecher describes as a "natural progression by which strikes move toward civil war," the striking miners confronted the coal operators, the mine guards, the state militia, the state courts, the governor, and even the UMWA. The Paint Creek-Cabin Creek strike became one of the most protracted and bloody labor-management conflicts in American history. More important, however, the southern West Virginia miners discovered unionism and the power of collective action.[1]

The UMWA officials responded enthusiastically to this moment. Here was their chance to organize the second most productive coal-producing state in the nation, one that for a decade had threatened the union's hold on the organized midwestern coal fields. The UMWA pledged its full support of the strike, enacted levies to finance it, and sent in its top officials, including Vice-President Frank Hayes and Mother Mary Jones, to the state.[2]

The coal operators also responded aggressively to the walkout, as they were determined not only to defeat the strike but to annihilate the union's foothold on Paint Creek. They withdrew recognition of the union, imported strikebreakers from New York and the South, and brought in 300 Baldwin-Felts detectives to break the strike.[3]

The Baldwin-Felts guards built iron and concrete forts that they equipped with machine guns throughout the strike districts and then evicted the striking miners from their company houses, destroying $40,000 worth of the miners' furniture in the process.[4] The guards then began to intimidate the strikers in their tent colonies at Holly Grove and Eskdale. The guards often prevented the strikers from leaving the tent colonies and prohibited them the use of company bridges, forcing the strikers and their families to wade through waist-high streams. They also stopped the strikers and their families from using the passenger trains into Charleston. The son of a miner who attempted to do so was kicked in the face and warned: "Get off, you damn son of a bitch, or I will shoot your brains out"; he was then literally thrown off the train.[5]

The mine guards also began to kill the striking miners both individually—for example, a black miner chasing his stray cow—and collectively. On June 5, the guards launched a surprise attack upon one of the tent colonies, wounding several people. Thereafter, according to Fred Mooney, "Every day or two they would sneak into the hills and sprinkle the canvass cities with showers of leaden pellets, caring not if their bullets hit men, women, or children."[6] Later during the strike, the mine guards rigged a train, called the "Bull Moose Special," with iron-plate siding and machine guns, and then at night, with its lights turned out, and with coal operator Quinn Martin and Kanawha County Sheriff Bonner Hill aboard, they drove the monster through the valleys, machine-gunning the people in the tent colonies on the sides of the hills.[7]

Covering the strike for a local Socialist newspaper, Ralph Chaplin, the future Wobbly editor, poet, and song writer, and "right-hand man" to William ("Big" Bill) Haywood, captured the miners' hatred for the mine guards in his poem "Mine Guard":

You cur! How can you stand so calm and still
  And careless while your brothers strive and bleed?
  What hellish, cruel, crime-polluted creed
Has taught you thus to do your master's will,
Whose guilty gold has damned your soul until
  You lick his boots and fawn to do his deed—
  To pander to his lust of boundless greed,
And guard him while his cohorts crush and kill?
Your brutish crimes are like a rotten flood—
  The beating, raping, murdering you've done—
  You psychopathic coward with a gun:
The worms would scorn your carcass in the mud;
  A bitch would blush to hail you as a son—
You loathsome outcast, red with fresh-spilled blood![8]

UMWA officials became less enthusiastic about the strike. Their aspirations diminished as the power and brutality of the coal establishment revealed itself and as the magnitude and expense of the strike placed a serious financial burden on the UMWA national office. Hayes explained that because in West Virginia "the coal companies own practically all of the land, the houses and the stores . . . it [was] necessary for the coal miners' union to feed, clothe and house the strikers upon the beginning of the struggle." The destitute conditions of the West Virginia miners, caused by "extremely poor" work in the past year, forced the union to spend large amounts of money out of a treasury already "depleted" from sponsoring strikes the two previous years.[9] Threatened with bankruptcy, and apparently believing that total victory was impossible, the union officials attempted to compromise with the operators. Twice in conjunction with the governor, they proposed settlement terms to the coal operators who, intent on total victory, rejected both proposals. Following the second request, a coal operator in McDowell County wrote a colleague that "the operators have disregarded the proposal . . . it looks very much as though it [the strike] will be fought out to the bitter end."[10]

The union officials, however, were reluctant to fight. Intimidated by the brutality of the coal companies and bound by traditional union procedures, UMWA officers were unable to provide the necessary leadership in what was fast becoming an extraordinary labor-management conflict. Jones recalled addressing a group of striking miners with one of the union representatives, when the coal diggers began screaming, "Organize us, Mother." When she agreed, the official quickly stated, "You can't organize those men because you haven't the ritual." Jones retorted, "The ritual hell, I'll make one up!" The union officer then said, "They have to pay fifteen dollars for a charter." The fiery angel of the miners responded, "I will get them their charter. Why those poor wretches haven't fifteen cents for a sandwich. All you care about is your salary regardless of the destiny of these men."[11]

Two months into the strike, a small, soft-spoken, but quick-tempered, twenty-four-year-old miner from Cabin Creek, Frank Keeney, ventured into District 17 headquarters in Charleston. He requested union aid in arranging mass meetings in the strike district. When the officials refused, Keeney "proceeded to read the riot act to the union officers" and assumed unofficial charge of the struggle—the rank-and-file miners had taken over the strike, and a new leadership had emerged among them.[12]

Unable to get the necessary support from the UMWA, Keeney and other miners turned to the Socialist party—and the Socialists helped. They supplied the strikers with speakers and conducted mass meetings to raise financial and moral support. They organized strike committees to bootleg guns and information into the strike zone and to smuggle information and pictures out. They publicized the outrages against the miners in their newspapers. The radicals encouraged the miners to fight and sanctified the violence under the name of direct action. As a result of this aid and urging, a number of the rank-and-file miners, including Keeney, joined the Socialist party and voted the Socialist ticket.[13]

The Socialists' justification for violence was important. By the time Keeney took control of the strike, it had assumed the dimensions of a life-or-death struggle, and the miners were prepared to fight for their existence. The miners joined the National Rifle Association in order to obtain government surplus guns at low prices. They formed squads of "minutemen." According to Mooney, the "minutemen" were "ready to go to the relief of strikers when they were in danger. The hour of the day or night was of no consequence. All that was necessary to obtain the help of these minutemen was to get word to them and they were off."[14]

Armed and organized, the striking miners unleashed their rage upon the Baldwin-Felts guards. They hid in the hills and sniped at individual guards, and squads of miners attacked companies of Baldwin-Felts men. In one instance, miners surrounded a camp of guards during the night, cleared away the underbrush, and silently waited till dawn. When the guards awoke and began preparing breakfast, the miners opened fire, killing thirteen to fifteen of them. At Mucklow, in reaction to the "Bull Moose Special" incident, the miners ambushed a company of mine guards. At least sixteen people were killed after several hours of fighting. Both sides were prone to bury their dead without letting the other know their casualties, and therefore the total number of dead is unknown. One observer estimated that the miners killed as many as 150 of the guards.[15]

The reactions of the miners to the deaths of the mine guards were indicative of their hatred for them and of the cruel dimensions that the strike had assumed. Joy spread through the tent colonies when a miner announced: "Good news, boys, Gaujot [a mine guard] is dead." While most of the miners expressed happiness, one simply commented, "Just another rat gone." On the coffin containing the body of a Baldwin-Felts guard whom they had killed, the miners attached a sign that read:

> Gone to Hell.
> More to Go.
> Damned Thugs.

On a train carrying wounded miners and mine guards to a hospital following a gun battle, a railroad conductor found a miner spitting tobacco juice in the eyes of a dying mine guard and heard the coal digger say,

"You'd better die, you son-of-a-bitch, for if you don't, I am going to kill you the minute we get out of the hospital."[16] Jones, who, according to a biographer, "never before and never again would . . . be so blatant in her appeals to violence," held up the bloody coat of a wounded mine guard and screamed: "This is the first time I ever saw a goddamned mine guard's coat decorated to suit me." The coat was then cut into pieces, which the miners wore as souvenirs.[17]

Obviously, a self-righteous enthusiasm for violence against the enemy was no longer the prerogative of the coal operators and their hirelings. Nor were the power advantages, which a willingness to use terror for political advantage often brings, now restricted the the operators and the guards. "Industrial history affords no more prolonged, cruel fight," wrote the editors of *Coal Age*, "than has been carried on during the last year in the southern coal fields of West Virginia."[18] The operators had encouraged violence for years; the situation became more deadly when the miners made it a sword that cut both ways.

The miners' violence involved more than the ruthless slaying of mine guards; as Chaplin noted, "These men are not in this struggle to fight; they are in it to win." The miners blew up the tipples of operating mines and the trains carrying coal that had been mined by scabs. They met trains that were bringing strikebreakers to the strike zone and forced the potential scabs to evacuate—an action that often pitted black strikers against black strikebreakers and immigrant strikers against immigrant strikebreakers.[19] The solidarity of black and white, Protestant and Catholic, immigrant and native miners was unbreakable.

The militancy of the miners caught the attention of radicals throughout the country. With Haywood and Max Eastman on the platform, a speaker from the Industrial Workers of the World (a Wobbly), Joshua Wanhope, told a mass meeting of radicals in New York, "There is hope for the laboring class as long as there are in it, as shown in West Virginia, men who, seeing that it is a case of killing or being killed, are willing to take guns and do a share of the killing."[20]

The solidarity of the miners inspired them. "Solidarity is something more than a word in Kanawha County," wrote Chaplin; "it is a tremendous and spontaneous force—a force born in the hot heart of the class struggle." Inspired himself, Chaplin wrote the song that became the national anthem of the American labor movement—"Solidarity Forever":

> When the union's inspiration through the workers'
> blood shall run,
> There can be no power greater anywhere
> beneath the sun.
> Yet what force on earth is weaker than the

feeble strength of one?
But the union makes us strong.[21]

The militant solidarity involved the women of the company towns as well as the men. It was the women who prevented the "Bull Moose Special" from making a return trip through the valleys by tearing up the railroad tracks. During a gun battle, when Maud Estep saw her husband killed, she grabbed his gun and shouted, "Make every shot count boys, do not waste your ammunition. My man is dead; his head is shot off. Give them hell." Then she began shooting. Mobs of women attacked strikebreakers with broomsticks and, according to a veteran of the labor war, "beat the hell out of them." "In West Virginia," a journalist from San Francisco reported, "women fight side by side with men."[22]

While women suffered from the same injustices of the company town (e.g., insecurity of housing, coal scrip, and the company store) as their husbands, they suffered even more from its peculiar physical conditions. The women had to try to keep the house clean in towns bathed in coal dust because the coal companies failed to provide the needed sanitary facilities. This task was compounded by the need to go to the creek and carry buckets of water several hundred yards because the company refused to install running water. "It's hard to get a house clean," complained a miner's wife, "when you're tired out from carrying water." The lack of running water forced the women to spend at least two days a week in the cold mountain stream washing clothes. For one woman, "the worst thing that could happen in the company town" occurred "when the clothes line broke." The men recognized the plight of the women in the company town and, once unionized, sought to ease it. A miner later wrote the *UMWJ* that "the women have to carry all their water a long distance in pails, to do their washing and for other purposes and it is a greater burden than most people might think. The coal miners of West Virginia should start an agitation and keep it up until this is remedied."[23]

Superstition and company policy prevented women from working in the dangerous West Virginia mines, but the women encountered a "subtler kind of danger at greater psychological distance . . . the ever present danger of her losing her main source of security, her husband." Therefore, a woman in a company town knew that she, too, was "potentially expendable, not physically like her husband, but psychologically."[24] The high rate of mine fatalities, in addition to increasing the women's hatred of the company, reinforced their sense of unity and oneness. A former resident of a company town recalled:

> Not a week passed but that tragedy touched some home. When a housewife chanced to glance through the window and see a group of

> miners bearing an improvised stretcher between them, she spread the alarm. In a twinkling women were on the porches, wiping hands on aprons, calling to one another. And before the grim-faced bearers . . . were halfway to the doctor's office, a multitude of folk trailed in their wake. Anxious, distraught women and children, uncertain of the fate of their loved ones, demanded to know the identity of the victim. When the dreaded news was revealed, the women gathered around their hysterical sister and offered comfort.[25]

The women possessed the same hostility for the coal companies as their husbands. Herein might lay part of the great appeal of Jones. The miners accepted and listened to her not because she drank and cursed like a man, but, as Chaplin explained, "She might have been any coal miner's wife ablaze with righteous fury." In this and all later strikes in southern West Virginia, the women strongly supported and fought for the union. Mooney later reflected, "I am incapable of describing the courage displayed by the heroic women who passed through the strikes."[26]

The story of the Paint Creek–Cabin Creek strike is more than a story of fighting and cruelty and the hardships that the men and women endured in the tent colonies. In these colonies they sang, danced, and played games; as much as they could, they enjoyed life. Chaplin commented, "In spite of the 'heart throb' articles in some of the daily papers, these people are not objects of pity. They are doing pretty well in their tents. . . . The fact that many of the strikers seem to rather enjoy the situation makes some of the local respectables furious with rage. It isn't just what one would expect of a striker to see him holding his head high and walking around as if he owned the whole valley."[27] This was the thrill of rebellion. In reference to mass strikes, David Montgomery has noted that "anyone who has experienced such moments knows the joy of assertion and comraderie and the extraordinary ingenuity which well up from those taking the risks and sacrifices of the struggle." An imprisoned Wobbly said it best sixty years ago: "In the struggle itself lies the happiness of the fighter."[28]

This was also the "gaiety, the festival atmosphere" that "marks so many of the mass actions . . . even when colored by anger and bitterness." It stems from the break with the "day-to-day boredom, monotony, subservience, and limitation of individual possibilities of which most people's daily lives are made up." The excitement also results from the "people's understanding that they can initiate and control action and themselves make the decisions about their lives." The union, to the southern West Virginia miners, had become, not a formal institution, but an intense, emotional unity. Although he had been a member of the

UMWA since 1902, Mooney recalled that the Paint Creek–Cabin Creek strike was "my first real introduction to unionism."[29]

Beliefs in what unionism should represent accompanied that explosion of union spirit on Paint and Cabin creeks. Early in the strike, the miners had presented the operators with their demands. Bread-and-butter issues, such as an increase in pay, ranked low in their order of importance; their chief concerns were recognition of the union and abolition of the mine-guard system, demands that the miners connected with the broad issues of freedom and justice.[30] The miners carried banners at their strike meetings which declared:

> Mountaineers are always free.
> Out of the State with the Baldwin murderers.
>
> No Russia for us. To hell with
> the guard system.
>
> Governor, Why don't you send the Baldwin
> bloodhounds out of the state.
>
> We have been sacrificed to the gods of greed.[31]

During one meeting, Jones told the striking coal diggers, "Your banners are history; they will go down to the future ages, to the children unborn, to tell them the slave has risen, children must be free."[32] This emphasis on children was one of the principal ways miners made clear that their motives were not narrowly selfish.

Beginning with this strike, and in every later strike in southern West Virginia, this emphasis on children was maintained; the victory was not for themselves, but for their children. The wife of a striking Paint Creek miner declared, "All that we ask is that our children be given a chance." The journalist covering the strike for the *Milwaukee Leader* reported that "the fight of the West Virginia miners is a struggle to preserve the coming generation from being despoiled by the profit hungry mining industry."[33]

No one was more eloquent about this point than Jones. "It is freedom or death, and your children will be free," she told the striking miners. "We are not going to leave a slave class to the coming generation." The miners agreed with her. When Jones shouted that "the next generation will not charge us for what we have done; they will charge and condemn us for what we have left undone," the miners began screaming, "That is right!"[34]

It was around the image of children that bread-and-butter issues revolved when they appeared. "As a believer in God's word I appeal to you to protect your little children by giving them a reasonable living," declared a union official during a strike in Raleigh County. It was around the image of childen that the union revolved. "I think the United Mine Workers is the best organization on earth for the poor children of any mining town," wrote another miner; "I am proud that I am a mine worker." It was around the image of children that all subsequent strikes in southern West Virginia would, at least partially, revolve. "I say to you that any man in this gathering today," a miner-preacher shouted to a group of miners during a walkout, "who does not join this strike and stand by it, even until death—for the sake of the children, is not worthy to call himself a Christian because he is not willing to stand up for the Kingdom of heaven. . . . The children are the Kingdom of Heaven." The coal operators were the enemy, and the miners saw them as threatening not only their immediate needs, but also their basic rights as men as well as the rights of society now and in the future. They were striking for their children and for brotherhood and justice. "We must redeem the world," Jones told the striking miners of Paint and Cabin creeks.[35]

A lull in the fighting occurred when the governor declared martial law and sent the entire state militia into the strike zone. The miners initially welcomed the troops, hoping that they would disarm both sides and permit a peaceful end to the strike. They even raised an American flag in honor of the arrival of the soldiers.[36] Those attitudes swiftly changed, however, as the miners became aware that the state militia was not there to restore the peace, but to break the strike. In his general study of the use of martial law in American history, constitutional historian Robert Rankin wrote that "in the history of the United States, martial law has never been used on such a broad scale, in so drastic a manner, nor upon such sweeping principles as in West Virginia in 1912–1913 during the Paint Creek trouble. Here is the climax of the use by the state of its power to declare and to carry into effect martial law with all its force."[37]

Instructed by the governor and their commanding officers, the soldiers arrested at least 200 striking miners and their leaders without warrants and detained them in makeshift jails, called bullpens, without the right of habeas corpus. In flagrant disregard of the U.S. Supreme Court decision in *Ex Parte Milligan*, the soldiers then established military tribunals and court-martialed over one hundred civilian miners. There was no pretense of balanced justice in this unconstitutional proceeding—the coal operator and Baldwin-Felts guards who were on the "Bull Moose Special" were never even questioned, while a striking miner was sentenced to five years in prison for telling an army officer to "go to hell." With true military efficiency, the miner was arrested on Tuesday, court-martialed on Wednesday, and was in prison on Thursday.[38]

This application of martial law caused a national uproar and, according to Rankin, "resulted in the most comprehensive discussion concerning the use of martial law and the manner of its exercise that has ever occurred in the history of the United States." The *New York World*, for example, editorialized, "West Virginia does what the United States can not do, it suspends the civil law in time of peace. . . . More than the welfare of one monopoly-ridden State is involved in this tyranny. It menaces the peace of every State. . . . The American people will not be denied trial by jury. They will not submit to despotism."[39]

Martial law and the military occupation of Paint and Cabin creeks halted the fighting during the winter months of 1912 and 1913 as the miners were overwhelmed by the power of the state militia. The miners, however, had not conceded defeat. Everywhere Chaplin went in the area under martial law, he heard the miners saying, "Just wait until the leaves come out," a reference to spring, when the leaves of the woody hillsides would provide the protection the miners needed to fight the state militia. Chaplin caught the intent of the miners' expression in his poem "When the Leaves Come Out":

The hills are very bare and cold and lonely,
  I wonder what the future months will bring?
The strike is on—our strength would win, if only—
  O, Buddy, how I'm longing for the spring!

They've got us down—their martial lines enfold us;
  They've thrown us out to feel the winter's sting.
And yet, by God, those curs can never hold us,
  Nor could the dogs of Hell do such a thing!

It isn't just to see the hills beside me
  Grow fresh and green with every growing thing;
I only want the leaves to come and hide me,
  To cover up my vengeful wandering.

I will not watch the floating clouds that hover
  Above the birds that warble on the wing;
I want to use this gun from under cover—
  O, Buddy, how I'm longing for the spring!

You see them here, below, the damned scab-herders!
  Those puppets on the greedy owners' string;
We'll make them pay for all their dirty murders,
  We'll show them how a starving hate can sting!

They've riddled us with volley after volley,
  We heard their speeding bullets zip and ring,
But soon we'll make them suffer for the folly—
  O, Buddy, how I'm longing for the spring![40]

Before the leaves came out, the international union capitulated. In April 1913 UMWA President John White and the newly elected governor of West Virginia, Henry D. Hatfield, approached the coal operators with another compromise settlement. Leaving out the rank-and-file miners' most important demands, particularly the ones asking for complete recognition of the union and the abolition of the mine-guard system, White and Hatfield asked the operators to grant (1) a nine-hour work day; (2) the right to select a checkweighman; (3) semimonthly pay; and (4) no discrimination against union miners.[41]

Although the operators had rejected earlier efforts to negotiate an end to the strike, by now they wanted to settle. The mines had been closed for nearly a year, and they had been unable to break the strike by intimidating the miners and were unable to import strikebreakers. And they, too, probably realized that the leaves were about to come out. Consequently, when the UMWA officials and Hatfield presented the new list of settlement terms, the operators readily accepted.[42]

The union officials, elated at ending the strike, were content with the compromise settlement. They apparently feared opposition to the agreement from the rank-and-file miners, however, for rather than submit the proposed contract for a referendum, as was usual in contracts involving union recognition and wage disputes, they called a convention of *selected* miners' delegates to vote on the settlement. Although Hatfield had sixteen of the delegates arrested upon their arrival in Charleston—the men known to oppose the compromise—the convention proceeded as planned, with Hatfield and union officials urging that the delegates accept the compromise. Following three days of "heated debate," the delegates ratified the contract, and the union officials announced that "the miners place all confidence in Governor Hatfield and leave it to him to see that they will be protected."[43]

The rank-and-file miners were outraged with the actions of the district union officials and with the compromise settlement. They claimed that the settlement was worthless since their most important demands were not included and the terms that were in it were already recognized by law, but not enforced. Further angering the miners was their belief that the coal operators could not have held out much longer and would have been forced to grant all their demands. The miners on Paint and Cabin creeks flooded the national office with petitions that assailed Hatfield and the contract. One local union complained that the settlement was "dangerous to the welfare of the union of coal miners in West Virginia and is hereby branded by this body as traitorous and delusive and not in any way in keeping with our peculiar demands." Because they wanted to stay on strike, the members of this local explained that they had been "constantly hounded by the officials of our union together with certain representatives of Governor Hatfield to go from our tents and seek employment." The petition asked that the contract be rescinded and concluded that the "actions of our officials have been absolutely contrary to the will and judgment of the Cabin Creek miners and productive of universal dissatisfaction." The members of another local union declared that they were "disgusted at the action of our officials in forcing upon us a method not accepted by us" and claimed that they would "ignore every act" of the officials until new ones were obtained.[44]

To force the rank and file to accept the settlement, Hatfield issued a "36-hour ultimatum," ordering the miners to return to work within that time or to face deportation from the state. To enforce this decree, the governor sent soldiers into the coal fields to question each worker and then to escort him back to the mines. The coal diggers, with little alternative, accepted the governor's dictates.[45]

Hatfield's military rule knew few limits. When two local Socialist newspapers pointed out that his actions were despotic and urged the miners to remain on strike, the governor recalled part of the state militia and sent the soldiers to Charleston and Huntington to smash the presses and incarcerate the editors.[46]

The suppression of the two newspapers had sudden and dramatic national repercussions, as the U.S. Senate now felt compelled to investigate the conditions surrounding the Paint Creek–Cabin Creek strike. In March 1913, Senator John Kern of Indiana had introduced a resolution in the Senate calling for a congressional investigation of the strike and, for the first time in American history, a senatorial investigation of the acts of a state government. Both the West Virginia coal operators and senators who advocated states' rights vigorously opposed the resolution and seemed ready to defeat it when Hatfield suppressed the newspapers. The Senate passed the Kern resolution shortly afterward by a "viva voce vote."[47]

After months of hearings, U.S. Senate investigators released a report that denounced the governor, the military authorities, and the coal operators. Citing numerous violations of the U.S. Constitution, West Virginia state laws, and human decency, they pointed out that miners were court-martialed and detained in jails "while the civil courts were open," in clear violation of the Supreme Court's decision in *Ex Parte Milligan*. They also showed that the miners were denied due process of law. For example, "no attempt was made to try them before a jury," although "the offenses for which the parties were tried . . . were offenses which could have been punished under the civil law and in the common-law courts of the state." The committee attacked Hatfield for his "arbitrary arrests outside the martial-law zone" and his selection of members of the drumhead courts. According to the senators, "The military tribunal deemed itself bound alone by the orders of the commander-in-chief, the governor of the state, and in no respect bound to observe the Constitution of the United States or the constitution or the statutes of the State of West Virginia." Shortly after the release of the report, Senator Kern told a convention of the American Federation of Labor meeting in Indiana, "This investigation has made an end to military law in this country."[48]

While the U.S. Senate committee investigated, the coal diggers rejected the settlement and renewed the strike. Although two months had elapsed since the UMWA convention had ratified the contract, the rank-and-file miners had never accepted the compromise settlement. Following the ratifying convention, the miners had staged wildcat strikes, had fought pitch battles with mine guards, dynamited mines, and had burned tipples in protest against the settlement. Meanwhile, union locals expelled district officers or asked for the recall of district and national union officials who had negotiated the compromise.[49]

The miners also prepared to renew the strike. They purchased materials and their own land on which they could establish tent colonies, built commissaries to procure and distribute food, and organized the strikebreakers who had been brought in during the initial phase of the strike. They also conducted their own conventions and held "monster" mass meetings in which they announced they would strike if their original demands were not granted.[50]

When reports of the impending strike reached Charleston, Hatfield and union representatives rushed to the coal fields and asked the miners to remain on the job, while they attempted to negotiate a new contract. Claiming that the UMWA officers did not represent them, the miners walked off the job. The UMWA hierarchy, embarrassed to have to recognize a strike that they had declared settled, excused the reversal by claiming that the operators had abrogated the settlement by "bad faith" in blacklisting the leaders of the initial phase of the strike, and capitulated to the rank-and-file. With the U.S. Senate daily exposing the gubernatorial and military actions during the martial-law periods of the first phase of the strike, the coal operators could not depend upon the governor to issue yet another proclamation of martial law. Consequently, they quickly granted all of the miners' original demands. The Paint Creek–Cabin Creek strike was won, and the miners of Kanawha and New River coal fields had a recognized union.[51]

The victory involved more than the winning of strike demands. It identified and solidified class lines, showed the miners the power of collective action, and taught them that unity could work, not only against the companies, but also against the state and even their own union. It demonstrated that they could initiate and control action and make decisions about their lives.

Immediately following the strike, an explosion of wildcat strikes erupted throughout the Kanawha and New River coal fields. The miners conducted these strikes to secure the dismissal of nonunion personnel and company spies and to obtain wage scales, checkweighmen, and the closed shop. The editor of *Coal Age* exclaimed that "petty strike fever" had stricken the southern West Virginia miners and noted that "all of these strike contentions have been of a somewhat trivial nature."[52] To the miners, the wildcat strike was a source of power and a means of imposing their will upon a hostile coal company over issues that touched them most immediately and deeply.

The Paint Creek–Cabin Creek strike also created a situation in which violence was accepted as a necessary, justifiable, and, at times, pleasurable tool to be used against the operators and mine guards who had adopted the use of terror earlier. Shortly after the strike, a miner wrote the *UMWJ* about the killing of a mine superintendent who had "always kept the miners under an iron rule." The slaying was justified because "as long as they believe in keeping the miners as they would slaves, and as long as they believe in that infamous guard system, peace and harmony is in doubt."[53] Terror now worked both ways.

The betrayal of the strike by the national union caused the miners to rely more heavily on the cadre of rank-and-file leaders who, sharing their experiences, had led the way to victory at Paint and Cabin creeks. Men such as Keeney, Mooney, Bill Blizzard, Lawrence Dwyer, and William Petry, who could command the respect and loyalty of the miners, were to give leadership for a truly rank-and-file working-class movement in 1915 and 1916.

These new leaders and the Paint Creek–Cabin Creek experience prompted repeated efforts among the miners to gain control of District 17 from its conservative, corrupt, and undemocratic officials. They formed a "rump" district union—the West Virginia Mine Workers—and remained outside the UMWA, actions that forced the national office to dismiss the incumbent officers and hold new elections. In November 1916 Keeney was overwhelmingly elected president of District 17, and Mooney became its secretary-treasurer. The rank and file had taken over District 17.[54]

The Paint Creek–Cabin Creek strike was fought not for the narrow reasons of higher wages and shorter hours, but for justice, fraternity, and liberty, principles that were born of the miners' local experiences rather than any general ideology. The dynamics of the strike sprang not from the Socialist party, nor the UMWA, but from the rank and file. The support of the Socialists was important; they had shown the miners ways to beat the mine-guard system, especially in circulating information, and they provided the miners with an ideological justification foı violence and an understanding of class injustice. However, the miners did not find in socialism the broad principles that they espoused;[55] their idealism was still in search of an ideology.

## NOTES

1. G. H. Edmunds, letter to the editor, *UMWJ*, Aug. 15, 1912, 1, 3; Evelyn Harris and Frank Krebs, *From Humble Beginnings: West Virginia State Federation of Labor, 1903–1957* (Charleston, W.Va., 1960), 73; Jeremy Brecher, *Strike!* (San Francisco, 1972), 236. For the labor unrest that the strike prompted throughout southern West Virginia, see, for example, George Wolfe to Justin Collins, Mar. 19, 1914, Justin Collins Papers, West Virignia University Library, Morgantown, W.Va., and *Beckley* (W.Va.) *Post Herald*, Aug. 2, 1975.

2. *UMWJ*, June 21, 1915, 5, Oct. 24, 1912, 4, Sept. 19, 1912, 1, and a UMWA circular, "Notice of Assessment," dated Feb. 10, 1913, John Mitchell Papers, Catholic University of America, Washington, D.C.

3. Testimony of Brant Scott, U.S. Congress, Senate Committee on Education and Labor, *Conditions in the Paint Creek District, West Virginia*, 63rd Cong., 1st sess., 3 vols. (Washington, D.C., 1913), 1:500–1 (hereafter cited as *Paint Creek Hearings*); testimony of Lee Calvin, ibid., 640–41; testimony of George Williams, ibid., 2:1854–65; testimony of Quinn Martin, ibid., 1:937; *UMWJ*, Jan. 16, 1913, 1, and Mar. 13, 1913, 1–3.

4. Fred Barkey, "The Socialist Party in West Virginia from 1898 to 1920" (Ph.D. diss., University of Pittsburgh, 1971), 110; *Coal Age* 4 (July 12, 1913): 66; *UMWJ*, July 4, 1912, 8; Ralph Chaplin, "Violence in West Virginia," *International Socialist Review* 13 (Apr. 1913): 731–32; testimony of Gianiana Seville, *Paint Creek Hearings*, 1:478–80.

5. Testimony of Nina West, *Paint Creek Hearings*, 1:465–68; Chaplin, "Violence in West Virginia," 731–32; *UMWJ*, July 11, 1912, 1, Mar. 20, 1913, 1, Sept. 19, 1912, 1.

6. *UMWJ*, May 16, 1912, 2, 5; Fred Mooney, *Struggle in the Coal Fields*, ed. Fred Hess (Morgantown, W.Va., 1967), 16; *UMWJ*, Dec. 26, 1912, 2; Selig Perlman and Philip Taft, *History of Labor in the United States, 1896–1932*, 4 (New York, 1966), 330.

7. *UMWJ*, Feb. 13, 1913, 1, 8, and testimony of Lee Calvin, *Paint Creek Hearings*, 1:640–42.

8. This poem is excerpted from Ralph Chaplin, *When the Leaves Come Out* (Cleveland, 1917), 31. For Chaplin's career with the Huntington *Socialist and Labor Star*, see David Corbin, *The Socialist and Labor Star* (Huntington, W.Va., 1971), 3–4, and Ralph Chaplin, *Wobbly, the Rough and Tumbling Life of an American Radical* (Chicago, 1948), chap. 11 ("The Kanawha Miners' Strike"). All the songs and poems by Chaplin cited in this chapter were either written during or inspired by the Paint Creek–Cabin Creek strike.

9. *Milwaukee Leader*, May 29, 1913; Thomas Cairnes (president of District 17) to John White, Apr. 24, 1912, District 17 Correspondence Files, UMWA Archives, UMWA Headquarters, Washington, D.C.; *UMWJ*, Oct. 24, 1912, 4, and Sept. 19, 1912, 1. The levies mounted to over $600,000. "Peace in West Virginia," *Survey* 30 (Sept. 13, 1913): 709 The union's retreat may have been in line with Brecher's thesis that unions act as moderating influences on rank-and-file militancy during mass strikes. Brecher, *Strike!*

10. *Paint Creek Hearings*, 1:520–24; *UMWJ*, Sept. 19, 1912, 1, and Oct. 24, 1912, 4; P. H. Kelly to Ren [J. A. Renahan], Sept. 17, 1912, Collins Papers.

11. Mary "Mother" Jones, *Autobiography of Mother Jones*, ed. Mary Field Parton (Chicago, 1925), 155.

12. Mooney, *Struggle*, ed. Hess, 27, and Jones, *Autobiography*, ed. Parton, 152.

13. Barkey, "Socialist Party in West Virginia," 113–19; Chaplin, *Wobbly*, 1; *Charleston Labor Argus*, Jan. 2 and 30, 1913; *Huntington Herald-Dispatch*, June 5, 1913. For an example of Socialists' encouragenment of direct action, see the speech of Harold Houston excerpted in *Paint Creek Hearings*, 3:2258–61. David Corbin, "Betrayal in the West Virginia Coal Fields," *Journal of American History* 65 (Mar. 1978): 987–1010.

14. Testimony of James Pratt, *Paint Creek Hearings*, 2:1760, and Mooney, *Struggle*, ed. Hess, 29, 31–32.

15. *Washington Times*, Aug. 4, 1912; the clipping in item 1941760, Document File 1942834, Record Group 94, Records of the Adjutant General, National Archives, Washington, D.C.; Barkey, "Socialist Party in West Virginia," 110; *New York Times*, Feb. 11, 1913; Perlman and Taft, *History of Labor*, 333; Chaplin, "Violence in West Virginia," 729–35; Mooney, *Struggle*, ed. Hess, 26; *Coal Age* 3 (Jan. 18, 1913): 119; Harold West, "Civil War in the West Virginia Coal Mines," *Survey* 30 (Apr. 5, 1913): 44.

16. Mooney, *Struggle*, ed. Hess, 17–18, 32–33, and *Coal Age* 1 (Sept. 14, 1912): 70–71.

17. Dale Fetherling, *Mother Jones, the Miners' Angel* (Carbondale, Ill., 1974), 86–87, 104, and *Coal Age* 2 (Sept. 14, 1912): 364.

18. *Coal Age* 3 (Apr. 26, 1913): 650.

19. Chaplin, "Violence in West Virginia," 735; *Paint Creek Hearings*, 2:1805–7; Mooney, *Struggle*, ed. Hess, 29–31; Barkey, "Socialist Party in West Virginia," 115; Harris and Krebs, *From Humble Beginnings*, 76–77.

20. *New York Times*, May 28, 1913.

21. Chaplin, "Violence in West Virginia," 730; Chaplin, *When the Leaves Come Out*, 28; Edith Fowke and Joe Glazer, *Songs of Work and Protest* (New York, 1973), 12–13.

22. Interviews with Frank Blizzard, Dry Branch, W.Va., Hobert Moss, Coalburg, W.Va., and John McCoy, Alum Creek, W.Va., summer 1975; Cora Older, "The Last Day of the Paint Creek Court Martial," *Independent* 74 (May 15, 1913): 1086–87; Mooney, *Struggle*, ed. Hess, 36–37.

23. Anna Rochester, *Labor and Coal* (New York, 1931), 90–91, and an interview with Mrs. Clyde Toliver, Sabine, W.Va., summer 1975. For the lack of running water and other problems involved in cleaning house in company towns, see U.S. Department of Labor, Women's Bureau, *Home Environment and Employment Opportunities of Women in Coal-Mine Families*, Bulletin no. 45 (Washington, D.C., 1925), 54–56; "Women and Child Labor," *Monthly Labor Review* 21

(Aug. 1925): 333–34; *Life and Labor Bulletin* 3 (Oct. 1925): 1; C. M. Walker, letter to the editor, *UMWJ*, Sept. 18, 1918, 12.

24. Lewis Field, Reed Ewing, and David Wayne, "Observations on the Relations of Psychological Factors to Psychiatric Illness among Coal Miners," *International Journal of Social Psychiatry* 3 (Autumn 1957): 143.

25. Jack Rogers, "I Remember that Mining Town," *West Virginia Review*, Apr. 1938, 203–5.

26. Chaplin, *Wobbly*, 120, and Mooney, *Struggle*, ed. Hess, 12. For examples of women's later support of the union, see Mrs. J. B. Crabtree, letter to the editor, *UMWJ*, Mar. 1, 1921, 5; Mrs. T. J. Parsley, letter to the editor, ibid., Apr. 15, 1921, 15; Maggie Workman, letter to the editor, ibid., Mar. 15, 1925, 12.

27. Testimony of Maud Fish, *Paint Creek Hearings*, 2:471; Mooney, *Struggle*, ed. Hess, 38; Chaplin, "Violence in West Virginia," 729.

28. David Montgomery, "Spontaneity and Organization, Some Comments," *Radical America* 7 (Nov.–Dec. 1973): 71. The Wobbly speaker is quoted in Danie Stewart, Introduction to Corbin, *The Socialist and Labor Star*, iii.

29. Brecher, *Strike!*, 237, 247, and Mooney, *Struggle*, ed. Hess, 15.

30. For the listing of the miners' demands, see chap. 2, or Harris and Krebs, *From Humble Beginnings*, 73.

31. *Paint Creek Hearings*, 3:2262, 2282. The miners also expressed these attitudes in their letters to the *UMWJ*. For example, Hubert Kirk wrote, "I hope the time will come when there will not be such a thing as a scab or a guard in the country, and then we will be free people." Letter to the editor, *UMWJ*, Mar. 2, 1913, 2.

32. Jones is quoted in *Paint Creek Hearings*, 3:2264, 2268–69.

33. *Milwaukee Leader*, May 26, 1913.

34. *Paint Creek Hearings*, 3:2268.

35. A. S. Riffle, Circular to Local Union Members, ca. Dec. 1919, Straight Numerical File, item 205194–50–49, Record Group 60, General Records of the Department of Justice; Gilbert Canfield, letter to the editor, *UMWJ*, Feb. 1, 1922, 7; William Shepard, "The Big Black Spot," *Collier's Weekly*, Sept. 19, 1931, 12–14; *Paint Creek Hearings*, 3:2264, 2268–69.

36. Robert Rankin, *When Civil Law Fails: Martial Law and Its Legal Basis in the United States* (Durham, N.C., 1939), chap. 6 ("The Use of Punitive Martial Law in West Virginia"), and Barkey, "Socialist Party in West Virginia," 115.

37. Rankin, *When Civil Law Fails*, 85.

38. Alpheus Mason and William Beany, *American Constitutional Law* (Englewood Cliffs, N.J., 1954), 451–55; Perlman and Taft, *History of Labor*, 332–33; Older, "Last Day of the Paint Creek Court Martial," 1086–89; testimony of S. F. Nanta, *Paint Creek Hearings*, 1:188–90.

39. The quote from the *New York World* is excerpted in "Sifting West Virginia's Wrongs," *Literary Digest* 46 (June 7, 1913): 1259–60. Rankin, *When Civil Law Fails*, 136. For a sampling of the newspapers' criticism of martial law in West Virginia, see ibid., and "The Constitution in a Labor War," *Literary Digest* 46 (Apr. 5, 1913): 756–68. For a popular protest against this abuse of martial law, see Straight Numerical File, item 165095–50, Record Group 60, General Records of the Department of Justice, and item 1941760, Documentary File 1942834, Record Group 94, Records of the Adjutant General. The president of the West Virginia Bar Association also assailed this abuse of martial law. See William G.

Matthews, *Martial Law in West Virginia*, Address before West Virginia Bar Association, July 16, 17, 1913 (Washington, D.C., 1913).

40. Chaplin, "Violence in West Virginia," 127, and *Wobbly*, 125–27. Also see Mooney, *Struggle*, ed. Hess, 38.

41. Wyatt Thompson, "How Victory Was Turned into Settlement," *International Socialist Review* 14 (July 1913): 12–17, and *UMWJ*, Apr. 24, 1913, 1, and May 15, 1913, 1–2.

42. Thompson, "Victory into Settlement," 12–17, and *UMWJ*, Apr. 24, 1913, 1, and May 15, 1913, 1–2.

43. *UMWJ*, Apr. 24, 1913, 4, and May 1, 1913, 1; *Charleston Gazette*, May 10, 1913; Thompson, "Victory into Settlement," 12–17; *Wheeling* (W.Va.) *Majority*, May 1, 1913. For the praise of Hatfield and support of the compromise by officials of District 17, see *Huntington* (W.Va.) *Herald-Dispatch*, May 28, 1913.

44. Thompson, "Victory into Settlement," 12–17; *Socialist and Labor Star*, May 2 and 9, June 6, 1913; *Wheeling* (W.Va.) *Majority*, Apr. 17 and 24, May 1, 8, and 15, 1913; Mooney, *Struggle*, ed. Hess, 39. The miners were correct in claiming these settlement terms were already recognized by law. For the right to a nine-hour day and the right to select a checkweighman, see *The (Mining) Code of West Virginia, 1906* (St. Paul, Minn., 1906). For prohibition of discrimination against union miners, see W.Va. State Legislature, *Acts, 1899*, 163. District 17 Correspondence Files at the UMWA Archives contain several petitions from local unions attacking the settlement and Hatfield. The quotes in this paragraph are from two: a petition from Miami Local to the International Executive Board, June 7, 1913, and a petition from Crown Hill Local to the International Board, June 7, 1913.

45. *Huntington* (W.Va.) *Herald-Dispatch*, Apr. 24, 1913; *New York Call*, Apr. 29, 1913; Wyatt Thompson, "Strike Settlement in West Virginia," *International Socialist Review* 14 (Aug. 1913): 87–89.

46. *Charleston Gazette*, May 1, 1913; *Socialist and Labor Star*, May 2 and 9, 1913; Leslie Marcy, "Hatfield's Challenge to the Socialist Party," *International Socialist Review* 13 (June 1913): 882; *Huntington* (W.Va.) *Herald-Dispatch*, May 9 and 10, 1913.

47. *Huntington* (W.Va.) *Herald-Dispatch*, May 28, 1913, and "U.S. Senate Decides to Investigate West Virginia," *Current Opinion* 55 (July 1913): 3–4.

48. U.S. Congress, Senate, *Digest of Report on Investigation of Paint Creek Coal Fields of West Virginia*, 63rd Cong., 2d sess. (Washington, D.C., 1914), 1–9, and *Miners' Herald* (Montgomery, W.Va.), Oct. 3, 1913.

49. *Paint Creek Hearings*, 2:1361–62; *New York Call*, July 25, 1913; *Huntington* (W.Va.) *Herald-Dispatch*, June 4, 1913; Mary Marcy, "Unions Repudiate Debs's Escort Haggerty," *International Socialist Review* 14 (July 1913): 22–23.

50. *New York Call*, June 16–July 6, 1913; *Socialist and Labor Star*, June 21, July 4, 1913; *Charleston Gazette*, June 25–July 7, 1913; *UMWJ*, June 19, 1913, 1.

51. UMWA circular, June 23, 1913, Mitchell Papers; *Milwaukee Leader*, June 6, 1913; *Socialist and Labor Star*, June 21, July 4, 1913; *Charleston Gazette*, May 10, 13, June 25, 28, 1913; *New York Call*, June 27, 31, July 3, Aug. 16, 1913; *UMWJ*, June 19, 1913, 1, 5, July 24, 1913, 1, 4; "Peace in West Virginia," 709. The contract did not include abolition of the guard system. The state legislature, however, had passed the Wertz bill, which theoretically outlawed the use of mine guards, and the miners apparently were satisfied. Time would prove that state politicians had again duped the miners (see chap. 5). For the details of the Wertz

bill, see *UMWJ*, Feb. 27, 1913, 1; W.Va. State Legislature, *Acts, 1913*, 173–74.

52. *Coal Age* 4 (Aug. 9, 1913): 213, 4 (Aug. 23, 1913): 285, 4 (Sept. 20, 1913): 433, 4 (Sept. 27, 1913): 469, and W.Va. State Department of Mines, *Annual Report, 1914*, 16.

53. Joe Bruttaniti, letter to the editor, *UMWJ*, Nov. 9, 1916, 9.

54. Frank Keeney and Lawrence Dwyer, "A Public Statement Issued by the West Virginia Mine Workers," Apr. 14, 1916, Keeney to John White, May 10, 1916, White to Keeney, May 16, 1916, all in District 17 Correspondence Files; *Charleston Argus-Star*, Mar. 30, 1916; David Corbin, "'Frank Keeney Is Our Leader, and We Shall Not Be Moved,'" *Essays in Southern Labor History*, ed. Gary Fink and Merl Reed (Westport, Conn., 1977), 145–46, 153; *UMWJ*, Feb. 15, 1921, 1–3; *Charleston Gazette*, Sept. 2, 1921; Mooney, *Struggle*, ed. Hess, 41–42.

55. Corbin, "Betrayal in the West Virginia Coal Fields." Thus, one month after the strike, the *Socialist and Labor Star*, a paper that had been helpful to the miners, editorialized to its black readers, "We do not want the amalgamation of the races nor do you." Aug. 8, 1913.

CHAPTER V

# Conspiracies and Control

> In a word, I say to you, that your remedy, is to fight. . . . Your remedy, I repeat, is to fight, and fight boldly, aggressively, defiantly, and unitedly. . . .
>
> Working in harmony as a trained army, you can successfully resist the avowed purpose of the United Mine Workers to confiscate your property.[1]
>
> Z. T. Vinson

"The armed revolution in West Virginia [the Paint Creek–Cabin Creek strike] was partially successful," a speaker told a gathering of southern West Virginia operators, "because there was no real co-operation upon the part of the state's mine owners to resist [the UMWA]."[2] This speech, delivered three months after the end of the Paint Creek–Cabin Creek strike, reflected the alarm of the coal operators in the wake of the conflict. They realized that the year-and-a-half struggle had not been an ordinary labor-management dispute. *Coal Age*, the journal of the coal operators, declared that the "so-called strike on Cabin and Paint Creeks was in reality an armed insurrection, formulated by agitators hired by the union, and afterwards reinforced by the socialists." The strike turned the coal establishment's hatred and fears of unionism into paranoia, and the operators prepared for the fight. U.S. Senator Nathan Goff of West Virginia defended the actions of the governor and militia and the court-martialing of the miners, declaring in the halls of Congress, "You cannot fight a war with kid's gloves on."[3] When they adopted this position, the operators themselves were transformed from individual producers of coal into a class.

The coal operators had obviously experienced the "panic-fear" that George Rudé called "one of the most constant elements contributing to certain states of collective mentality." Rudé noted that "panic-fear" grips socioeconomic groups when there is a "threat, real or imaginary, to three matters of vital moment—to property, life and means of subsistence."[4]

The Paint Creek–Cabin Creek strike presented such a threat to the coal establishment, and the coal operators reacted with the kind of panic that created class solidarity and that produced the cooperation and uniformity that they needed to keep the union out of West Virginia and to convince themselves, their miners, and the nation that their interests were not brutally selfish ones, but were motivated by the common good.

As a result of the Paint Creek–Cabin Creek strike, for the next decade the coal operators perceived the UMWA as a monstrous force whose intention was not merely unionization but the destruction of the southern West Virginia coal industry and eventual domination of the United States. The operators of Mingo County declared that the UMWA "is unlawful per se, revolutionary in character, and a menace to the free institutions of this country." For the control of the southern West Virginia coal fields, the chairman of the Logan County coal operators' association exclaimed, UMWA officials "are willing to sacrifice the lives, not of themselves, but of their members in a treasonable attempt to override all constituted State and Federal authority."[5]

These alarmist perceptions of the UMWA and its vile intentions were more than, as some writers have claimed, simply public efforts to discredit the union movement. The operators expressed these same views in private. In letters to each other, the coal operators referred to the UMWA and union miners as "reds," "radicals," and "anarchists." Justin Collins expressed a common view of the operators when he wrote to the National Association of Manufacturers that unionism "in the end, through political clap-trap and demagogism, breeds socialism, communism and hence—national decay."[6]

The operators' paranoid fears of unionism were exacerbated by the realization that in the year that the Paint Creek–Cabin Creek strike ended, the UMWA enacted into its constitution a clause stating that the miners were entitled to the "full social value of their product." This provision caused one coal operator to contend that the UMWA "is not a labor union. . . . Its demands are not in accord with unionism, but are socialistic pure and simple." The Socialist party, he explained, "had captured and took charge of the UMW of A and changed its constitution to read . . . 'We are entitled to all of it [profits], without one cent going to the mine owner for the plant, the tipple, the mine, or lessar.'" The chairman of the Williamson Coal Operators' Association, Harry Olmsted, pointed out that the UMWA "definitely abandoned the trade union movement" in 1913, when it "changed its constitution, and declared its purpose to be that of securing for its members the 'full social value of their product.'" But worse, according to Olmsted, was that "their whole object is now and has been since 1912, to cripple all the protecting powers of government, so that their armies can march unmolested into the territory of non-Union mines and shoot down the workman and destroy the mining plants at will."[7]

The southern West Virginia coal operators had practical, economic reasons for opposing every aspect of unionism. They disapproved of being bound to union-negotiated wage scales that would be based, not on local conditions, but on conditions in competing coal fields. Union-wage scales also meant the elimination of one of their major advantages over other coal-producing states and would hinder their ability to compete in the national coal market. Furthermore, they were highly concerned about the inefficiency of union labor. "Union labor is very inefficient," explained an official of Island Creek Coal Company, "a fact which is well recognized and well known in the industry." The operators complained that under union contracts, coal companies could not discharge the less productive workers. With the union's grievance procedure, they claimed, every punitive action on the part of the company would "end up in Charleston [District 17 headquarters]." The coal company, under a union contract, could not offer bonuses and higher pay to the best workers. Hence, unionism stifled efficiency and individual initiative.[8]

Each specific union policy or demand carried a larger, more sinister dimension, for every demand was simply an integral part of a master plan. A Logan County operator explained:

> It is the purpose of the United Mine Workers, first to unionize the mines, and then increase their demands for their labor, and keep on in a progressive increase until at last it will not be worthwhile for anyone to own the mines, and then take possession of the mines themselves. . . . It is a demand of the United Mine Workers to get control of a necessity of life. In a word, to establish a soviet government. Lenin and Trotski demanded no more.[9]

Anything connected to the UMWA was viewed with alarm. Collective bargaining was "extortion" and therefore "unlawful and should be prevented." A Kanawha County coal operator explained that he opposed the "checkoff" because "we are furnishing them [UMWA] munitions of war to club us with all the time." In a letter to the U.S. attorney general, another coal operator charged that the "checkoff" was itself a "conspiracy . . . the acceptance [of which] means to turn over the management to them."[10]

Around 1910 the miners' union pushed for state laws that would require the certification of miners. The union hoped to decrease mine accidents by reducing the number of unskilled miners. The union also supported certification as a substitute for the failure of both the mine inspectors to enforce existing mine laws and the coal operators to heed them. In response to the UMWA efforts, the state legislatures of Indiana and Illinois passed the needed safety legislation.[11] But to the southern West Virginia coal establishment, the proposals carried sinister intentions. Collins claimed that this was a step toward unionization that would "lead directly to union dominance in West Virginia." Officials of U.S. Coal and Coke at Gary, West Virginia, declared that certification "was doubtless fostered by the miners' union with the intention of throwing a legal barrier around the miner and his job, and thus rendering it unlawful for the coal operators to place at work as miners men who had not seen at least two years' service in the mines in Pennsylvania."[12]

Similarly, Collins believed mine explosions were not accidents but union schemes to organize and destroy the coal industry in southern West Virginia. "Once union miners had caused the explosion," he wrote to a friend, "the union could run to the state legislature and ask for 'drastic legislation' against the coal operators." "The union with all their following, and demagogues throughout the country, would like to see all kinds of legislation enacted which will practically put the operator out of business." He wrote to another coal operator that "you will observe with what wonderful regularity [mine explosions] occur about the time the West Virginia legislature goes in session, and with what vigor and insistence the labor union lobby and all the labor union politicians insist upon licensed miners."[13]

The southern West Virginia coal operators panicked in 1913–14 when the railroads and coal operators of other states pushed for lowering the freight-rate differentials to help neutralize the competitive advantage of the superior and cheaper West Virginia coal. This conspiracy began, according to the southern West Virginia coal establishment, when the UMWA "gained control of the Ohio legislature as well as certain officials in Ohio." Subsequently, the Ohio legislature passed "unwise" coal-mining legislation that crippled the state's coal industry. The legislature then forced the railroads in Ohio to reduce their freight rates. The railroad companies in Ohio, the West Virginia coal operators explained, "realizing the dangers confronting them," then made "secret agreements" with the coal and railroad companies of other states to overcome their difficulties by raising the West Virginia rates.[14]

Hence, according to the southern West Virginia operators, the proposed changes in freight-rate differentials constituted an "illegal combination of railroads and coal operators and coal miners of other states." The purpose of these proposals was to force "the West Virginia Railroads, against the will and judgement of their operating officers, to increase freight rates on coal to the Great Lakes for the purpose of keeping

West Virginia coal out of the Western market." Another spokesman declared that the real intent was "to destroy the coal industry of this state and to hopelessly cripple West Virginia's railroads."[15]

The greatest conspiracy of all, however, was the alleged agreement between the coal operators that belonged to the Central Competitive Field and the UMWA to unionize southern West Virginia. This plot dated back to the creation of the Central Competitive Field agreement system in 1898, when the midwestern coal operators extracted from the union a promise—the southern West Virginia operators claimed a signed agreement—to organize the West Virginia coal miners. In its mildest form, the West Virginia operators saw the agreement as an attempt to "impose such terms and conditions [upon the southern West Virginia coal industry] as would benefit the operators of the Central Competitive field." In its most severe form, the operators declared that the agreement was designed to eliminate coal production in southern West Virginia totally. The president of Red Jackett Consolidated Coal and Coke, Mingo County, testified before the U.S. Senate that the purpose of the "infamous conspiracy of 1898" was "to destroy the business of these operators and to deprive the laborers in those fields of the opportunity to work." The operator explained that by the late 1890s the midwestern coal operators and UMWA officials

> discovered that those mountains and valleys of southern West Virginia ought never to have been settled and developed; that that region should have remained in a state of nature and that inasmuch as the mistake had been made of not preventing its development that mistake ought to be corrected by destroying what was already there. As there was no system of birth control when these industries were born, they accordingly determined to kill the thing they say should never have been brought into the world.
>
> And, Mr. Chairman, as stated, a part of the effort to stifle and to kill this coal industry of the South—including southern West Virginia—has been the effort to organize that territory.[16]

The Paint Creek–Cabin Creek strike was the culmination of that agreement. The southern West Virginia coal operators charged that the UMWA and Central Competitive Field operators, acting together, not only called the strike and supplied organizers and strike benefits to the striking miners, but they also supplied the guns and sent in outside miners to increase the number of strikers. The idea of this joint conspiracy was first mentioned in 1907. For the next five years, the coal operators frequently referred to it in conjunction with the union. With the Paint Creek–Cabin Creek strike, nearly every coal operator in southern West Virginia cited it as one of the major reasons for opposing the union.[17]

Between 1913 and 1922, the southern West Virginia coal operators felt surrounded by conspirators and potential invaders who wanted to deprive them of their property. "The businessman never gets his day in court," explained a Boone County coal operator who claimed that trade unionists had already captured most of American society and were attempting to socialize the nation. "The churches seeking popularity," the operator charged, "cater to socialist doctrines. The newspaper seeking popularity caters to the class which will give it the most circulation, the politician seeking for the same popularity caters to that class."[18]

The nonunion southern West Virginia coal fields, the operators claimed, were the nation's safety valve against labor's domination of the United States. If the UMWA organized southern West Virginia, the coal operators contended, "they can tie up the wheels of industry." "Shall the United Mine Workers get possession of the remaining nonunion mines and thus, like the United Mine Workers of Great Britain, be able to freeze and starve the people of the United States, to stop all industries unless they can have their way?" a coal operator asked.[19]

The Logan County Coal Operators' Association declared that if the UMWA unionized the southern West Virginia mines, "they will have a hundred and ten million people of this country by the throat." Therefore, they claimed, unionism "is not a matter of industrial democracy; it is a question of industrial autocracy. It is a demand of the United Mine Workers, or, rather their leaders, to get control of a necessity of life." "Such a centralized power having control of such a thing as our fuel supply," a coal operator stated to a U.S. Senate committee, "is un-American and dangerous in the extreme."[20]

Consequently, the great majority of the southern West Virginia coal operators refused to deal with the UMWA. Kanawha County coal operators declared their intentions of breaking the union in their coal fields and the contract they had been "forced" into signing, while the nonunion coal operators steadfastly opposed recognition of the UMWA. At the conclusion of one U.S. Senate investigation of mining conditions in southern West Virginia, Frank Walsh, former chairman of the U.S. Commission on Industrial Relations and co-chairman of the War Labor Board, made a plea for negotiations between the operators and the union; the operators' attorney quickly responded: "The committee knows our attitude. We will not have anything whatever to do, under any circumstances with the United Mine Workers of America or their representatives. We have no settlement to make and no conference. We stand absolutely on our rights."[21] Their rights were their property and prosperity and their view of the public good. A settlement or even a conference with UMWA officials would constitute surrender to a radical, traitorous force with which they were at war. The UMWA was an organization that they still claimed was alien to the southern West Virginia coal fields.

The coal operators still maintained that their miners did not want the union. Many of the miners "frankly state they will not work under the union; they will abandon mining rather than do so, and will seek other occupations." UMWA officials, a Logan County operator explained, "have been very jealous of the cordial relationship between the mine owner and the employees ever since we have had a field here. They know that this is a field which is a large producer, and that the employee is well treated, and that everything is cordial between capital and labor, and just as long as that exists, they can not form their organization here in these fields." Another coal operator told a U.S. Senate committee that his miners would "mob" any UMWA organizer who came into the area.[22]

The operators then offered various explanations for why their miners had struck and were joining the union. A coal operator in Kanawha County argued that his miners, who were really "good, honest, straight-forward citizens," joined the "anarchists and radical element" in the Paint Creek–Cabin Creek strike "because of the fearful word 'scab.'" A Boone County operator asserted that union organizers influenced miners through their wives and children. "The miner's children will say [to the miner] that the other children 'won't play with me; they call me a scab,'" the operator explained. He further elaborated: "They also influence the wife. She finds that the [other] women will not associate with her, and she says 'Women will not associate with me and they call me a scab's wife.'" The operator concluded that "there is this pressure put on him and it is absolutely impossible for him to resist."[23] A coal operator at Crumpler claimed that the union movement among his miners was "due . . . to the unflagging industry of its [UMWA] efficacious propagandists." A Raleigh County coal operator later charged that UMWA organizers were a "bunch of tough agitators, big strong rascals," who stirred up his miners "with bootleg whiskey."[24] Mingo County operators contended that UMWA organizers used terrorist tactics to recruit miners; they would destroy company property and then "threaten to kill" the miners who refused to join the union. Similarly, a Boone County operator testified that "the professional organizer unionizes a field by saying to the miners: 'If you do not do as we tell you and sign up we will punish you.' . . . These workmen cannot stand that terrorism."[25] The coal operators were determined to keep that terrorist, alien force out of their company towns.

The first need was to combine forces. "There is nothing more certain than that the United Mine Workers will take your properties out of your hands," a speaker told the southern West Virginia operators, "unless you organize among yourselves a force strong enough to resist them." He then advised a policy of containment: "You must have an efficient organization, opposing them at every point and every angle, and all the public officers will protect both you and your property from assault." The speaker finished with the major point of his address: "If you organize and prepare for war, you will not have labor troubles."[26]

Following the Paint Creek–Cabin Creek strike, coal operators' associations sprang up throughout southern West Virginia. Employers' associations have commonly functioned as "bargaining associations" in order "to deal collectively with unions." They have, however, also appeared as "militant associations" in order "to oppose collective bargaining and to discredit and defeat unionism."[27] While most of the coal operators' associations in other states appear to have been bargaining associations, those formed in southern West Virginia were militant associations.

In September 1913, two months after the Paint Creek–Cabin Creek strike, coal operators from southern West Virginia met in Huntington and formed the Operators' Protective Association, with E. E. White, a Raleigh County operator, as chairman. The purpose of the meeting and of forming the association was to take "concerted action for the protection of the property of all the coal operators in West Virginia from destruction by Socialists, otherwise known as the United Mine Workers of America." In pursuit of this purpose, the Operators' Protective Association established a "million dollar defense fund" to fight the union. A central committee of twenty operators was established to supervise the fund and to distribute it as needed.[28]

Coal operators' associations were not novel to southern West Virginia in 1913. In 1903, the operators in the Kanawha coal field had formed the Kanawha County Coal Operators' Association, which signed with the UMWA following the Paint Creek–Cabin Creek strike. According to a Boone County operator, the Kanawha County operators did not realize then that "radicals and anarchists" controlled the miners' union. In 1907 Collins spearheaded the formation of the Tug River Coal Operators' Association. "Unless we adopt measures and keep at it constantly to prevent the union from getting hold of us," Collins wrote to Tom Felts, "they will succeed in doing so. I do not intend to have any union and I think the time to start preventing it is right now." When that association signed a contract with the UMWA, Collins fumed. "We paid in our money to fight the union," he wrote to his superintendent, "this being the sole purpose of the assessment and payment." Collins then advised his superintendent to take the company out of the Winding Gulf Coal Operators'

Association and to align with the newly formed Pocahontas Coal Operators' Association.[29]

Despite his disgust with the Winding Gulf Association, Collins was somewhat pleased that coal operators' associations, such as the Pocahontas Association, were appearing throughout southern West Virginia to combat the union movement. "Most of the operators of the state," he explained, "seem to be waking up to the fact that unless they look out . . . [for] union labor, that they are the last."[30] Following the Paint Creek–Cabin Creek strike, more and more operators recognized the need for united action against the UMWA. The operators in Mingo County organized the Williamson Coal Operators' Association; its purpose, according to its secretary, was that the operators "might better fight the union." The association established special committees to guard against union organizing efforts.[31] In January 1913 the operators in the Logan coal field formed the Guyan Valley Coal Operators' Association. Speaking for the association, George Jones, the general manager of Lundale Coal Company, Amherst, Logan County, explained that "we oppose the unionization of the Guyan coal field. We are against the union and expect to do everything in our power to prevent its coming into our mines. There have been some IWWs, Bolshevists, and these men have sown seeds of dissension, and we have picked these men out and discharged them."[32] Writing about the "new harmony" among the coal operators, a superintendent wrote Collins that "a far greater spirit of uniting together for one common cause exists than was [even] exhibited at the meeting held in 1913 when one million dollars was pledged for a defense fund."[33]

The coal operators' associations used various methods to block the union movement. They established lobby committees to urge Congressmen to pass legislation restricting labor unions. The Smokeless Coal Operators' Association established an office in Washington, D.C., to lobby more effectively. The associations created publicity bureaus to promote their cause and to discredit the UMWA and its leaders. The Williamson Coal Operators' Association bought a local newspaper, from which it delivered anti-union tirades and attacked district and national UMWA officials.[34]

The most concerted effort at defense was the strengthening of the mine-guard system. Shortly after the Paint Creek–Cabin Creek strike, the West Virginia legislature, probably in response to the national publicity on the activities of the Baldwin-Felts guards during the strike, seemed to outlaw the mine-guard system with the passage of the Wertz bill, which read, in part: "it shall be unlawful for any . . . deputy or deputies to act as, or perform any duties in the capacity of guards or watchmen for any private individual, firm or corporation . . . or to represent, in any capacity, as officers of the law, any individual, person, firm or corporation."[35]

Neither the coal operators nor the mine guards had anything to fear from this law—and they knew it. The secretary of the West Virginia Coal Association, Jessie Sullivan, pointed out to concerned coal operators that the Wertz bill conveniently did not contain a penalty clause for violators. Consequently, the coal companies continued to employ privately hired and privately paid guards to serve as public officers until 1933, and the state did not prosecute the companies. The state ignored the ruling of the West Virginia State Supreme Court of Appeals in *West Virginia vs. Tomlin* that held that where a statute forbids an act affecting the public but is silent as to penalty, the violation of the statute is punishable by common law.[36]

With the state unwilling to enforce the law prohibiting the use of mine guards, Baldwin-Felts detectives continued to keep UMWA organizers out of the company towns and to spy upon the miners. The guards now joined the work force as miners, made friends with the local coal diggers, and learned their private opinions. Miners who complained about living or working conditions or who spoke in favor of the union were immediately reported to company officials and were subsequently discharged, evicted, and blacklisted. In areas where the UMWA had gained a foothold, the miner-spy often joined the union and kept Felts and appropriate coal operators informed about the organization's activities. The coal operators did not deny their use of spies but rather defended it. The secretary of the Williamson Coal Operators' Association declared, "We claim that we have the right to employ secret service men, or detectives, to protect our interests. We want to know what our men are doing; what they are talking about. We want to know whether the union is being agitated."[37]

The Logan County Coal Operators' Association formed its police bulwark against unionism from a local source: it simply hired the county sheriff, Don Chafin, and his deputies. The treasurer of the Logan County Coal Operators' Association testified that the association had paid Chafin $2,725 a month to work for it. The deputy sheriffs, like the county sheriff, were paid by personal checks from the coal operators. In 1921 the association paid out $61,517 to these deputies for protection against the union. One coal operator justified the companies' paying the deputies by noting that "unions are now controlled by radicals, Industrial Workers of the World and Bolshevists, who are not in sympathy with American ideals and who are seeking to destroy industry."[38]

The anti-union activities of the Logan County sheriff and his deputies became a national disgrace and part of the region's folklore. Two deputies were stationed at each railroad depot and on each train to prevent union organizers from entering Logan County. All strangers, upon coming into the county, were searched and forced to identify themselves. When the clerk for the West Virginia Department of Mines, Toby Heizer, arrived in Logan to induct some miners into a fraternal organization, he was accosted by several deputies and ordered to identify himself. He refused to do so, and the deputies dragged him off the train, beat him, and forced him to leave the county.[39] A newspaper reporter from Washington, D.C., after a visit to Logan County, wrote: "Everywhere one goes down in this county he hears the name of Don Chafin, high sheriff of Logan County. One can see that he has struck terror into the hearts of the people in the union fields. Although a State officer they do not trust him. Every kind of crime is charged to him and his deputies. He is king of the 'Kingdom of Logan.' He reigns supreme by virtue of a State machine backed by the power of the operators."[40]

Chafin was hated by UMWA officials. When visiting Charleston on one occasion, Chafin walked defiantly into District 17 headquarters. The District vice-president, William Petry, without saying a word, reached into his desk drawer, grabbed a .22-caliber pistol, and shot Chafin four times. A local policeman, rushing to the scene of the shooting, heard the union official lament, "That's what happens when a man carries a toy pistol. That goddamned son-of-a-bitch is liable to get well. I should have had my old 'forty-four.'"[41] Chafin did live, and he continued his anti-union policies and brutality.

The American Civil Liberties Union declared that Logan County had become a "national scandal"—as indeed it had. A U.S. Senate investigating committee denounced the Logan deputy sheriff system as "vicious and un-American," declaring that it was "contrary to the genius and spirit of our institutions." A UMWA official told a U.S. Senate committee: "I would just love to see this committee go down to Logan County without anybody knowing who they were." Senator Burton K. Wheeler responded, "They [Logan deputy sheriffs] would probably throw the Senate committee in jail." The president of the American Federation of Labor, Samuel Gompers, declared that Logan County was the "last remains of industrial autocracy left in the United States."[42]

By 1913–14, everything about the southern West Virginia company towns reflected the operators' distorted, paranoid attitudes toward the UMWA. Hundreds of mine guards and deputy sheriffs patroled the roads and railroads and roamed the towns on foot and on horseback, carrying shotguns, rifles, pistols, blackjacks, and clubs, while they searched for union organizers and union miners. Gatling guns, stored in the basements of the company stores, were immediately placed on top of the store and coal tipple at the first sign of labor strife. The company towns had become martial societies.

Indeed, they were closed, martial societies. All strangers and visitors were searched and questioned before entering the company town, and many were often prohibited from doing so. Miners who lived and worked in the towns needed passes simply to leave or re-enter the town. Unfriendly newspapers, not only pro-union ones, but any that dared to criticize, even mildly, the coal companies, were kept off company grounds. Not only were free speech and public meetings forbidden, but the miners were not allowed to congregate in groups of more than two. Mail continued to be scrutinized, read, and sometimes censored by the company-store postmasters. As an added measure of protection, the companies, around 1913–14, began to enclose their towns with barbed-wire fences in an effort to wall out the UMWA. The Elk River Coal and Lumber Company not only constructed a twelve-mile fence around its town of Widen, but it also cleared away all the trees and undergrowth for several hundred yards from the fence, so that the guards and company officials could spot anyone approaching the town.[43]

Beneath the coal operators' professed assurances of the allegiance and contentment of their miners, however, lay their awareness that the miners felt that the system was unjust and that the miners wanted the union. The slightest hint of miners' discontent caused panic among the operators and sent them, once again, demanding federal troops to crush the rebellion.[44] The Paint Creek–Cabin Creek strike and the brutal efforts to suppress it had embarrassed them, for it revealed to the nation not only the structure of order and power in southern West Virginia, but also that the operators had lost the allegiance of their coal diggers and that their entire structure of order and power was in danger.

To preserve what was theirs, the southern West Virginia operators turned to a more subtle form of defense against the union: building a model company town. This was the era of model town building in southern West Virginia, and it was during this period that the operators constructed expensive and elaborate recreational facilities for their miners, installed social and cultural institutions to "uplift" the miners, and brought in outside school teachers to instruct the miners' children and welfare workers to teach the miners' wives. Coal operators also encouraged more cooperative attitudes from company store clerks, and new emphasis was placed on the role of the mine foreman.

These efforts, expenditures, and elaborate constructions represented more than the coal companies' concern for the physical well-being and mental improvement of the miners and their families. They were more than sedatives to make the exploitative and oppressive conditions of the company town less abrasive and more acceptable to its inhabitants; they went beyond welfare capitalism. They constituted efforts to break down the militant solidarity of the miners' working-class culture by inculcating new cultural values and norms and by molding the social and economic behavior of their work force along new lines, ones more acceptable and beneficial to the coal industry. These were attempts to regain the respect and loyalty of the miners while preserving the company town. "Is it not true," a coal operator asked, "that too many coal operators treat their employees as belonging to what they term the 'mining class'?" The operator exhorted, "Because their occupation is different, does not make them worthy of less consideration . . . it places a larger responsibility on those possessed of greater advantages." The new order was not designed to eliminate class differences, but to make them acceptable to the miners. Kinder and more benevolent, indeed paternalistic, concern on the part of the operator was essential to this plan.[45]

The operator was to become a father figure toward his employees; he was to be their provider, protector, and dispenser of justice. An elderly coal operator recalled that his miners could have anything they wanted—"All they had to do was come to me. To use the expression of the Middle Ages, I was high justice, the middle and the low. As we did in the army, we would say to a man, 'Do you take company discipline or do you want a court-martial?' A sensible man would always say, 'I'll take company discipline.' "[46]

Their miners had erred in striking and joining the UMWA, but now under the careful eye and teachings of the coal operators, the prodigal sons would be kept in hand. If their altruism and guidance failed, the operators had a device for keeping their potentially wayward children in line—the yellow-dog contract, by which operators forced prospective employees to disavow union membership and to agree not to join a labor union during their employment. The purpose of the contract is illustrated by the following conversation between a southern West Virginia coal operator and a U.S. senator:

Coal operator: We give jobs to all of those people but we do not want to give jobs to men who are members of an organization that came down there with the avowed purpose of destroying our business.

Senator: But you want to be a judge of the kind of an organization that they shall join.

Coal operator: I think that that is something that employers probably have to do. If they do not I do not see who is going to decide it for them.[47]

The operators' attempt to decide what was best for their miners applied not only to economic affairs (e.g., unionism) but to all aspects of life, including politics. The following dialogue between a U.S. senator and C. A. Cabell, a coal operator on Cabin Creek, is revealing:

Senator: Did they ever try to hold their Socialist meetings there [operator's company town]?
Cabell: No. If they had, I would not have allowed it.
Senator: That would be a political meeting, would it not?
Cabell: Well, I reckon it would, from a Socialist point of view; I suppose if properly conducted and all that, for the proper purpose, it would be a political meeting and nobody would object to it; but I think I would be the better judge of that, being right amongst the people.[48]

The new order also involved new roles for the wives of the coal operators. A Kanawha County coal operator explained that "there are dozens of tactful things a successful [operaror's] wife" could do to "ease" the tension between the social classes in the company town. He advised that the women modify their dress "so as to prevent envy and sullenness in the miners' women as they pass in their daily walks about the camp." "Richly dressed women parading through a mining camp have about the same effect as a bird of gorgeous plummage feeding among the sparrows—they always want to peck at it." He further urged that they "enter the life of the [coal] camp," where they could perform the services of a "professional uplift worker." "The Sunday school, the church and the moving picture," he declared, "are places that should have a plentiful sprinkling of officials' wives."[49]

Company store clerks were also drafted into the service. The company store was no longer to serve simply as an institution to exploit the miners, but it was to promote better relations between the company and its employees. "Old creeds and old faiths are being discarded; they do not conform with new experiences," a company pamphlet to its store clerks read as it advised a "better consideration of the need and requirements" of its miners. "The trouble in the past," the handout explained, was "a failure of cooperation where it should have had it." Instructing the store clerks to give better service to the miners, the pamphlet announced that the company was seeking "new ends by new methods based upon broader, more altruistic foundations."[50]

The coal companies placed great stress on the role of the company official who was closest to the rank-and-file miner—the mine foreman. "There is nothing that I know of that will more quickly take the life out of union interference," a mine superintendent wrote a foreman, "than to make your men satisfied." With such a belief, the coal companies carefully instructed their foremen on how to treat their miners. Maintaining that the "disposition of the foreman to his men" was probably the most important element in keeping the miners satisfied and in "holding them in their places," one company official advised foremen to be "forceful" and "leave no room for argument," but not to "cuss" or "snap" at their men. "The all-round successful foreman," the official wrote, "is the one who is able to win and hold the respect and confidence of both his men and his employers."[51]

Collins advised a beginning foreman that "your people should love and respect you and they will not do so unless you make them earn their money, keep discipline and good order without any Sunday School business and treat all with fairness and consideration." He finished his instructions by stating that "you must always bear in mind that place that is paved with good intentions, and that the success of the institution is immeasurably more yours than ours."[52]

Most foremen accepted their new roles and burdens. One of them wrote to *Coal Age* six months after that Paint Creek–Cabin Creek strike that "before the present seeds of discontent have been nurtured into full-grown rebellion," it was the duty of the mine foreman "to disabuse the minds" of the miners "of these visionary fallacies [union-radical ideas] and put them back on the bed-rock of sound economics."[53]

The coal operators' concern with the foreman's relationship with the miners did not stop with advice. The Baldwin-Felts detectives who were hired to guard against union agitators were also instructed to spy upon the foremen. The foremen were dismissed for misconduct as quickly as the miners were for unionizing. When a mine guard–spy reported a three-day brawl between a superintendent and foreman, the coal operator immediately fired the company officials involved and instructed the new ones to "clear out and straighten up" the work force. Collins wanted his foremen respected and loved, but not too loved. Upon a spy's report, Collins fired another foreman for drinking with the miners. According to Collins, such behavior "has tended to make bosses good fellows."[54]

The coal companies used their foremen in other ways to suit their new purposes. One company increased the number of foremen in order to provide more company supervision of the miner at work and hence to reduce the "miner's freedom" on the job. Several coal companies considered the idea of more internal promotions "from the rank and file of the laboring classes" in an effort to foster ambition and ideas of individual social mobility among the miners. The coal companies were willing to go a long way in trying to break down the solidarity of the miners, but not that far, so nothing came of those plans.[55]

The new order also involved bringing in welfare workers to teach the miners new values and to infuse new aspirations. Shortly after the Paint

Creek–Cabin Creek strike, welfare workers were found in over 70 percent of the company towns in southern West Virginia. One of them explained that the trouble with the miners' life-style was that it was "too simple" and "too socialistic"; and, of course, "socialism leads to anarchy." Because of the lack of newspapers and department stores, the welfare worker continued, "There is little to want." Therefore, their duty was to teach the miners and their families "to want"—"to want because their neighbors have."[56]

Another social worker explained that a coal company operator "had given little thought or consideration to [the miner's] welfare and uplift until a strike [Paint Creek–Cabin Creek], followed by bloodshed." The strike, according to this welfare worker, "made the miner, his life, conditions and surroundings the topic of the hour." Subsequently, the coal company for which she worked brought her and two other social workers into the town with instructions "to bring something of happiness, contentment and economy into the homes of the community." "We saw at once," the social worker explained, "that the situation would demand tact, patience and, above all else, no outward affiliation with the company itself, for the operator, as an operator is sincerely hated."[57]

The social workers in almost three-fourths of the company towns in southern West Virginia helped the miners and their families build playgrounds and form social clubs. They showed the miners how to "upgrade" their homes and how to save money. They also taught the miners' wives to sew, cook, can food, and clean house.[58] One welfare worker explained that "the problem of more wholesome and attractive food and greater efficiency and economy in cooking . . . is essential for contented miners' families." A representative of the coal industry claimed that as a result of such instructions, "these charming little misses [miners' daughters] will know how to organize homes even though reared in an unorganized field."[59] Company officials probably recognized, too, that better diets would decrease the miners' dependency on alcohol for calories and that better health would make them more productive miners. But foremost, the operators hoped that these services would greatly change the miners' life-style. Upon hiring two social workers, an official for U.S. Coal and Coke at Gary, McDowell County, explained that his company was "making an attempt to encourage a higher standard of living through training the women and children in the art of housekeeping."[60] Higher standards of living would teach families "to want because their neighbors have." The welfare director of the Logan Mining Company explained that in teaching the women to clean house, "it would have been much easier to have given the town a thorough cleaning and to have made cleanliness compulsory but we believed that the voluntary method would give better results in the long run." "So far as the company is concerned," the welfare director declared, "it has at a comparatively trifling cost secured a reduction in labor turnover and an improvement in the morale of the working force."[61]

The improved physical condition of the company towns was the most obvious sign of the new times. During the Paint Creek–Cabin Creek strike, the editor of *Coal Age* blasted the southern West Virginia coal operators for allowing the deplorable conditions in their company towns. West Virginia operators, the editor declared, "have held the whip hand and have successfully fought down every rebellion on the part of their employees. Such actions as this on the part of the coal companies aggravated the acute labor situation of the field to the breaking point." The editor, therefore, advised better and more humane living and working conditions. He even suggested the possibility of union recognition: "The very union so many coal operators have been fighting will eventually prove the bulwark of their defense against such dangerous anarchistic bodies as the Industrial Workers of the World."[62] The West Virginia operators could not understand his final point because they did not see the difference between the UMWA and these more radical organizations. They did, however, recognize the need for better living and working conditions.

A coal operator from southern West Virginia during the strike lamented that the miners' houses were "mere shacks, rough boards thrown together." He stressed the need for change because under present physical conditions "no attempt is made toward instilling any idea of improvement or betterment."[63] Improved housing conditions, as well as the use of social workers, the operators believed, could make the miners individually competitive.

The result was an era of building new model towns and of facelifting older towns in southern West Virginia. At Fireco, Raleigh County, for example, the coal operator, Colonel William Leckie, described by *Coal Age* as a "genial Scotchman, who combines business sagacity with a tender heart," installed the "modern conveniences" of baths and electric lights, a barbershop, and recreational facilities, such as a movie theater and a dance hall. Leckie and his associates, according to *Coal Age*, "have planned for the spiritual, social and physical welfare of the employees of the company." A few coal companies even brought in "sociological experts" to advise them on how to make conditions more suitable to their workers.[64]

As a means of further improving company-miner relations, while at the same time stimulating social competitiveness, one coal operator suggested the possibility of allowing the miners to own their homes. "When a miner has something he can call his own, something he can improve," the operator argued, "then he is going to appreciate his job more and will take better care of it."[65] This proposal was carrying the new order too far. The operators, although they were losing a considerable amount of money on company housing (within a three-year period, the Kanawha operators alone lost $422,894 on company housing[66]), still felt the need to retain this form of power. A few coal operators admitted as much. A Fayette County operator bluntly told a U.S. Senate committee that he used the housing contract because "I want to avoid dealing with the United Mine Workers of America." Company housing also served to minimize the everyday type of complaint from the miner. A Logan County miner explained that he and his fellow workers believed that the company checkweighman was cheating them. So they went to the company office "and asked them . . . [to] come out . . . to weigh our coal." Instead of weighing his coal, "they give me a house notice and put me out."[67]

Some of the operators, with an uncanny ability for rationalization, maintained that company housing was for the miner's, not the company's, benefit. The miner, they claimed, "infinitely prefers to sacrifice the academic civil liberty of harboring agitators for the very practical civil liberty of not having his home dynamited."[68]

The coal operators had the problem of trying to establish better relations with their miners and to instill individual competition without sacrificing company housing—a valuable control against agitators, complaints, and strikes. One way that they attempted to inculcate a sense of proprietorship and pride in residence was by encouraging and awarding miners for growing gardens. In addition to supplying miners with seed and fertilizer, they brought in agricultural experts to lecture the miners and their families on the best ways to raise gardens. Then, once a year, the coal companies presented cash awards (it was a twenty-dollar gold piece at Weyanoke, Raleigh County) to the miners with the most beautiful and most productive gardens. Many companies also presented cash awards to the miners with the most well-kept and improved houses.[69]

Encouraging the miners to grow gardens and to upgrade their houses was certainly an effort on the part of the company to occupy their employees' spare time, the misuse of which was the cause of labor trouble; one coal operator declared: "There are two schools of miners, those who expend their surplus time and energy in fighting and those who use it for gardening."[70] Prizes for gardens and homes also had, undoubtedly, other purposes. At a minimal cost to the company (e.g., a twenty-dollar gold piece per year), the operators could impart in the miner a pride in his yard and his house, although he owned neither. It was, furthermore, a cheap means of upgrading the physical environment of the town, and, most important, it stimulated neighborly (social) competition in a rigidly ordered social structure.

In contrast to the small, wooden, monotonously similar, and usually stark miners' houses situated along the company road at the foot of the mountains stood the coal operator's house. Made of brick or cut stone, these elaborate and costly white-columned structures consisted of ten to twenty rooms, were two or three stories tall, and were generally placed on the mountainside overlooking the company town.[71] The operator's house was what E. P. Thompson calls "theater." Thompson explains that "a great part of politics and law is always theater"; in establishing a social system, ruling classes, rather than depending on daily exhibitions of force, attempt to endorse and secure the proper order by a "continuing theatrical style."[72]

"Theatrical style" was the dramatic display of wealth and power, such as the operator's house, which, according to veterans of the southern West Virginia coal fields, gave the operators the symbolic power of overlooking their miners, while it forced the miners to look up—literally—to the coal operators. The symbolism was reinforced when the miner wanted to see the coal operator; he had to climb up the hill as well as enter through the back door.[73]

"Theater" also involved conspicuous displays of dress—dress to show rank and station. While the coal operators wanted their wives to dress inconspicuously, they themselves were careful to maintain a show of rank to the miners. J. G. Bradley, president of Elk River Coal and Lumber at Widen, but a resident of Dundon, had three different types of clothes that he wore, depending on the town he was in. The "patchy ones" he wore in the town in which he lived. The "moderately decent ones" he wore in other commercial towns. He saved his best clothes to wear in his company town of Widen.[74]

"Theater" was also the "social lubricant of gestures," acts designed to "make the mechanisms of power and exploitation evolve more sweetly." The prizes for homes and gardens, as well as awards to the most productive miners, were certainly such "social lubricants." Falling into this category were the parties that the coal operators conducted for the war veterans among their work force and the turkeys that they gave their miners every Christmas.[75] In addition, there were the various amusements the operators provided.

In upgrading their towns, the operators included numerous recreational facilities, including swimming pools, tennis courts, playgrounds, ball fields, and movie houses, and they sponsored activities, such as Boy Scout clubs and baseball teams.[76] These amusements, while occupying the leisure time of the miners and their families,[77] also served to establish better relations between labor and management. During the final month of the Paint Creek–Cabin Creek strike, Ira Shaw, the industrial secretary of the YMCA, lectured at the West Virginia Coal Mining Institute on the benefits of his association in a mining town. Calling attention to the lack of leisure-time activity in the southern West Virginia company towns, Shaw explained that "no man is likely to become dangerous to society so long as he is working. It is misuse of leisure that in the final analysis works ruin to our laboring men." It led, according to Shaw, to the use of alcoholic stimulants and to labor agitation. He therefore proposed that the West Virginia coal operators construct YMCAs in their company towns. The YMCA "has been working its way among miners for several years without failure," Shaw declared, and "it succeeds in enlisting practically every man in the operation in its program, and aims to meet all the social and educational needs of the community." "The miner will react upon his surroundings . . . according to the motive of the operator," Shaw declared; "all the miner needs is opportunity and friendly leadership, and this latter is all-important."[78]

During the next five years, YMCAs sprang up throughout the coal fields. The coal company at Decota, Kanawha County, constructed an elaborate one. Three months later, the president of the coal company announced that the YMCA was promoting "better character" among his miners. The YMCA, he explained, "has raised the old men as well as the young from low and injurious amusements to those of a higher and more cultivated character." Furthermore, it had reduced drinking by 50 percent.[79]

Company officials at Ramage, Boone County, announced their intention to construct a YMCA, explaining that "there has been a large amount of dissatisfaction among the miners of West Virginia; a sentiment of resistance has been carefully fostered by agitators and the strikes along Cabin Creek and Paint Creek." Therefore, the coal company decided to build the YMCA, not simply for amusement, but to create a "sentiment" among the miners: "The important thing is not to build anything but to create a sentiment. After all, it is a harder task to mold a sentiment than to erect a hall." On the day the YMCA building was dedicated, with the president of the coal company, the governor of West Virginia, and the former UMWA president-turned-coal operator, T. L. Lewis, seated on a platform and an audience of miners seated on the ground, the main speaker, I. M. Taggart, talked of the virtues of the organization and of the need for reconciliation between coal company and coal miner. "There has always been more difficulty in getting employees and employers together for the purpose of discussing a difference than there ever was in finally settling it. Neither of the parties in a controversy can approach a problem in an intelligent manner if they have been denouncing or endeavoring to take an unfair advantage of one another."[80]

A coal company in Fayette County established a YMCA, called "Blue Triangle House," for the women in its company town. Lessons in cooking, canning, and house cleaning were taught there, and the director of the "Blue Triangle House" explained that he tried to avoid "the image of an uplift organization." The chief aim was to develop miners' daughters "along the same lines as girls in college and in the cities," and he believed that the YMCA for females would produce a "happier, healthier womanhood," and in consequence, "result in a better [mining] town."[81]

Pleased with the efforts in the southern West Virginia coal fields, the Industrial Department of the YMCA invited Bradley (the coal operator at Widen, where there was an elaborate YMCA) to speak at its Conference on Human Relations in Industry at Silver Bay, New York. The announced purpose of the conference was to discuss ways "to have employees and welfare workers to come together in hope that they might reach a better understanding of each other's problems." The conference leaders probably realized that they had made a mistake when the southern West Virginia coal operator, although a builder of a YMCA, introduced his speech by stating that he employed only nonunion labor, was "opposed to the United Mine Workers of America," did "not believe in the union," and would "have nothing to do with it," and that he believed in the "armed-guard system and the company store."[82]

The elaborate and costly movie theaters, such as the 250-seat auditorium at Widen and the $25,000 movie theater at Colliers, also had purposes other than to entertain miners. One operator stated, "The miner seems to be more easily reached through the eye than the senses, and this makes the pictures a powerful moral and educational agent."[83] The movie became another instrument through which the operators attempted to inculcate new values and ideals.

The operators were careful of the movies that they showed to their miners—control of the movie theater gave them the right to censor movies. A Kanawha County coal operator explained that he intended "to give [the miners] pictures having an educational value, avoiding the lurid, unreal wild-west type." This operator's singling out of western movies reflected a general belief among coal company officials that violence on the screen, especially in the form of western movies, was a source of working-class violence.[84] A Boone County operator, who was concerned about "all these here niggers carrying high-powered rifles," explained to a U.S. Senate investigating committee that "those little boys who are about 10 years old; he wants to be known as a bad man, and he goes to the rotten moving-picture shows that pretends to represent the western bad man, and he gets a pistol, and he goes off and wants the people to believe that he is bad. He gradually grows up to be a gunman." "As a result of the moving-pictures, do you think?" a senator asked. "As a result of the moving-picture shows," the operator responded.[85] Another coal operator explained, "This powerful moral and educational agent would teach the miners to 'dress in better style' and teach them 'better manners.'"[86]

Another powerful "moral" and "educational" agent—and one the coal operators used heavily in attempting to institute sociocultural change—remained the school system. Since the opening of the southern West Virginia coal fields, the coal companies had provided space for school grounds, built schools, equipped them, and subsidized school budgets.[87] Following the Paint Creek–Cabin Creek strike, this interest in and support of education increased.

The coal companies made vigorous efforts to improve the quality of the public schools. They sponsored and supported the movement for the consolidation of the archaic, smaller public schools and sought more competent teachers for the schools. Because of low salaries and the lack of "convenient and desirable homes," all of the counties in southern West Virginia, despite the coal companies' support, lacked interested, qualified teachers. Few good teachers applied for positions, and many of the better ones either left the area after a brief time, for higher salaries in other states, or took employment in the mines, where the pay was also higher.[88]

Following the Paint Creek–Cabin Creek strike, in an effort to attract better teachers, the coal companies increased their subsidies of teachers' salaries. Furthermore, the coal companies built houses for the school teachers. These houses, called "teacherages," were equipped with bedroom suites, kitchen-dinette furniture, and some living room furniture. The "teacherages" were rent-free.[89]

That the Winding Gulf Collieries hired R. A. Riggs illustrated the companies' concern for competent instructors. Riggs was teaching in Curtisville, Pennsylvania, when the company decided to recruit him, because, according to the superintendent, he was a "first class man." Discovering that the county salary was only $75 a month, Riggs said he would not take the position for less than $125 a month. The company agreed to subsidize the fifty-dollar difference. Riggs then decided to bring his family with him to Raleigh County and demanded suitable housing. The company agreed to this; it built him a "suitable dwelling house" of six rooms, with steam heat, electricity, and bathrooms. The teacher then signed a contract with the coal company that required him to teach for eight months and work for two months in the company store and gave him a two-month sabbatical to "do such research and investigation work as is necessary in connection with the educational features."[90]

A retired principal and school teacher at Widen, Parker G. Black, recalled his initial employment interview with the company superintendent. According to Black, the superintendent stated: "Mr. Black, we want good men. I will pay you twice what you are now getting. . . . You will be in complete charge of the school, and will have a free hand . . . I do not want to be bothered. I do not know anything about running a school, but I do know when a man makes a major mistake, and if you make one, I'll sure as hell tell you so." The superintendent proceeded to tell Black to "make your own budget and buy your own equipment" and to employ teachers at fifteen dollars a month beyond the state pay, and "we [the company] will pay the extra."[91]

This heightened concern with education had several sources. As the coal companies consolidated and became better organized, they brought in better trained personnel to administer the corporations, and these people often were strongly interested in education. In 1914, for example, Cabin Creek Consolidated Coal Company hired Josiah Keely as general manager. Keely had degrees from West Virginia University and Harvard University, had served as principal of a West Virginia preparatory school in Morgantown, had been treasurer of the West Virginia Teaching Association, and was a member of the West Virginia State Board of Examiners.[92] Company officials like Keely undoubtedly stimulated the company's interests in education.

The coal companies, which still blamed the miner for mine accidents, believed a more intelligent work force would decrease accidents and increase the production of coal. Therefore, they made early efforts to Americanize the immigrant as well as the illiterate southern black and native miners. In pursuit of this goal, they created special schools, apart from public schools, to teach their workers to read and write.[93]

Most important, education could help preserve the socioeconomic order by eliminating what many company officials considered to be a major cause of the Paint Creek–Cabin Creek strike and the union movement—ignorance. "We have, on the average, a more illiterate class of people working in the mines," a coal operator told a U.S. Senate committee that investigated the Paint Creek–Cabin Creek strike, "than almost any other occupation. This is taken advantage of by a great many people;

in other words, this ignorance is taken advantage of . . . by the disturber of the peace [the UMWA organizer]."[94]

A coal company official at Crumpler, West Virginia, accounting for the rise of unionism in his area, declared that "one beholds a mad stampede of unreasoning destruction and barbarous cruelty." He explained that "the essential difference between the primitive man [the radical] and the civilized man is that one is ruled by impulse, the other by reason." He suggested two ways to reduce the growth of radicalism: "The exclusion of the morally uneducated and mentally undeveloped immigrant [but] at the present stage of development, this would work a serious hardship on American industries." He therefore recommended a second way, "the 'moral education' of this primitive type, both native and foreign." "Certainly, this sounds like a difficult proposition," he continued, "but when we remember that man is endowed by nature with reason, which enables him to anticipate his needs and make a law unto himself, it is clear that he has risen above the primitive level; and the morally educated man has overcome the blind instincts of the primitive man."[95]

Several observers have recently called attention to the idea that once a society has become industrialized, school systems become less agents of change and more agents of maintaining the social structure.[96] This situation appeared to be true for the coal field schools, too. The coal operators were certainly aware of the vital link between schooling and socioeconomic stability. While hiring a new school teacher, Collins wrote his superintendent that "nothing will contribute more to the success of our coal plant." Collins's superintendent wrote to the candidate for the position: "We believe by getting the proper man at the head of [school] affairs, we could and would secure a good many benefits, directly and indirectly, to the general business of our company. We want a man to lead the children in the proper paths, both morally, educationally, socially, and physically, and in fact, become a leader as it were of the social work within our camp."[97]

In the coal field school, the miners' children were taught the social roles that suited the company and were molded to assume the economic tasks of their parents—only to perform them better. In 1914, the school at Glen White, Raleigh County, in addition to courses in reading, writing, and arithmetic, instituted courses in the properties of mine gasses, mine ventilation, the geology of coal, mining methods, and care of mine-safety lamps. Fifteen years later, the West Virginia State Board of Education added a course in coal mining to the public school curriculum; the National Coal Association commented that "the board is to be complimented for their step making for the betterment of miner and mining conditions."[98]

Courses were also added to teach new values and aspirations and to stimulate competitive individualism. The black schools in McDowell County, for example, instituted a "school savings bank" program. Each child was given one dollar to place in the bank and a bankbook. Each month thereafter, a bank clerk came to the school and stamped the children's books to show the accumulation of money through interest. Thus the miners' children learned the capitalistic values of thrift, investment, and unearned increment. According to the state Superintendent of Black Schools, this program would teach the miner's child "a new point of view, and thus inspire new ideals, new hopes, and greater desires."[99]

Behavior that could not be molded along the lines desired by the company was simply banned from the company towns; such was the case with the miners' continual use of intoxicants. Since the early days the coal operators had been worried about the effect of alcohol on the miners' work. By 1913 the operators' concern with the miners' use of alcohol assumed new dimensions: they felt intoxicants were responsible for much of the miners' violent behavior toward the coal companies and mine guards. They claimed liquor made miners susceptible to the appeals of union organizers and that moonshine was an important cause of strikes; during the Mingo County strike, one coal operator explained that "the liberal presence of very potent moonshine spirits . . . is fully as much responsible for the present troubles as is any question of unionism."[100] The coal companies initiated programs to reduce the miners' consumption of intoxicants. The Cabin Creek and Carbon Fuel Company bought out a local saloon (the owner agreed to sell only after the company's mine guards knocked his teeth out and broke his ribs) and replaced it with a YMCA. Writing about the company's actions, *Coal Age* remarked a few months later that "some well informed men went so far as to say that if the operators had begun this work five years ago in the entire field, the [Paint Creek–Cabin Creek] strike would not have occurred."[101]

Several coal companies, following the Paint Creek–Cabin Creek strike, initiated what one company called a "Clean Out the Drunks" campaign. Drunkenness, as well as unionizing, became a reason for dismissal, eviction, and blacklisting. The coal companies pushed for and obtained, in 1914, statewide prohibition in West Virginia. "I am not a prohibitionist," one southern West Virginia operator announced, "but . . . the coal states, at least, should be dry. I believe the operators are unanimous on this question." Upon passage of the legislation, *Coal Age* editorialized: "The mining industry of West Virginia is to be congratulated on the enactment of a statewide prohibition law. . . . If this measure is faithfully carried out, there is sure to be a noticeable improvement in the morale of men in the mining communities. Nothing has been more active in the destruction of character and prosperity in coal camps than the unrestricted sale of intoxicating liquors."[102]

When the coal companies discovered that their miners had begun crossing the borders into Kentucky and Ohio to buy alcohol, they campaigned for national prohibition. Fearing this legislation could not be obtained, the president of a West Virginia coal company declared: "In case of failure of the present movement to establish [national] prohibition, we respectfully urge that a five-mile dry zone be established around coal mines."[103]

The southern West Virginia coal operators flaunted the new order—theirs was not an effort to be hidden. "The paternalistic plan of the operators providing everything in cooperation with the employee," a pamphlet by the Pocahontas Coal Operators' Association exclaimed, "is so much a part of the life of these people that it is doubtful if it can be improved by outside influences [i.e., the UMWA] or even by a government edict." Rivaling George Baer's classic statement of the God-given, labor-leading talents of the anthracite coal operators, the association's pamphlet exalted: "The men who brought the capital to these hills, to replace the isolated, lonely log cabins with regular employment, earnings, the modern conveniences, comforts and enjoyments of civilization, are the natural labor leaders." With the new order, the miners would "begin to think clearly and for themselves" and, consequently, "will take their problems to the office and get their solution from the MANAGEMENT." The pamphlet concluded that "there is no reason or excuse for denying that most of the development of these communities is due to the initiative and foresight of the coal operators. It is to their credit that they saw fit to provide for their employees the advantages of modern life, both economical and social."[104]

This line of reasoning not only illustrates the operators' views toward their miners and the new—for them, natural—order, it also reflects the transition of the operators' thought and attitudes toward the company town. They no longer defended, or apologized for, the company town as a necessity for feeding, clothing, and housing thousands of workers in a sparsely populated mountainous wilderness; they were proclaiming the company town as a positive good!

The coal establishment now believed, or at least asserted the idea, that the company town was not merely the only way to provide for their workers, but that it was the best way. The coal companies made the most of their contributions to the school system in the state. They claimed that their workers had better, roomier houses at lower rents than did unskilled city workers. But this was not all; the coal companies now praised each and every aspect of the company town. Company stores were now the best means of providing their workers with the best quality of goods at the lowest prices. Because of the company store, the "miners and their families do not suffer by comparison with city stores." Coal scrip was viewed as "a great convenience" for the coal company and the coal miner. It relieved the company of the "expense of much bookkeeping and the losses of uncollectable bills." Consequently, the coal company could pay higher wages and charge lower rents for the houses and lower prices for goods in the company store. Unlike workers in large cities, the coal companies pointed out that their employees never suffered from a "fuel shortage." On the contrary, "the miners receive all the coal [for fuel] they can use," and at a minimal price. "No matter what the market price of coal is, the [small] charge to the miner remains the same."[105]

Their miners, the coal companies claimed, enjoyed cheaper and purer water than workers in commercial cities. "Water is furnished absolutely free in approximately 99% of the mining towns of West Virginia." Stating that they made bacteriological tests on their water supply every month, they declared, "It might be truthfully said" that the water supply is "more pure and wholesome than that to be found in any of our large cities."[106]

The company-town system, the companies averred, permitted the miners to enjoy a better sewage-disposal system than they would if they worked in commercial cities. "The presence of a central ownership and maintenance makes possible the installation of expensive sewage systems which are beyond the means of individuals and small unorganized communities. It also insures a systematic regulation of sewage disposal and enforcement of rules promoting the general health of the community."[107]

In contrast to the congested, working-class ghettos of the urban north, the coal companies proudly pointed out that the miners' children had plenty of time and space in which to play. "Boys and girls are to be found the year round using swings, slides, and other [recreational] apparatus." The miners' children also enjoyed the best of playground equipment, furnished, of course, by the coal companies: "The very best apparatus is installed by many of the companies at their expense."[108]

Not only did the miners have the best, but the most beautiful: "What is more romantic or picturesque than a garden?" the coal companies asked: "Here lovers meet under a full moon. . . . Here those who love nature get inspiration from the beautiful growing plants . . . [and] we are safe in saying that the most beautiful gardens in West Virginia are to be found in coal mining communities."[109]

The southern West Virginia coal establishment did not have a writer-philosopher-propagandist as eloquent and profound as George Fitzhugh was for the Southern planters and slavery, but they had writers, such as Phil Conley, who were as prolific and insistent as Fitzhugh. In pamphlets, books, letters, and speeches, Conley constantly called attention to, and praised, the benefits that the coal industry had bestowed upon the state and its people and asserted the positive good of the company town. "The change from isolated mountainous conditions to well-regulated towns, is phenomenal," he wrote. "The log cabins of one and two rooms have been replaced by well-built comfortable homes of four, six and eight rooms that have modern conveniences. . . . Educational and social conditions are now afforded where before opportunities for development were scarce." Conley also recognized the potential the coal industry offered: "The world is at the door of these mountain people, who for generations were shut off by barriers of isolation. Unlimited opportunities are afforded the native mountain men and the foreign employees to advance to responsible positions."[110]

The thrust of Conley's writings was directed at proclaiming the virtues of the company town and the advantages that it offered to the miners in contrast to "independent" towns. "Established industry had learned," Conley wrote, "just as the Southern slave-owner had learned many years before, that a valuable employee must be treated properly." Therefore, the coal operators provided the miners with "comfortable houses and garden plots, churches, schools, YMCAs, and other community recreational facilities." As a result, a coal company town was a "better place in which to live than the majority of small independent towns." According to Conley,

> There are more home conveniences such as electricity, water, well-built, comfortable homes, board walks, fenced yards and gardens, and the majority of them are more sanitary.
>
> There are more advantages socially, found in the community buildings, picture shows, and parks.
>
> The schools, many of the churches, Sunday schools and other educational and religious activities are superior to those to be found in the usual small town of equal population.
>
> Athletics for children and adults are provided for in ball parks, children's playgrounds, and recreational buildings. The ordinary small town of like number of inhabitants has not the facilities to compare with those to be found in the mining towns.[111]

Conley recognized that not all towns nor all parts of the town were ideal and picturesque, and he explained why:

> It is the minority, the more ignorant people, colored, white and an occasional foreigner in this field, who are discontents, radicals in regard to labor relationships, who are agitators and trouble-makers. It is this class with their low standards of living, and ignorance, that characterize a mining town as uninhabitable both to the outsider and the respectable mine employee.
>
> This class is content with the poorest sort of housing and living conditions, proven every day by the fact that the better houses given them are shown every sort of disrespect and defacement.

As for why these degenerate radicals were permitted to continue their harmful effects upon the town, Conley blamed, of course, the union:

> There is hardly a company official who does not earnestly desire to remove every unsightly house from his operation, provided the element contented to live there go with them.

But Conley explained:

> It is impossible under the regulations of the miners' union for a company to get rid of this element; discharge for radicalism or trouble-making is never allowed by the union leadership. The best the company can do is to segregate them, "at the lower end of town where their dirt doesn't bother anyone else." But their poisoned minds and low standards contaminate the whole town and lower the morale of the community.

This radical minority, Conley explained, was responsible for all of those derogatory newspaper and magazine articles and the government reports about the miserable conditions of the company town in southern West Virginia:

> These are the people who by their poor living and attitude, exploited by publicity agents, visited by investigators, have created the idea in the mind of the public that all coal miners, as a class, are unintelligent, poor livers and ignoramuses.[112]

Conley was right in pointing out that journalists and other observers had overlooked (and still do) the good points and material benefits of the company towns in favor of detailing the seamier, more depressed conditions. But what Conley ignored, or was perhaps incapable of understanding, was that, to the miners, the institutions that the company built and the services that the company supplied, regardless of their expense and benefit, could not escape the taint of company control. Withdrawal of the use of the school and church during strikes was an obvious example. The health services that the company supplied gives another.

The coal companies provided their miners with good health-care facilities and services. Several of the companies had built hospitals for their miners. U.S. Coal and Coke at Gary, McDowell County, had an annual "Negro Health Week," in which all the children of black miners were examined and treated for medical defects. The company bought dentistry equipment valued at over $40,000 and hired thirteen dentists and dental hygienists who visited the county schools three times a year to care for the children's teeth.[113]

Foremost in medical service was the company doctor. Almost every company town in southern West Virginia had a qualified, licensed medical doctor who, at minimal cost to the miners (usually a dollar a month), was always available to the miners for any needed medical service. Every miner's wife either had a doctor or a competent midwife, usually the former, present when she gave birth.[114]

The miners appreciated the doctors' services, but they did not appreciate the doctor, who was a company official and, therefore, controlled by the company. The superintendent at Davy, McDowell County, provided the miners with an extreme example of company control when he fired the company doctor who married a woman whom the superintendent had not chosen. Most doctors were more obedient. During strikes, they denied medical services to the miners and their families. Ignoring the traditional right of privileged communication between doctor and patient, company doctors often served as spies for the companies, informing officials of the private opinions of their patients. At least one company doctor was a Baldwin-Felts agent.[115]

The company doctor sided with the coal company in most affairs; indeed, he was often the company's mouthpiece. For example, Dr. E. F. Heikell, an employee of Continental Coal Company, told a Rotary Club meeting that the miners themselves were to blame for the "deplorable health conditions that prevail among the helpless, hopeless, non-union miners and their families." "Until we get laws," the company doctor declared, "that will give us authority to make a well man go to work and provide for his family we are not going to change present conditions."[116]

The miners' resentment of company doctors led to open hostilities against both the company and the doctor. Two months after the conclusion of the Paint Creek–Cabin Creek strike, 500 miners at Paint Creek Collieries struck because a company had failed in its promise to dismiss a company doctor who had spied on them during the strike.[117]

Despite the intentions and efforts to institute a new order, the operators' philosophy and fears of unionism allowed them to go only so far. As one miner wrote, "In our town we have many good things, there are different kinds of churches and good schools, but there is another thing of much more importance . . . the coal operators have intentionally overlooked—and this is our freedom."[118] The new order had failed; the union movement expanded.

## NOTES

1. Z. T. Vinson, *Advocating Co-operation and Organization of West Virginia Coal Operators*, Address before West Virginia Mining Institute (Huntington, W.Va., n.d.), 5–6.

2. Ibid.

3. *Coal Age* 4 (Aug. 9, 1913), 205. Goff is quoted in Dale Fetherling, *Mother Jones, the Miners' Angel* (Carbondale, Ill., 1974), 99.

4. George Rudé, *The Crowd in the French Revolution* (New York, 1972), 221–23. Eugene Genovese detailed how the threat of slave revolts produced uniformity and conformity among the southern slaveholding aristocracy. *Roll, Jordan, Roll* (New York, 1974), especially 596.

5. Operators' Association of the Williamson Field and the Logan Coal Operators' Association, *The Issue in the Coal Fields of Southern West Virginia: Statement to President Harding* (n.p., n.d.), 14, 23 (hereafter cited as *Issue in the Coal Fields*).

6. George Wolfe to Justin Collins, Oct. 11, 1914, and Collins to the National Association of Manufacturers, Nov. 19, 1925, Justin Collins Papers, West Virginia University Library, Morgantown, W.Va.; William Warner to Gov. John J. Cornwell, Aug. 19 and Sept. 22, 1917, and N. S. Blake to Cornwell, June 4, 1917, John J. Cornwell Papers, West Virginia State Department of Archives and History, Charleston, W.Va.

7. Operators' Association of the Williamson Field, *Statement Made before the Sub-committee of the Committee on Education and Labor*, Report filed July 14, 1921 (n.p., n.d.). 14; William Coolidge, *Brief in Behalf of Island Creek Coal Company* (Boston, 1921), 4; *Issue in the Coal Fields*, 16; Non-Union Operators of Southern West Virginia, *Statement to the United States Coal Commission*, Report filed January 12, 1923 (n.p., n.d.), 14.

8. Bituminous Operators' Special Committee, *The United Mine Workers in West Virginia*, Statement submitted to the U.S. Coal Commission, August 1923 (n.p., n.d.), 80–83; U.S. Congress, Senate Committee on Education and Labor, *Conditions in the Paint Creek District, West Virginia*, 63rd Cong., 1st sess., 3 vols. (Washington, D.C., 1913), 1:2060 (hereafter cited as *Paint Creek Hearings*); testimony of Langdon Bell, U.S. Congress, Senate Committee on Interstate Commerce, *Conditions in the Coal Fields of Pennsylvania, West Virginia, and Ohio*, 70th Cong., 1st sess., 2 vols. (Washington, D.C., 1928), 2:1851–57 (hereafter cited as *Conditions in the Coal Fields*); testimony of William Coolidge, U.S. Congress, Senate Committee on Education and Labor, *West Virginia Coal Fields: Hearings . . . to Investigate the Recent Acts of Violence in the Coal Fields of West Virginia*, 67th Cong., 1st sess., 2 vols. (Washington, D.C., 1921–22), 2:1967 (hereafter cited as *West Virginia Coal Fields*); testimony of William Wiley, ibid., 962; William Wiley, *Facts about the Armed March* (Charleston, W.Va., 1921), 12, 37, 40–42. See also Bituminous Operators' Special Committee, *Comparative Efficiency of Labor in the Bituminous Industry under Union and Non-Union Operations*, Evidence and Testimony Submitted to the U.S. Coal Commission,

Sept. 10, 1923 (n.p., n.d.), and Bituminous Operators' Special Comm., *UMW in West Virginia*, 80–83.

9. Testimony of William Coolidge, *West Virginia Coal Fields*, 2:908–11. See also Coolidge, *Brief in Behalf of Island Creek Coal Company*, 12–13; *Issue in the Coal Fields*, 5–14.

10. Coolidge, *Brief in Behalf of Island Creek Coal Company*, 15; testimony of Quinn Martin, *West Virginia Coal Fields*, 1:555; C. M. Fenton to Attorney General, June 8, 1914, File 50, Record Group 60, General Records of the Department of Justice, National Archives, Washington, D.C.

11. UMWA National Convention, *Proceedings, 1911*, 58, and William Graebner, *Coal-Mining Safety Legislation in the Progressive Era* (Lexington, Ky., 1976), 125.

12. Collins to Tom Felts, Feb. 6, 1908, Collins Papers, and Frank Kneeland, "The Moving Picture in Coal Mining," *Coal Age* 5 (June 27, 1914): 1037.

13. Collins to Tom Felts, Feb. 6, 1908, and Collins to I. T. Mann, Jan. 15, 1909, Collins Papers.

14. J. W. Dawson, *The Greatest Crisis that Ever Confronted the State of West Virginia*, Address before the West Virginia Board of Trade (n.p., n.d.), 2–5, and J. C. McKinley, *The Coal Crisis*, Address before the West Virginia Board of Trade (n.p., n.d.), 8–9.

15. Dawson, *Greatest Crisis*, 2–5, and McKinley, *Coal Crisis*, 8–9.

16. Clarence Bonnett, *History of Employers' Associations in the United States* (New York, 1956), 453–60, and testimony of Langdon Bell, *Conditions in the Coal Fields*, 2:1834–35.

17. Operators' Assoc., Williamson Field, *Statement before the Sub-committee, 16*–59; Coolidge, *Brief in Behalf of Island Creek Coal Company*, 3–4; *Issue in the Coal Fields*, 10–12; Wiley, *Facts about the Armed March*, 12; testimony of Quinn Martin, *West Virginia Coal Fields*, 2:555. The operators began to raise this issue on the eve of the Paint Creek–Cabin Creek strike; see Neil Robinson, "West Virginia on the Brink of a Labor Struggle," *Coal Age* 2 (Sept. 14, 1912): 361–62.

18. Wiley, *Facts about the Armed March*, 33.

19. Testimony of William Coolidge, *West Virginia Coal Fields*, 1:908; *Issue in the Coal Fields*, 21; Coolidge, *Brief in Behalf of Island Creek Coal Company*.

20. Logan District Mines Information Bureau, "Why Coal Production from Non-Union Fields Constitutes a National Safeguard," *Coal Facts* 5 (Dec. 30, 1921), and testimony of Langdon Bell, *Conditions in the Coal Fields*, 2:1882–84.

21. Testimony of Quinn Martin, *West Virginia Coal Fields*, 2:552–53, and ibid., 1045. Such general statements of conspiracies were common, but the operators never questioned that the UMWA was the spearhead of the effort. "This hostile force of great power and influence is . . . trying, as we believe, and as we allege, and as we proved, to destroy us." Testimony of Langdon Bell, ibid., 1557.

22. Testimony of Walter Thurmond, ibid., 1:540; Bituminous Operators' Special Comm., *UMW in West Virginia*, 6; *Conditions in the Coal Fields*, 1:643–50.

23. Testimony of Quinn Martin, *West Virginia Coal Fields*, 1:556, and Wiley, *Facts about the Armed March, 25*–26.

24. Ernest Bailey, letter to the editor, *Coal Age* 4 (Jan. 3, 1914): 3; Bituminous Operators' Special Comm., *UMW in West Virginia*, 39–40; Laurence Leamer, "Twilight of a Baron," *Playboy*, May 1973, 170.

25. Operators' Assoc., Williamson Field, *Statement before the Sub-commit-*

*tee*, 8–9, and Wiley, *Facts about the Armed March*, 26.

26. Vinson, *Advocating Co-operation*, 17.

27. Gordon Watkins and Paul Dodd, *Labor Problems* (New York, 1940), 814–16; Frederick Ryan, "The Development of Coal Operators' Associations in the Southwest," *Southwestern Social Science Quarterly* 14 (Sept. 1933): 133–44; Bonnett, *Employers' Associations*, 470.

28. *Appeal to Reason* (Girard, Kans.), Oct. 11, 1913; *Coal Age* 1 (Sept. 27, 1913): 469; Vinson, *Advocating Co-operation*, 21; *Miners' Herald* (Montgomery, W.Va.), Sept. 27, 1913.

29. Testimony of William Wiley, *West Virginia Coal Fields*, 1:523; Collins to E. E. White, Mar. 24, 1915, to Wolfe, Apr. 10, 1915, also Wolfe to Collins, Apr. 14, 1915, and Collins to Wolfe, Apr. 15, 1915, Collins Papers.

30. Collins to Wolfe, Apr. 10, 1915, Collins Papers.

31. F. Hinrichs, *The United Mine Workers of America and the Non-Union Coal Fields* (New York, 1923), 140, and *West Virginia Coal Fields*, 1:237.

32. Jones is quoted in Arthur Gleason, "Private Ownership of Public Officials," *Nation* 110 (June 12, 1920): 725. Also see *West Virginia Coal Fields*, Report, 7.

33. Wolfe to Collins, Apr. 13, 1915, Collins Papers.

34. West Virginia Mining Congress, *Circular*, July 7 and July 8, 1913, found in Collins Papers; *Coal Age* 20 (Dec. 30, 1920): 1344; *West Virginia Coal Fields*, 1:237.

35. West Virginia Legislature, *Acts, 1913*, 173–74. For public response to the publicity, see the petitions and letters in Item 165095, File 50, Record Group 60, General Records of the Department of Justice.

36. Wolfe to Collins, Feb. 2 and Feb. 27, 1913, Collins Papers; Committee on Coal and Civil Liberties, "Coal and Civil Liberties, Report to the U.S. Coal Commission" (typewritten report, submitted Aug. 11, 1923), U.S. Department of Labor Library, Washington, D.C., 14–20.

37. Winthrop Lane, "Labor Spy in West Virginia," *Survey* 48 (Oct. 22, 1921): 110–12; *New York Call*, Mar. 9, 1913; *UMWJ*, Sept. 13, 1917, 15; *New York Times*, June 17, 1913; Richard Hadsell and William Coffey, "Baldwin-Felts Detectives in the Southern West Virginia Coal Fields," manuscript in preparation (copy in the author's collection), 6–14. In October 1917, Felts billed Collins for services that included: "One inside man at District 29 Headquarters, $100; One union man who travels our mines, $150; Five plainclothes men, $625; One confidential inside man to report on employees, $150; Two deputy sheriffs, $250." Wolfe to Collins, Oct. 10, 1917, Collins Papers.

38. Bituminous Operators' Special Comm., *UMW in West Virginia*, 87–89; *UMWJ*, May 15, 1922, 4, and June 15, 1922, 7; *West Virginia Coal Fields*, Report, 7; *Report and Digest of Evidence Taken by a Commission Appointed by the Governor of West Virginia in Connection with the Logan County Situation* (Charleston, W.Va., 1919), 9; Elliott Northcott (U.S. attorney) to the Attorney General, Dec. 18, 1922, File 205194–50–30, Record Group 60, General Records of the Department of Justice. Chafin apparently became an employee of the association when he first became county sheriff in 1917; he had a total of 457 deputies. In addition, he had "scores" of "special constables," who held badges and who were authorized to carry pistols. Howard Lee, *Bloodletting in Appalachia* (Parsons, W.Va., 1969), 88–89.

39. Northcott to Attorney General, Dec. 18, 1922, File 205194–50–30, Record Group 60, General Records of the Department of Justice, and Gleason, "Private Ownership," 725.

40. *Washington Star,* Sept. 7, 1921.
41. Petry is quoted in Howard Lee, *Bloodletting in Appalachia*, 92–93. Also see *UMWJ*, Oct. 1, 1919, 16.
42. "Free Speech in Logan," *Survey*, Apr. 15, 1923, 79–81; *West Virginia Coal Fields*, Report, 7; testimony of Van Bittner, *Conditions in the Coal Fields*, 1:1268. Gompers is quoted in *Charleston Gazette*, Sept. 2, 1921.
43. Committee on Coal and Civil Liberties, "Coal and Civil Liberties," 14–15; testimony of C. A. Cabell, *Paint Creek Hearings*, 2:1498; Wolfe to Collins, 1918, and Collins to Wolfe, July 6, 1918, Collins Papers. For the use of the post office in the company stores to censor the mails, see William Warner to Attorney General A. Mitchell Palmer, Nov. 25, 1919, File 1-74-16-130-83 (205194-50-33), Record Group 60, General Records of the Department of Justice. For further efforts to seal off the company towns from union organizers, see Collins to Raymond DuPay, June 2, 1913, and L. R. Taylor to Collins, June 16, 1913, Collins Papers. Many of the commercial, independent towns were just as closed to outsiders. The mayor of Mt. Hope, for example, openly remarked that "no union man gets any protection around here." *Appeal to Reason* (Girard, Kans.), May 17, 1913.
44. When a brief strike occurred in the Kanawha District shortly after the Paint Creek–Cabin Creek strike, the secretary of the Kanawha Coal Operators' Association immediately wired the U.S. Attorney General to request troops. "If some action is not taken promptly by the Federal Government," he wrote, "we will have another war." D. T. Evans to U.S. Attorney General J. C. McReynolds, June 2, 1914, File 50, Record Group 60, General Records of the Department of Justice.
45. M. P. H., letter to the editor, *Coal Age* 13 (Apr. 20, 1918): 756–57. In the United States, these practices are merely depicted as "welfare capitalism" or employer paternalism. E. P. Thompson has pointed out that in nineteenth-century England "what is generally seen as gentry paternalism" was an attempt by ruling social classes to mold working-class behavior by establishing a cultural hegemony. This may well have been the case in the United States, or at least in the southern West Virginia coal fields in the post–Paint Creek–Cabin Creek era. "Patrician Society, Plebian Culture," *Journal of Social History* 7 (Summer 1974): 382–405. For a general discussion of "welfare capitalism" throughout the United States, see Irving Bernstein, *The Lean Years: A History of the American Worker, 1920–1933* (Boston, 1960); Stuart Brandes, *American Welfare Capitalism, 1880*–1940 (Chicago, 1976); Thomas Brooks, *Toil and Trouble: A History of American Labor* (New York, 1964), chap. 10; David Brody, "The Rise and Decline of Welfare Capitalism," in *Change and Continuity in Twentieth Century America: The Twenties*, ed. John Braeman et al. (Columbus, Ohio, 1968), 160–71. Bernstein argues that "welfare capitalism" was primarily a defense against union encroachment, and Brandes claims that "welfare capitalism" included other objectives, such as to increase productivity and to reduce labor turnover. The system that developed in southern West Virginia certainly included all these objectives, particularly the reduction of geographic mobility. See, for example, Wolfe to Collins, Feb. 13, 1914, Collins Papers. Also the operators believed that such efforts would increase company profits: see Collins to E. C. Berkley, June 15, 1916, Collins Papers.
46. Quoted in Leamer, "Twilight of a Baron," 168.
47. Testimony of Langdon Bell, *Conditions in the Coal Fields*, 2:1832–62.
48. Testimony of Cabell, *Paint Creek Hearings*, 2:1535.

49. Josiah Keely, "Successful Wives in Coal Camps," *Coal Age* 11 (Apr. 7, 1917): 591–92.

50. This pamphlet is reprinted in *Conditions in the Coal Fields*, 2:1844–45.

51. Wolfe to E. E. White, Sept. 10, 1925, and Collins to E. C. Berkley, June 15, 1916, Collins Papers. Collins's letter shows that the concern with foremen was also to increase corporate profits. *Coal Age* 8 (June 22, 1918): 1168.

52. Collins to Berkley, June 15, 1916, Collins Papers.

53. Ernest Bailey, letter to the editor, *Coal Age* 5 (Jan. 3, 1914): 9.

54. Wolfe to Collins, Jan. 9, 1914, E. C. Payne to Felts, Mar. 11, 1912, Felts to Collins, Mar. 9, Nov. 12, 1912, Collins to Felts, Mar. 11, 1912, Collins Papers.

55. Carter Goodrich, *The Miner's Freedom* (Boston, 1925), 124; D. H. Perdue, letter to the editor, *Coal Age* 13 (Apr. 13, 1918): 709; Keith Dix, *Work Relations in the Coal Industry: The Hand Loading Era, 1880–1930* (Morgantown, W.Va., 1977).

56. Bituminous Operators' Special Committee, *The Company Town*, Report Submitted to the U.S. Coal Commission (n.p., 1923), 28, and Margaret Jordan, "A Plea for the West Virginia Miners," *Coal Age* 6 (Dec. 5, 1914): 914–15.

57. Bituminous Operators' Special Comm., *Company Town*, 28, and Jordan, "Plea for West Virginia Miners," 914–15.

58. Jordan, "Plea for West Virginia Miners," 914–16; Isabella Chilton, "Self-Improvement for a Company Town," *Survey* 41 (Mar. 15, 1919), 873; *Coal Age* 5 (May 2, 1914): 738; Isabella Chilton Wilson, "Welfare Work in a Mining Town," *Journal of Home Economics* 11 (Jan. 1919): 21–23. Many of these efforts were probably designed to reduce workers' extravagance, which employers often viewed as a cause of wage demands. I have not found any information directly connecting this goal to the southern West Virginia operators' efforts. Other social and labor historians have suggested that the idea of reducing workers' extravagance was the case in other areas and industries. See Brandes, *Welfare Capitalism*, 33, and Herbert Gutman, "Work, Culture, and Society in Industrializing America, 1815–1919," *American Historical Review* 78 (June 1973): 560.

59. Quoted in Brandes, *Welfare Capitalism*, 58–59.

60. *Coal Age* 5 (Feb. 2, 1914): 738.

61. Chilton, "Self-Improvement," 873, and Wilson, "Welfare Work," 21–23.

62. "The West Virginia Strike," *Coal Age* 3 (Apr. 26, 1913): 650.

63. Letter to the editor, "Sociological Conditions in West Virginia," *Coal Age* 2 (Nov. 23, 1912): 733.

64. *Coal Age* 6 (Aug. 22, 1914): 311–12, 12 (Sept. 12, 1918): 1217, 5 (Feb. 14, 1914): 295. Experts in housing and town building were also brought in to talk to operators' conventions and meetings. See, for example, R. H. Hamill, "Design of Buildings in Mining Towns," *Coal Age* 11 (June 17, 1917): 1045–48. For descriptions of the model towns built during this era, see Winthrop Lane, *Civil War in West Virginia* (New York, 1921), chap. 3 ("Some Good Mining Camps"); Thurmond, *Issue in the Coal Fields*, 92–93; Betty Cantrell, Grace Phillips, and Helen Reed, "Widen, the Town J. G. Bradley Built," *Goldenseal* 3 (Jan.–Mar. 1977): 2; "Fireco—A New Mining Town in West Virginia," *Coal Age* 12 (Sept. 12, 1918): 1217. The coal companies advertised their model company towns to attract miners. See, for example, *UMWJ*, Jan. 1, 1917, 25, Jan. 11, 1917, 30, Jan. 25, 1917, 14, 15, 16, May 31, 1917, 11, June 7, 1917, 15. Typical was the advertisement of Spruce River Coal Company that read "Miners Wanted: The Ideal Min-

ing Town of the State. YMCA, Three Schools, Splendid Churches, Shower Baths, Playgrounds, Baseball Parks . . . , ibid., Nov. 1, 1917, 32.

65. Letter to the editor, "Sociological Conditions in West Virginia," *Coal Age* 2 (Nov. 23, 1912): 733.

66. Hinrichs, *Non-Union Coal Fields*, 60; Philip Conley, *Life in a West Virginia Coal Field* (Charleston, 1923), 36; Bituminous Operators' Special Comm., *Company Town*, 10.

67. Testimony of J. D. Boone, *Conditions in the Coal Fields*, 2:1745–46. Also see U.S. Department of Labor, Bureau of Labor Statistics, *Housing by Employers*, Bulletin no. 263 (Washington, D.C., 1920), 21, and Arthur Gleason, "Company-Owned Americans," *Nation* 110 (June 12, 1920): 794–95.

68. Bituminous Operators' Special Comm., *Company Town*, 33.

69. *Coal Age* 6 (Aug. 15, 1914): 264, 273, 6 (Aug. 22, 1914): 311–13, 11 (July 28, 1917): 166, 11 (Nov. 17, 1917): 845, 856; testimony of Robert Gross, *Conditions in the Coal Fields*, 1:1928.

70. "Moonshine's Lively Part in Mingo Troubles," *Literary Digest* 70 (Sept. 17, 1921): 37–39.

71. Mark Gillenwater, "Cultural and Historical Geography of Mining Settlements in the Pocahontas Coal Fields of Southern West Virginia" (Ph.D. diss., University of Tennessee, 1972), 87–90, and W. P. Tams, *The Smokeless Coal Fields of West Virginia* (Morgantown, W.Va., 1963), 74. For descriptions of some of these mansions, see J. T. Peters and H. B. Carden, *History of Fayette County* (Fayetteville, W.Va., 1926), 563, and *Charleston Gazette*, Nov. 4, 1977.

72. Thompson, "Patrician Society," 390.

73. Interview with Mr. and Mrs. Carl Hazzard, Mullins, W.Va., summer 1975.

74. Thompson, "Patrician Society," 389–90, and Cantrell, Phillips, and Reed, "Widen," 2.

75. Thompson, "Patrician Society," 390, and interview with Sydney Box, Glen White, W.Va., summer 1975.

76. Lane, *Civil War*, chap. 3; *Coal Age* 6 (Nov. 28, 1914): 878–80, 11 (Sept. 12, 1918): 492–93; "Activities in a Modern Mining Town," *West Virginia Review*, June 1925, 329; testimony of Josiah Keely, *Conditions in the Coal Fields*, 2:2100.

77. The operators viewed leisure time as a cause of labor unrest. One of them explained that union agitators had been successful in his area because "time hung heavily on the hands of the men and they seemed to scent trouble where there was none." *Coal Age* 6 (Nov. 28, 1914): 878.

78. Ira Shaw, "Welfare Work among Miners," *Coal Age* 4 (July 5, 1913): 21–22. The use of YMCAs as a form of social control was not novel to the southern West Virginia coal fields. Railroad companies had previously used YMCAs to "reduce the traffic at local saloons" in their work camps. In Pennsylvania YMCAs at coal-mining camps cashed miners' paychecks to keep them from going into towns and cashing them in saloons. The secretary of the Industrial Department of the YMCA, Charles Townson, represented the association's general purpose in industrial matters as improving "working, living and leisure conditions; increas[ing] happiness and contentment; greater efficiency in production; better relations between employer and employee." "The Association," he wrote, "creates an atmosphere of friendliness and confidence, which helps to prevent misunderstandings and to make possible the adjustment of differences when they do arise." C. Howard Hopkins, *History of the YMCA in North America* (New York, 1951), 122, and Charles R. Townson, "Industrial Program of the Young Men's Christian

Association," *Annals of the American Academy of Political and Social Sciences* 103 (Sept. 1922): 134–37.

79. "The Cabin Creek YMCA, Decota," *Coal Age* 4 (Nov. 15, 1913): 741; Fred Ridge, "An Important Factor in Modern Industry," ibid. (Nov. 29, 1913): 33; testimony of C. A. Cabell, *Paint Creek Hearings*, 2:1440; C. A. Cabell, "Building a Model Mining Community," *West Virginia Review*, Apr. 1927, 210.

80. *Coal Age* 4 (Nov. 28, 1914): 873–75, and *McDowell* (W.Va.) *Times*, Oct. 30, 1914.

81. Helen Hutchinson, "What the YMCA Is Doing in West Virginia," *UMWJ*, Feb. 1, 1921, 16, Aug. 1, 1921, 4.

82. Quoted, ibid., Sept. 15, 1922, 5.

83. Lane, *Civil War,* 33; "Theater for Miners," *Dramatic Mirror* 80 (Aug. 7, 1919): 1227; "Cabin Creek YMCA," 741–42; Conley, *Life*, 16.

84. Testimony of Langdon Bell, *Conditions in the Coal Fields*, 2:1845, and "Cabin Creek YMCA," 742.

85. Wiley, *Facts about the Armed March*, 28.

86. Conley, *Life*, 16. Also see Kneeland, "Moving Picture in Coal Mining," 1036–37, and *Coal Age* 5 (Feb. 14, 1914): 295.

87. For a detailed description of the coal companies' early support of schools, see chap. 3.

88. W.Va. State Superintendent of Free Schools, *Annual Report, 1914*, 72–78, and *Annual Report, 1912*, 49; F. C. Cook, superintendent of McDowell County schools, in W.Va. State Superintendent of Free Schools, *Biennial Report, 1903–4*, 199; *Biennial Report, 1906*, 210–11; *Biennial Report, 1910*, 4.

89. Logan District Mines Information Bureau, "Why Coal Producers Aid West Virginia Schools," *Coal Facts* 4 (Nov. 4, Nov. 30, 1921); W.Va. State Superintendent of Free Schools, *Biennial Report, 1920*, 16; Hinrichs, *Non-Union Coal Fields*, 71; W. S. Rosenhelm, "Billion Dollar Coal Field," *West Virginia Review*, Oct. 1924, 20–21; Josiah Keely, "The Cabin Creek Consolidated Coal Company," ibid., June 1926, 348.

90. Collins to Wolfe, May 23, 1916, Wolfe to Riggs, May 15, 1916, Wolfe to Collins, May 15, 1916, Wolfe to Riggs, May 15, 1916, Contract, Winding Gulf Collieries Co. and R. A. Riggs, July 4, 1916, Collins Papers.

91. Cantrell, Phillips, and Reed, "Widen," 4.

92. For the consolidation of coal companies in the Kanawha coal field, see Elizabeth Goodall, "History of the Charleston Industrial Area" (M.A. thesis, West Virginia University, 1937), 41–43. For Keely's background, see "Who's Who in Coal Mining," *Coal Age* 6 (Aug. 29, 1914): 344.

93. Graebner, *Coal-Mining Safety*, 161–64; Sam Reynolds and W. H. Reynolds, "Human Element in Coal Element," *Coal Age* 1 (May 11, 1912): 1021, and 4 (July 12, 1913): 61; testimony of William Coolidge, *West Virginia Coal Fields*, 2:911. In the quest to reduce mine accidents, the miners supported the effort to Americanize [teach English] to immigrants. *UMWJ*, Nov. 15, 1919, 8.

94. Testimony of John Laing, *Paint Creek Hearings*, 2:1658–59.

95. Ernest Bailey, letter to the editor, *Coal Age* 6 (Aug. 22, 1914): 1217. Also see M. P. H., letter to the editor, ibid., 13 (Apr. 20, 1918): 756–57. For a similar position, see West Virginia Engineer, "Sociological Conditions in West Virginia," ibid., 2 (Nov. 23, 1912), 733.

96. See Martin Carnoy, *Education as Cultural Imperialism* (New York, 1974); Michael Katz, *Class, Bureaucracy, and Schools* (New York, 1971), and his *The*

*Irony of Early School Reform* (Cambridge, Mass., 1968); Joel Spring, *Education and the Rise of the Corporate State* (Boston, 1972); David Cohen and Marvin Lazerson, "Education and the Corporate Order," *Socialist Revolution* 8 (Mar. 1972): 47–72; Robert Wiebe, *The Search for Order* (New York, 1967), especially 66–69.

97. Wolfe to Collins, May 15, 1916, and Collins to Wolfe, May 23, 1916. Also see L. C. Epperly to Collins, July 3, 1924, Collins Papers.

98. *Coal Age* 3 (Nov. 21, 1914): 838, and National Coal Association, *Bulletin* no. 959 (Aug. 24, 1929): 3.

99. State Superintendent of Black Schools, "Report," in W.Va. State Superintendent of Schools, *Annual Report, 1916*, 16–20.

100. Leamer, "Twilight of a Baron," 170; "Moonshine's Lively Part in Mingo Troubles," 36; Oscar Cartlidge, *Fifty Years of Coal Mining* (Charleston, W.Va., 1936), 70–73; *Charleston Daily Mail*, Apr. 8, 1921; testimony of Walter Wood, *Paint Creek Hearings*, 2:1363; testimony of Edward Bragg, ibid., 1:570–71.

101. "Cabin Creek YMCA," 741–42; Ridge, "An Important Factor in Modern Industry," 33; testimony of Wade Perry, *Paint Creek Hearings*, 3:2236–38.

102. West Virginia Superintendent, "Payday Drinking," *Coal Age* 4 (Oct. 4, 1913): 478–79; "Prohibition in West Virginia," ibid. 5 (Mar. 7, 1914): 416; "Prohibition and Labor in West Virginia," ibid. 13 (July 25, 1918): 181; testimony of Charles Pratt, *Paint Creek Hearings*, 2:2121; testimony of Edward Bragg, ibid., 1:570–71.

103. *Coal Age* 13 (Aug. 29, 1918): 408. Also see ibid. 13 (July 25, 1918): 181.

104. Pocahontas Coal Operators' Association, "Paternalism," a pamphlet in District 17 files, UMWA Archives, UMWA Headquarters, Washington, D.C. During the 1902 coal strike, Baer, president of the Reading Railroad and leader of the anthracite operators, wrote that "the rights and interests of the laboring men will be protected and cared for—not by the labor agitators, but by the Christian men to whom God in his infinite wisdom has given control of the property interests of this country." Joseph Rayback, *A History of American Labor* (New York, 1966), 211.

105. On company aid to schools, see Logan District Mines, "Why Coal Producers Aid Schools"; testimony of Walter Thurmond, *West Virginia Coal Fields*, 1:538–39; Conley, *Life*, 33–41. For the new position on company housing, see Non-Union Operators, *Statement to the U.S. Coal Commission*, 3; Logan District Mines Information Bureau, "Why Miners Do Not Own Their Own Homes," *Coal Facts*, vol. 3; Bituminous Operators' Comm., *Company Town*, 12–18. For the coal companies' new attitudes toward company stores, see Logan District Mines Information Bureau, "Company Stores Protect Mine Workers' Pocket Books," *Coal Facts* 7 (Dec. 17, 1921); West Virginia Coal Association, "Stores in Mining Towns," *West Virginia Review*, Jan. 1925, 1; Conley, *Life*, 43–47. For the abundance of fuel, see West Virginia Coal Association, "Fuel for Miners' Families," *West Virginia Review*, Aug. 1925, 1, and Conley, *Life*, 30–32.

106. West Virginia Coal Association, "The Water Supply in Mining Towns," *West Virginia Review*, Nov. 1924, 1; Conley, *Life*, 23–24; Bituminous Operators' Special Comm., *Company Town*, 19–20.

107. Conley, *Life*, chap. 3, and Bituminous Operators' Special Comm., *Company Town*, 21–22.

108. West Virginia Coal Association, "Playgrounds in Mining Communities," *West Virginia Review*, Dec. 1924, 1; "Activities in a Modern Mining Town,"

ibid., June 1925, 329; Non-Union Operators, *Statement to the U.S. Coal Commission*, 3.

109. West Virginia Coal Association, "Gardens in Mining Towns," *West Virginia Review*, Oct. 1924, 1. For other articles proclaiming the positive good of the company town, see West Virginia Coal Association, "Community Work in Mining Towns," ibid., June 1924, cover; Jesse Sullivan, "West Virginia's Greatest Industry," ibid., Apr. 1927, 232–33; West Virginia Coal Association, "West Virginia Coal and the Nation," ibid., Feb. 1929, 29; West Virginia Coal Association, "Hard Work and Skill in the Coal Business," ibid., Oct. 1928, 1; C. H. Mead, "The Winding Gulf Coal Fields," ibid., Apr. 1927, 211–12; George Wolfe, "Winding Gulf District," ibid., June 1925, 326–37; "Story of West Virginia's Famous Smokeless Coal Fields," ibid., June 1926, 290–99.

110. Conley's works include *Life; History of the West Virginia Coal Industry* (Charleston, W.Va., 1960); "An Enterprise That Costs the State $30,000,000 Annually," *West Virginia Review*, Sept. 1929, 450–51; and "McDowell County," ibid., June 1924, 18. Conley also wrote to government officials in behalf of the coal companies' company towns: see Conley to Secretary of Labor (James J. Davis), Sept. 23, 1922, and Conley to the Chief of the Children's Bureau of the U.S. Department of Labor (Grace Abbott), Sept. 28, 1922, both letters in file 20–23–8, Record Group 102, Records of the Children's Bureau. Conley also edited and published *West Virginia Review* between 1923–47, a "magazine devoted to stories and feature articles about West Virginia people, industries, traditions, and history." Conley, *Coal Industry*, 298. The quotes in this paragraph are from *Life*, 57–58.

111. Conley, *Coal Industry*, 83, and Conley, *Life*, 13, 55–56.

112. Conley, *Life*, 28.

113. W.Va. Bureau of Negro Welfare and Statistics, *Annual Report, 1925–26*, 9–11; "McDowell County, Coal Mining Being Almost Sole Industry, Provides Free Dentistry," *Coal Age* 18 (Oct. 7, 1920): 743–44; Brandes, *Welfare Capitalism*, 101. This attention to health services may support Spring's thesis that corporate concerns for employees' health stemmed mainly from their concern for productive and contented workers. Spring, *Education and the Corporate State*, 33–35.

114. Tams, *Smokeless Coal Fields*, 53, and Nettie McGill, *The Welfare of Children in Bituminous Coal Mining Communities in West Virginia*, Children's Bureau, U.S. Department of Labor publication no. 117 (Washington, D.C., 1923), 54–57.

115. Helen Norton, "Feudalism in West Virginia," *Nation* 133 (Aug. 12, 1931): 154; Wolfe to Collins, Dec. 20, 1915, and Sept. 27, 1927, Collins Papers; Sterling Spero and Abram Harris, *The Black Worker* (New York, 1974), 377–78.

116. Quoted in *UMWJ*, Mar. 15, 1930, 15. Also see Robert Sherrill, "West Virginia Miracle," *Nation* 158 (Apr. 28, 1969): 531.

117. *Coal Age* 4 (Sept. 20, 1913): 433. For other labor-management disputes involving company doctors, see George Korson, *Coal Dust on the Fiddle* (Philadelphia, 1938), 57–58; Judson Godfrey to A. Mitchell Palmer, Feb. 2, 1920, Straight Numerical File, item 205194–50–45, Record Group 60, General Records of the Department of Justice; W.Va. State Federation of Labor, *Proceedings, 1918*, 54.

118. Joe Bruttaniti, letter to the editor, *UMWJ*, Nov. 9, 1916, 9. Another miner asked for "malice or paternalism to none, but justice to all." Thomas Epps

to A. Mitchell Palmer, Jan. 26, 1920, Straight Numerical File, item 205194–50–135, Record Group 60, General Records of the Department of Justice. For a general discussion of the incompatibility of paternalism and capitalism, see Genovese, *Roll, Jordan, Roll*, 661–66; E. P. Thompson, "The Moral Economy of the English Crowd in the Eighteenth Century," *Past and Present* 50 (Feb. 1971): 82–96; Elizabeth Fox-Genovese, "The Many Fates of Moral Economy," ibid., 58 (Feb. 1973): 161–68.

CHAPTER VI

# "We Shall Not Be Moved"

> The public does not know that a man who works in a coal mine is not afraid of anything except his God; that he is not afraid of injunctions, or politicians, or threats, or denunciations, or verbal castigations, or slander—that he does not fear death.[1]
>
> John L. Lewis

Following the Paint Creek–Cabin Creek strike of 1912–13, the struggle to organize the coal fields took on broad and lofty meaning for the coal miners. The fight for the UMWA called for sacrifice, exhausting commitment, and devotion; it offered in return a feeling of righteousness and a "comprehension of human existence and its direction—in essence, the rewards of a religion."[2] "The church has had its place—its natural place—in our lives," a UMWA district leader in southern West Virginia told a gathering of his colleagues. "Our various fraternal societies have had their place. But, after all, the one organization that has done more for us than all the others combined, I say to you, is the union movement [UMWA]." The official continued: "The trade union movement is our very life because what the trade union movement does for us through its strength, power and influence determines the kind of life that we and our wives and little children are living in this country." But the union offered more than bread-and-butter benefits, he declared; the UMWA "not only determines what our material life shall be but it determines to a large degree our spiritual life as well. . . . The trade union movement gives us that existence that allows us to love our neighbor and worship God as we should."[3] The UMWA to the southern West Virginia coal miners was more than an economic institution; it was a crusade promising human dignity and religious purposefulness.

The spiritual significance and moral fervor that the southern West Virginia coal miners came to attach to UMWA was formed over a long period of time. The early miners had brought strong religious identifications with them into the coal fields. The native mountaineers had long found an emotional outlet and spiritual comfort in the campfire evangelist meetings of the Methodist and Baptist circuit riders.[4] In the midst of racism and oppression, southern blacks had found security and pride in the independence of the black church.[5] The European immigrants who had crossed an ocean in search of home and work found religion to be their one link with their homeland and culture. According to Oscar Handlin, "The more thorough the separation from the other aspects of the old life, the greater was the hold of the religion that alone survived the transfer. Struggling against heavy odds to save something of the old ways, the immigrants directed into their faith the whole weight of their longing to be connected with the past."[6]

This variety of religious beliefs created few social tensions and conflicts in southern West Virginia. From early frontier days, the native mountaineers were religiously mobile, joining the denomination of whatever minister crossed the mountains to bring them the gospel, and the multiplicity of religions in the Appalachian mountains had prevented religious intolerance.[7] The influx of tens of thousands of European immigrants and southern blacks strengthened this tradition of tolerance, for there were too many people belonging to too many different persuasions to allow any one group to dominate. For example, when one miner became intolerant of the increased number of Italians in his coal camp, he began spreading anti-Catholic propaganda. Shortly afterwards, he was attacked and knifed by a group of Italian-Catholic miners. The miner saw the light, as it were, and later reflected that "religious controversies concerning what form of worship one should adhere to are silly, asinine, and contemptible. . . . All that they have accomplished is a bloody record to which the atheist, agnostic, and unbeliever can refer."[8]

Traditional religious identifications were further strained by the industrial capitalism in the coal fields and, more important, by the company town. Social and religious historians have recognized the general disruptive impact of industrial capitalism upon American working-class religion.[9] Historians of the European immigration into the United States also have acknowledged that, while the religious traditions that the immigrants brought with them were quite strong, their traditional religious institutions were weakened or altered during settlement in a new land.[10]

In the southern West Virginia coal company towns, religious traditions met perhaps their greatest challenge. Gone were the circuit-riding ministers and campfire revival meetings that the native mountaineers had known; gone were the prestige and independence of the black church and the black minister that the southern migrant black miners had enjoyed; and gone were the power and prestige of the Protestant churches and the Roman Catholic church that the European immigrants had known. In their stead stood the company-sponsored and -controlled preacher, the company-built and -controlled clergy, all operating within the confines of the company-controlled town.

Investment in religion, the coal companies claimed, was simply a business proposition. "The coal companies," read an announcement by the West Virginia Coal Operators' Association, "are intensely interested in their communities and give every instance of developing the spiritual nature of their employees and their families. It is a good business proposition for any coal company to take an interest in and to assist in the building up of religious and community activities." The economic benefits that accrued through the sponsorship of religion included the stabilization of the mobile work force and the extolling of the work ethic to create better work habits among the miners. A publication of the company church at Glen White, Raleigh County, for instance, featured a businessman lecturing two miners for making homebrew. With an obvious reference to the dangers of mining coal while under the influence of alcohol, the businessman told the miners, "There can be no personal liberty when what a man does interferes with or makes hazardous the lives of others. Boys, what we need are real honest-to-goodness citizens. Men and women who hold their country in high respect, who obey the laws, who stand for the right conduct on the part of all."[11] These churches were dedicated to creating "honest-to-goodness" citizens in accord with the company's version of both honesty and goodness.

The coal companies' sponsorship of religion presented multiple problems for new miners. In the company town a miner's denomination became not a matter of personal preference or tradition, but was determined by the coal company, which selected the faith in the town where he was living and working.[12] Because of their extreme geographic mobility, many coal diggers belonged to between six and ten different denominations during their lives in the coal fields.[13] But even if a miner stayed in one town for a long time, which was unusual, his denomination could switch with a change in ownership of the company town, occasioned, for instance, by the death of a coal operator or purchase by another company.

Because the coal operator did not provide churches for all the religions represented in his company town, many of the migrant miners could not carry on their customary religious faith and practice. Sydney Box, a British immigrant, for example, had been a member of the Church of England, but could find no Anglican or Episcopal church in the southern West Virginia coal fields. Consequently, he was a Baptist while in Glen White, a Methodist in Glen Rodgers, and later, a Presbyterian.[14]

Catholics often found themselves without official churches and were forced to hold services in their company houses or in the Protestant churches when the latter were not in use. Many Catholics simply attended Protestant services or at least depended upon Protestant ministers to conduct baptisms and weddings.[15]

Despite such adjustments, the new miners tended to transfer their religious allegiances to the company church and to carry on traditional religious customs there. The minister, for example, although a company employee, was a prominent person in the community and enjoyed close social contact with the miners. Fred Mooney recalled that the company minister was always invited to a miner's house for dinner following Sunday morning services. Baptisms were community affairs; in Baptist towns, Saturday afternoon might be spent damming the local creek to raise the water level high enough for the ritual on the next day. Weddings were particularly joyous occasions as the miners took off a day from work and celebrated with an "all-day drunk." After performing the wedding services, the minister usually joined the miners in a nip or two. Saints' Days, which the miners called "Big Sundays," were also celebrated by drinking, as Protestant miners joined the Catholics in filling a bathtub with homebrew or moonshine and then helping them to drink it.[16]

What is most important is that these early miners attended the company churches and thus were never separated from the Holy Scriptures. Together, European immigrant, southern black, and native mountaineer learned and sang hymns, such as "I Shall Not Be Moved," that they later transformed into some of the most popular and militant songs in American labor history. These hymns, especially during the early years, also performed a valuable service to the miners in keeping them in close, personal contact with God. "There's a sermon in those songs," an elderly Mingo County miner stated, "so we sang them all the time"—and sing they did, not only in church, but also in groups in front of the company store and at work in the mines. "Only men of faith can sing in time of adversity," George Korson explained, "and of faith the miners had an abundance."[17]

By the time of the Paint Creek–Cabin Creek strike, however, the social contact between the miners and the ministers, and the miners' identification with the company church, had disintegrated. After two decades of life and work in the company towns, the miners had developed a working-class identity and now resented the political tasks of the company ministers, whose primary purpose was to provide a bulwark against the rising tide of unionism that was engulfing southern West Virginia. The Winding Gulf Coal Company, for example, decided that "it would be a good thing" to build a Catholic church and to hire a certain priest because "it was of great advantage" to another coal company to have had a priest during a previous strike where "the priest absolutely refused to allow his flock to have anything to do with the union."[18]

Once he was hired, the company preacher avoided the social and economic problems of his flock of miners. To do otherwise, to preach unionism, would result in the loss of his job because, according to a Logan County coal operator, it would "constitute a misuse of the cloak of religion."[19] It could also result in bodily harm, as happened on several occasions.[20] Simply stated, after the Paint Creek–Cabin Creek strike, there were no preachers of the social gospel in the southern West Virginia coal company towns.

Instead, the company ministers preached "other-worldliness," telling the miners to "be ye content with your wages" and that happiness and salvation lay in tolerating the hardships of this world. They gave divine blessings to the coal operators, who, they told the miners, were not only their best friends but were favored by God. For example, an announcement posted by a black church in southern West Virginia read:

> 8 P.M. The Hon. Mr. A____, the owner of this beautiful plant and the greatest Negro friend in West Virginia, will address the audience. 8:30 P.M. Mr. B____, the second man in the kingdom, the honored and highly respected superintendent of this town, the man we hope when the mantle falls from Elijah it will fall upon him as it did upon Elisha, will also address us.[21]

Some of the company preachers openly denounced the union as a "secret organization" that was "ungodly and wicked" and discouraged their congregations from joining it. Enraged by such anti-union sermons from the company clergy, a southern West Virginia miner wrote the *UMWJ*, complaining that "instead of preaching the Gospel of the Son of God, they preach the doctrine of union hatred and prejudice."[22]

The company ministers performed other duties for the company that further estranged them from the miners. During slack runs they served as labor agents, going south to recruit blacks, or worked in the company store. During strikes company ministers commonly served as mine guards. Company ministers in Logan County did not hesitate to join the army that Sheriff Don Chafin formed to repel the miners' armed march on Logan. At times company ministers themselves worked as strikebreakers. A miner wrote the *UMWJ* that during a strike at his mine "the leaders of the church, including the minister, went to scabbing." In their tent colony, however, the miners formed their own church, which, according to the miner, was an "organized church."[23]

During labor troubles the company church itself could also be used by the coal company. The churches were commonly "nailed shut" and marked "off limits" to the miners during strikes. During the Paint Creek–Cabin Creek strike, the company church at Pratt served as a munitions depot for the coal operators and their mine guards, and a Catholic church served as a makeshift jail for incarcerated strikers—without protest from the Catholic diocese.[24]

When the local company minister failed to provide the necessary spiritual opiate, the operators did not hesitate to call upon more powerful spiritual forces, bringing in nationally renowned preachers and evangelists to preach anti-unionism to the miners. Following the armed march on Logan, the Logan County Coal Operators' Association sponsored an appearance by Billy Sunday, who probably hated union organizers more than he did old Nick, if, in fact, he saw a difference. "I cannot believe that God had anything to do with the creation of these human buzzards [UMWA organizers]. I'd rather be in hell with Cleopatra, John Wilkes Booth and Charles Giteau than to live on earth with such human lice. . . . If I were the Lord for about fifteen minutes I'd smack the bunch so hard that there would be nothing left for the devil to levy on but a bunch of whiskers and a bad smell."[25]

The miners were not totally abandoned by the organized churches and ordained ministers in the coal fields. Churches and pastors in commercial towns in southern West Virginia often provided the miners with moral sympathy and spiritual support. Congregations of such churches passed resolutions supporting miners during strikes and set aside days of prayer to pray for their victory. A black minister who had been driven out of a company town because of his pro-union proclivities found refuge in the commercial town of Matewan, where he continued his ministry and began to hold union organizing meetings.[26]

At the same time, no one was more feared by the coal companies than a company minister who had become so repulsed by the companies' exploitation and oppression of his flock that he converted to unionism. Joining the armed march on Logan, the Reverend J. E. Wilburn announced to the miners: "Boys, I have refused up until this time to have anything to do with or take part in your union activities, but this morning I am laying aside my Bible until this community is made a safe place in which to live. I am ready to go to the front and when I get there the gunmen are going to know I am in action." Following the armed march, the Reverend Wilburn was jailed for killing Logan County deputy sheriff John Gore.[27]

Although the ministers in commercial towns were often sympathetic to the miners, their influence was negligible because even off company property, no one, not even an ordained minister, was immune from the power of the coal companies. When Free Will Baptists held their state convention in Logan, Chafin and his deputies broke into the building and forced the ministers from towns that had UMWA locals to leave the county. The company officials, however, were not always so nice, especially if a minister dared, even off company property, to denounce the company policies. When sixty-five-year-old Reverend Sam Betts set up tents and held a weekly revival meeting several hundred yards off the property of Winding Gulf Coal Company, he made the unfortunate mistake of declaring that "miners should have more for their work" and of denouncing the company superintendent as a "thief, liar and oppressor." Although the revival meeting was clearly outside company land, the company bookkeepers, store clerks, and mine guards attacked the evangelist and beat him until, according to the superintendent, he bled "like a sticked pig." They also broke several of his ribs. The minister was hospitalized for over a week, and the miners who protested against the outrage were immediately fired and replaced with "foreigners."[28]

Although the company ministers who turned against the company were removed immediately and the churches in the commercial towns were distant and had minimal influence in the daily affairs of the miners, both performed a valuable service. They reminded the coal diggers that the failure of religion in the company towns was not totally the fault of Christianity nor the ministry, but caused mostly by company policy. After a coal company had driven a company minister who turned pro-union from its company town, the UMWA local passed a resolution protesting "the actions of the company superintendent in denying a minister of the Gospel the right to hold religious service at the regular meeting place, the church, on company premises."[29]

The company church and company minister predominated, however, to the disgust of the southern West Virginia coal diggers. Angered over the company churches' indifference to their plight and by the company ministers' opposition to the UMWA, by the time of the Paint Creek–Cabin Creek strike the miners had revolted against both the churches and the ministers. "We are beginning to see the light for ourselves," a miner wrote, "and realize that the company preachers are selling us out to the bosses for a mere mess of the porage."[30] During the Paint Creek–Cabin Creek strike, a UMWA leader baited an audience of miners: "Let me tell you, them preachers are owned body and soul by the coal operators. Do you find a minister preaching against the guards?" "No," cried the crowd almost in unison, "they are traitors, moral cowards." Following the strike, UMWA locals passed numerous resolutions that condemned the local company churches, including the Catholic ones, for their neutral or pro-company stance during the conflict, and the miners publicly denounced their company ministers. This disrespect mounted, at times, to open hostility. In 1925, for example, a group of miners in the company town of Plymouth, Boone County, attacked and beat a company preacher following one of his anti-union sermons.[31]

The miners' contempt for the company church, the company minister, and the company religion was also displayed in everyday thought and action. In the company town, the preacher suffered a decline in prestige and influence as the miners showed him little respect and never looked to him for personal advice or counsel. This disregard was caused by problems of communication, and not only the realization that the miners and ministers were on different sides of the union-management conflict. Company ministers, for instance, often complained that they had to "talk down" to their congregations and that the miners "did not care for the staid, dignified services of the company church." The miners, on the other hand, grumbled that they could not understand the sophisticated sermons of their seminary- and college-trained company ministers, and they resented the company preachers' condescending attitudes toward them. After one church service, the miners led an arrogant, "eddicated" preacher, who had boasted that he could mine coal better than any miner, into the churchyard to ride a notoriously stubborn mule named Katie. After Katie had thrown the minister for the fourth time, the preacher began screaming and cursing at the mule and the miners. Thus, the minister delivered the first sermon that the miners truly enjoyed. More often the miners showed their disgust by falling asleep or gambling and drinking during company church services.[32]

As it had failed as a house of God, so the company church also failed as a social center. Mother Jones best expressed the miners' feelings about the company church when a local justice of the peace questioned her about holding a union meeting in a "House of God [a company church] with everyone carrying a gun." Jones quickly retorted, "Oh, that isn't God's house. That is the coal company's house. Don't you know that God Almighty never comes around to a place like this."[33]

The miners' growing contempt for company religion was vividly illustrated by their ceasing to attend company church services. Within a year after the Paint Creek–Cabin Creek strike, less than one-tenth of the miners on Cabin Creek attended church services, and less than one-fourth of the black miners in southern West Virginia attended church. By the 1920s, church attendance was so low in the Winding Gulf coal fields that less than one-half of the company towns retained churches and only two or three of those still held regular church services. Little wonder that in the early 1940s "Ripley's Believe It or Not" reported that the company town of Smithers, Kanawha County, was the largest town in the United States without a church![34]

The sincerity of the miners who did attend church is questionable, as gambling, drinking, and rowdiness commonly occurred during company church services. In the company town of Sharples, church services "had to cease," a coal operator testified, because they were continuously "broken up by the miners coming in and out in such a way to show their disrespect for the church."[35]

A Cabin Creek coal operator, who claimed his company church had become strictly a "female institution," may have been right when he told a U.S. Senate committee investigating the Paint Creek–Cabin Creek strike that the drop in church attendance among the coal miners was because the miners were "not as religious as they were in the old days."[36] But this reason fails to account for the miners' initial efforts to transfer their traditional religious beliefs and practices to the company church. Nor does it account for the continuing religious sentiment among the miners outside the church.

Although, as adults, UMWA District 17 leaders never attended church, they constantly salted their speeches and letters with biblical passages and references to the Deity. More important, District 17 officials found in the Scriptures justification (what E. P. Thompson called a "legitimising notion of right") for their roles as union activists. Mooney defended his reputation as an agitator with this explanation: "Solomon said that the seeker after knowledge would always encounter trouble. Solomon must have known that no man or woman can retain knowledge without that knowledge affecting the social system." When Frank Keeney was betrayed by the craft unionists in his fight for industrial unionism in the West Virginia State Federation of Labor, the UMWA District 17 president reflected that "Christ himself, who built the first organization for the union of men . . . was betrayed by one of his twelve men."[37]

The decline in church membership of workers occurred throughout the United States, although apparently not as severely elsewhere as in southern West Virginia. James Maurer, a noted national labor leader during the early twentieth century, contended that "if the workers had the same faith in the church that they have in the Bible, there would not be half enough churches in the country to hold them." Claiming that ministers' sermons were largely irrelevant to the needs of labor, Maurer asserted, "In the mills, mines and factories . . . the ministers will find among the toilers more of the true spirit of God than he will in many churches, but cannot hope to understand that spirit by discussing the evils of Sunday baseball."[38]

Miners at the Acme Mine, Cabin Creek, West Virginia, circa 1900.
West Virginia and Regional History Collection, West Virginia University Library, Morgantown

Frank Keeney (right) and Fred Mooney (left) after their election as president and secretary-treasurer, respectively, of District 17 of the United Mine Workers of America. UMWA Archives

Don Chafin was the "high sheriff" of Logan County, West Virginia. *UMWJ*

Mother Mary Jones and Frank Hayes, vice-president of the international UMWA, on the porch of the home where Jones had been confined since her participation and subsequent court-martial during the Paint Creek–Cabin Creek strike.
*International Socialist Review*

During the Paint Creek–Cabin Creek strike the coal operators ran the "Bull Moose Special" through one of the miners' camps; armed guards on board opened fire on the people in the camp.
*UMWJ*

Coal miners and union officials had no doubt that West Virginia was the "Russia" of the United States.
*UMWJ*

Mine guards, company spies, and hired gunmen made a mockery of the West Virginia state motto, *Montani Semper Liberi.*
*UMWJ*

Miners of the Cannelton Coal and Coke Company at Cannelton, West Virginia, in the 1920s.
West Virginia and Regional History Collection, West Virginia University Library, Morgantown.

UMWA President John L. Lewis (center) is shown here with some of the attorneys and defendants at the treason trials held at Charleston, West Virginia, after the armed march on Logan County.
*UMWJ*

This is the paradox of Jones. According to one of her biographers, Jones, who practically resided in West Virginia during these years, "was bitter at what she considered the shams of the church . . . the company-controlled houses of worship in many mining towns; and the clergy's failure to champion the workingman's revolt." Yet her speeches were filled with religious overtones and biblical references, as she looked, and asked the miners to look, to the heavens for guidance and wisdom. In one sentence Jones could prophesy revolution and urge "her boys" to buy ammunition and guns for protection and to kill the "God damn mine guards," and in the next sentence refer to her personal connections with the Almighty. "I have passed the scripture period of three score years and ten," she declared in one of her most violent speeches during the Paint Creek–Cabin Creek strike. "I am eighty-one, but I have a contract with God to see you boys through before I go."[39]

The coal miners' distaste for company religion and their nonattendance at the company church did not represent an absence of religious values, but the opposite; they held Christianity dear enough not to make a sham of it. David Cross, an elderly black miner in Wyoming County, realized that the company paid the minister, and he "didn't care for it." According to him, "If you're a Christian, you're suppose to act and believe like a Christian." Instead of attending the company church, Cross found spiritual contentment and secular guidance by reading the Bible. Fleet Parsons, an active UMWA organizer in Kanawha County, who "never professed a religion in his life" and claimed "preachers were phonies," rose before sunlight and read the Bible by firelight.[40]

The southern West Virginia miners, by the time of the Paint Creek–Cabin Creek strike, had separated the company church from religion. A Boone County coal operator made clear how total this division had become: "When I ask the miners to come out and take an interest in . . . [the company church] they tell me God is one question, and the church is another." A union leader expressed a similar position when he shouted to a gathering in Mingo County: "Why don't you [company] ministers go out and preach as Christ did? You don't because you are afraid of the high-class burglars and their dollars. You are gambling in Christ's philosophy, but you are not carrying out Christ's doctrine."[41]

Here is evidence not only of the miners' disassociation of church and religion, but also of a distinct conception of God and the philosophy and doctrines of Christ. That those independent religious sensibilities never took on a definite institutional form, never spawned a sect or denomination, was mainly due to the power of the coal operators. The preconditions, after all, were there. As Liston Pope explained, sect emergence "represents a reaction, cloaked at first in purely religious guise, against both religious and economic conditions. Overtly, it is a protest against the failure of religious institutions to come to grips with the needs" of the people.[42]

The miners drew much of their spiritual sentiment from their previous religious experiences. But the milieu that had given meaning and vitality to that sentiment was gone, and new social and economic experiences demanded a new institutional form of expression for those values. Unable to transfer them to the company church, as they apparently tried to do, the southern West Virginia miners vested this sentiment in the UMWA.

That the miners formulated a religion on and around this secular, and what has traditionally been viewed as simply an economic, institution need not be treated as an exceptional phenomenon. "Each society and each age," Erik Erikson has observed, "must find the institutional form of reverence which derives vitality from its world image."[43] In the world of these miners, the UMWA was the only institution that could serve this function. Like the church, the UMWA was a flesh-and-blood reality, existing in the real world and offering its adherents a better, more meaningful life on earth. Dignity, hope, self-respect, personal responsibility, and a place in the universe—these were elements in the worldly role that their own church had once provided and that the company church could not provide. "We older miners have suffered for want of unionism," wrote an elderly miner from Longacre in 1915. "We would suffer from hunger and thirst for a place where we could speak as men; we worked 12 hours a day and never saw a pay. Now under organized conditions we have our payday. We have gotten our freedom; we can speak where we please. . . . All miners should make the organization all that it can be made, for it has given us all we have in West Virginia."[44]

At the same time, the UMWA, like the church, performed the role of the "interfering community" that sets itself "against the world, sometimes says 'no' to what the world asks of it . . . in the name of another world." A ballad written by a southern West Virginia miner extolled the union as church:

> When you hear of a thing that's called union,
> You know that they're happy and free,
> For Christ has a union in heaven,
> How beautiful union must be.[45]

Like a church, the UMWA offered the miners a transcendence of life on earth. Another area coal digger wrote:

We will have a good local in Heaven,
Up there where the password is rest,
Where the business is praising our father,
And no scabs ever mar or molest.[46]

The theology of this secular church tapped a wellspring of creativity in the mining communities. "It is the lower classes," Ernest Troeltsch has pointed out, "which do the really creative work forming communities on a genuine religious basis. They alone unite imagination and simplicity of feeling with a nonreflective habit of mind, a primitive energy, and an urgent sense of need."[47] The miners' religion was straightforward, filled with simple dichotomies; there was the right side representing the forces of good, and the wrong side representing the forces of evil. There was the instinctive identification with the children of Israel that Thompson noticed generally in the "cults of the poor." "Brother Lee Hall," a miner from Sprigg, Mingo County, wrote, "has been a brother indeed—a Moses who is attempting to lead us out of bondage, through the wilderness and into the promised land."[48] Southern West Virginia miners found pride in the fact that the historic Jesus had also been a workingman who had performed "muscular labor," and they described Jesus as the first radical who had challenged a corrupt and tyrannical establishment in the name of good.[49]

These religious images fired a morality of stark contrasts that underlay the miners' sense of spiritual and class justice. Their religion did not become what Methodism was to Thompson's nineteenth-century workers, a "chiliasm of despair" and "psychic masturbation."[50] From 1912 to 1940, the miners were served by a meaningful, purposeful religion that gave them assurance and direction in both their individual and collective lives. Against its images, they were able to judge and condemn the socioeconomic order and to justify and motivate social activism.

To dwell upon these common traits is, as Eugene Genovese pointed out, to miss and misunderstand the larger and more profound complexities and subtleties, as well as the more meaningful context, of a people's religion—especially a religion of resistance.

> The historical Jesus need not be interpreted as a political revolutionary, in either a social or a national-liberation sense, as some radical critics have tried to do. The living history of the Church has been primarily a history of submission to class stratification and the powers that be, but there has remained, despite all attempts at extirpation, a legacy of resistance that could appeal to certain parts of the New Testament and especially to the prophetic parts of the Old. The gods must surely enjoy their joke.[51]

It is in this context that the miners' religion should and must be understood.

Basic to the miners' autonomous religious proclivities was their development of their own spiritual leaders—miner-preachers—separate from the company church.[52] The unordained miner-preacher was generally a man of little education but considerable mining experience who had suddenly "got the call to preach the word of God." The call could apparently strike any miner at any time, but it often came after several years of working in the mines. The Reverend G. W. "Preacher" Warner was forty years old, and he had been digging coal for twenty-five years when he "got the call." The Reverend Homer Santrock had been mining coal for over thirty years when he realized "God wanted me." Like the Apostle Paul, the Reverend Roy Lester was walking down a road, on his way to the mines, rather than to Damascus, when he had a "vision" and "got the call."[53]

Because they were not paid for their services—some of them actually refused payment because they considered it their duty to preach—the miner-preachers continued to work in the mines following their call. Consequently, the miner-preacher loomed conspicuous in the life, work, and culture of the miners where the mine pit often became the pulpit. The miner-preacher conducted prayer meetings for the miners before they entered the mines, praying for the Lord's protection from slate falls and explosions, and he often held another prayer meeting upon leaving the mines, "thanking the Lord for coming out alive." Religion, then, became an everyday, not a one-day-a-week, experience to the miner in southern West Virginia.[54] Because he worked and prayed with his fellow miners on a daily basis, the miner-preacher developed a personal identification and sympathy with his flock and an ability to address himself to their misery and hopes rarely achieved by the ordained, institutional minister.[55] As a result, it was the miner-preacher, not the company preacher, from whom the miners sought personal counsel in secular as well as spiritual matters.

Given his background as a coal miner and his constant communication with the other miners, it is little wonder that the miner-preacher became an active voice for social justice in the coal fields and a strong proponent of unionism. Walter Seacrist was a miner and lay minister in Holley Grove, on Paint Creek in Kanawha County. He not only preached unionism to the miners, but also served as vice-president of the West Virginia Mine Workers' Union and wrote some of the most militant union songs that have come out of the coal fields.[56] According to Keeney, Roy Combs, a miner-preacher in Mingo County, was one of the most effective

UMWA organizers in southern West Virginia. During Keeney's drive to organize southern West Virginia, Combs established several union locals and later served as a leader of the Mingo County strike, conducting numerous mass organizing meetings and delivering powerful pro-union sermons.[57]

The miner-preacher's undeviating faith in both God and the goals of the miners provided the miners with hope and inspiration during their darkest hours. Following the armed march on Logan, when hundreds of miners were in the Logan County jail waiting trial for treason, the Reverend J. J. Jeffries, a miner-preacher, formed a choir and led the incarcerated miners in hymns each night. After the singing, Jeffries conducted a prayer meeting. According to Mooney, Jeffries "prayed as one who believed in the effects of prayer and sang as though he enjoyed it."[58]

Not all of the miner-preachers could write militant labor songs or spend time in jail with their flock, but they all could summon unusual enthusiasm and courage for the cause of the union movement, as they combined the secular and the spiritual into uncompromising secular-spiritual goals. "I say to you that any man in this gathering today," a miner-preacher shouted to a group of miners during a walkout, "who does not join this strike and stand by it, even until death—for the sake of the children, is not worthy to call himself a Christian because he is not willing to stand up for the Kingdom of heaven. . . . The children are the Kingdom of Heaven." Nobody matched the eloquence of the miner-preacher who joined his colleagues on the armed march on Logan with the blessing: "Now I lay down my Bible and take up my rifle in the service of the Lord." Upon hearing that classic statement, Mooney exclaimed: "By God! That's what I call preaching."[59]

Although the miner-preachers provided the miners with guidance and inspiration, their influence should not be overestimated. The southern West Virginia miners, like most religious dissidents, placed greatest stress upon that "source of unquestioned authority" that "could undercut all other traditions and institutions"—the Bible.[60] Such reliance upon the Bible as a guide to spiritual faith and secular practice was essential for a society in which each person was a call away from priesthood, either as a lay clergyman, a miner-preacher, or a union leader.

That their reading of the Bible served as the "source of unquestioned authority" for the miners' religious perspective does not mean that theirs was simply a fundamentalist, literalistic religion, as Jack Weller and others have claimed. Weller has stated that southern West Virginians "have never appreciated anything but a simple literalistic belief in the Scriptures."[61] The southern West Virginia miners combed the Holy

Scriptures for passages that explained their particular predicament and offered understanding, hope, and salvation. To the Bible were sometimes added other religious authorities. Mooney explained that the Bible was a place where "knowledge could be obtained" and that it was a "book which laid down a routine of life." "I wanted to know the Bible," he wrote. "This desire became an obsession with me and I was not satisfied until I had read it several times." But Mooney also read "Dr. Talmedge, Dwight Moody, and Reverend W. S. Harris's *Hell before Death*," which "explained many things that I wanted to know such as industrial evils, panics, unemployment, who controlled finance and politics."[62] The Bible was the beginning but not the end of social and economic truths, and it could serve to stimulate political activism. On this basis, a socialist miner could advise his fellow coal diggers that they should "read the Bible more and the capitalist press less."[63]

In using the Scriptures for both justification of and motivation for union activism, the miners found the Bible to be a gospel of unionism and a handbook of social justice. Whit Collins of Logan County claimed that God favored the UMWA because the Bible "speaks of the union all the way through it. The Bible says everybody is got the right to live right and if a man don't work, he don't eat and live right." The Reverend Lester explained the compatibility of the "union and God" by relating that the "Bible claims that all men are equal and no one should be mistreated. The union's purpose was to eliminate inequality and mistreatment."[64]

Miners' letters often read like Sunday school lessons or ministers' sermons, with specific references to biblical passages that illustrated that the Word of God was the word of unionism. Classic was the miner who discovered the Lord's pro-union stance in Ecclesiastes 4:10–12: "Two are better than one, because they have a good reward for their labor. For if they fall, the one will lift up his fellow, but woe to him who is alone when he falleth; for he hath not another to help him up. And if one prevail against him, two shall withstand him, and a threefold cord is not quickly broken."[65] This stress on unity was omnipotent. It wielded together peoples of diverse ethnic backgrounds, races, nationalities, and religions into the dynamic spiritual collectivism that existed in the company towns. "Jesus is love," an elderly miner in Keystone, McDowell County, declared. "There's no need for anybody to tell me they are Methodists, Baptists, Catholics—Jesus is love."[66]

Love—community concord—was vitally needed in a world filled with blacklists, depressions, lockouts, strikes, murderous explosions and slatefalls, and mine guards. Mooney's fellow miners, themselves quite poor, sacrificed to aid him when he had been fired and blacklisted for his union activities. "This is the spirit of fellowship, love, and devotion that permeates the life of the union coal miners," Mooney reflected. "He will give until it hurts and then divide the rest." The southern West Virginia miners had grasped and practiced the essence of "Christ's doctrine and philosophy." "It behooves each one of us," read the constitution of the West Virginia Mine Workers' Union, "to follow a line of conduct that will bring about a closer friendship, one that is true to God and man."[67]

Numerous versions of Christ's parable of the Good Samaritan circulated among the miners. According to one, a young boy asked an "old Negro" which church he belonged to. The "old Negro" replied: "Bless you child, it's this way. There are three roads leading from here to town. A straight road, a road that goes round in sort of a circle and a road through the woods. When I go to town with a load of grain they don't ask, 'Uncle, what road did you come by?' but 'Is your wheat good?'"[68]

Another story told of a "poorly paid pastor of a country church in West Virginia" who became ill one winter.

> A number of his flock decided to meet at his house and offer prayer for the speedy recovery of the sick one and for material blessings upon the pastor's family. While one of the deacons was offering a prayer for the pastor's household, there was a loud knock at the door. When the door was opened a stout farmer's boy was seen.
>
> "What do you want?" asked one of the elders. "I've brought pa's prayers," replied the boy. "Brought pa's prayers? What do you mean?"
>
> "Yep, brought his prayers, and they are out in the wagon. Just help me and we'll get them in." . . .
>
> Pa's prayers consisted of potatoes, flour, bacon, apples, cornmeal, turnips and clothing for the sick one and his family. The prayer meeting adjourned in short order.[69]

These lay traditions are understandable in a society where institutional religious leaders were regarded as company "sucks," and survival depended upon brotherly and sisterly cooperation.

The southern West Virginia miners were never perplexed with Abraham Lincoln's problem of wondering whose side God was on. They never doubted that their God—the God of good and comfort—was pro-union. During the Paint Creek–Cabin Creek strike, a UMWA spokesman explained to his audience of coal miners that "you will eventually win . . . because your cause is just and you are advocating the principles for the betterment of mankind, and God holds the balance of power."

> Union miners, link together
> Keep the password in your mind,

the southern West Virginia miners sang in their version of "A Miner's Life," because,

> God provides for every miner,
> When in union they're combined.[70]

One UMWA leader established an even more direct connection between God and union by asserting, "The labor movement was not originated by man. The labor movement . . . was a command from God Almighty. He commanded the prophets thousands of years ago to go down and redeem the Israelites that were in bondage, and he organized the men into a union and they went to work. They got together and the prophet led them out of the land of bondage. . . . For the first time the worker was free."[71]

More important than their simple, albeit bold and blunt, declarations of the Lord's pro-union leanings was that the miners found the Lord's blessings in everything about them. In nearly everything they said and did the miners acknowledged the presence of God. "All our beautiful surroundings," read the constitution of the West Virginia Mine Workers Union, "tell us of God's greatness and goodness, of the universe of which we are children of His Creation."[72] The union was necessary for life as God intended. Accordingly, the practical benefits of unionism, the bread-and-butter policies, took on a higher meaning. When twelve southern West Virginia miners appeared before Jones to ask for the union obligation, she replied: "Now boys, you are twelve in number. That was the number Christ had. . . . You shall preach the gospel of better food, better homes, a decent compensation for the wealth you produce."[73] When the miners converted one of their favorite hymns into a militant union song, one stanza went:

> The union gives us beans, pork and potatoes,
> And we shall not be moved.[74]

It is no coincidence that this recognition of God, particularly of His blessings, was most pronounced and conspicuous during strikes, when they needed those blessings most. "The Lord has been on our side as far as the weather is concerned," wrote a miner during the Paint Creek–Cabin Creek strike. "It seems as though the Lord is with the striking miners in this isolated place," another miner declared. "Almost with the first day of the strike He began to bring forth one of the finest crops of sallet, shonnie, dandelion and other wild greens. Long before the season was over, the ground hog was nice and fat. The striking miners have been catching them in large numbers." During the Mingo County strike, a miner wrote, "Because of the Lord's blessings we have plenty of clothes and plenty to eat. . . . We strikers of Mingo County are sure to win."[75] The miners found God's presence in anything that enabled them to win a strike and recognition of the union. To the southern West Virginia miners, the God of their culture was the God of comfort.

Once the miners had accepted unionism, they expected the Lord to bless them—He was, after all, pro-union—and so their thankfulness was somewhat muted. What they were vocally grateful for was that God had awakened them to the meaning of unionism. Unionism to the southern West Virginia miner may have been divine, but the struggle for it took place in a flesh-and-blood world where men were corrupted by ignorance, fear, greed, and the sin of "worshipping of false idols" [for "false idols," read coal operators], and such human failings were the cause of nonunionism. In southern West Virginia for a brief period, Malcomb Ross noted that "sin is to be lukewarm for the [union] cause." Once awakened to unionism, the miners were prolific in praising and thanking God. "Thank the Lord," a Mingo County miner wrote during the struggle to organize Mingo County, "we have our eyes open and refuse slavery from now on." "Because of the Lord," another miner wrote during the same strike, "we have plenty of clothes and plenty to eat, so all we have to do now is to thank God for the saving of our souls."[76]

But the miners always knew that what the Lord gave to corruptible man, man could take away. "Many of us have sung the song like this: 'I thank God Almighty that I am free at last.' But if we allow the operators to sweet talk us against our obligation we will be ruined at last." No wonder that a miner who betrayed or hurt the union cause by sins of omission, such as failing to pay dues or to attend union meetings, was branded as Judas.[77]

In awakening the miners to unionism, God worked in mysterious ways. Miners often viewed mine guards, the militia, and the state police as worldly, pagan vehicles, like the Assyrians and Amelekites in ancient times, through which God was working to test the miners' faith and to show the miners the evils of nonunionism and of worshipping those "false idols," the coal operators. Long struggles were the tribulations of the faithful and the trials of conversion for the uncalled. After telling the story of how the "children of Israel" were able to unite and defeat the Amelekites after seven years of labor, a southern West Virginia miner explained to the *UMWJ*, "The fact is clearly demonstrated in my mind that it is only through a united effort and a rigid observance of the laws of God that we miners can be free." Incarceration was a divine test. A miner saw his time in jail as "only history repeating itself, for when Joseph was sold into Egypt, his brothers meant to do him harm but God meant it for good."[78]

Most mysterious were the apocalyptic experiences of a Fayette County miner who claimed that he had received "many visions from God." During one such vision, the miner wrote that "God spoke to me . . . and said if Socialism were adopted by all the world, everybody would go to heaven on horses of fire and chariots just like Enoch."[79] This miner's conversion experience was atypical, for the miners who looked for social salvation through socialism were few, and the number of miners who had such apocalyptic visions was probably even less. But this experience does illustrate that the miners did not need Karl Marx or socialist agitators to give them visions of a workers' paradise. They had developed their own, not by reading the *Communist Manifesto*, but by reading that "source of unquestioned authority"—the Bible. The God who had provided them with tangible earthly blessings also offered them hope and rest in a unionized heaven:

> We will have a good local in heaven,
> Up there where the password is rest,
> Where the business is praising our Father,
> And no scabs ever mar or molest.
>
> Our Savior is on the committee,
> He is pleading our cases alone,
> For ages He's been on committee,
> Pleading daily to God on His throne.
>
> The Bible up there is the Journal,
> And the members all know it is true;
> The contract up there is eternal—
> It was written for me and for you.[80]

The southern West Virginia miners did not need national UMWA organizers to tell them of the blessings of unionism. A strain of Christian perfectionism accompanied the miners' belief in the power of the union:

> When you hear of a thing that's called union,
> You know that they're happy and free,
> For Christ has a union in heaven,
> How beautiful union must be.
>
> There's nothing like union to me,
> It's a home of the happy and free,
> It's a heaven of rest for the miners
> How beautiful union must be![81]

This belief in a "better life to come" in a unionized heaven also gave the miners ideas and ideals to fight for on this side of the workers' paradise. With the union representing God's will—it was the church through which

God was working—the miners believed that once saved, they were capable of perfecting (unionizing) the world in God's image. Consequently, in attempting to unionize the southern West Virginia coal fields, the miners maintained that they were doing "God's work." "We are doing God's holy work; we are breaking the chains that bind us," a UMWA speaker shouted to a group of his fellow miners during the Paint Creek–Cabin Creek strike. "We are putting the fear of God into the robbers. All the churches here . . . couldn't put the fear of God into them, but our determination has made them tremble." "Coal miners, wake up and shoulder the cross," a miner appealed to his nonunion colleagues. "Are you willing to let the operators make conditions as they desire?" Another miner wrote: "God has given us this organization for he knew that the time would come when it would be our resort. According to the teachings of the word the poor are His chosen, and I feel that no man is working in the mines with the expectations of riches."[82]

The ideal of Christian perfectionism informed and sanctioned class struggle. "I want these men [coal operators and company ministers] to study these few lines from the Bible," a miner's wife wrote, quoting from the book of James—which Archibald Robertson had described as "this revolutionary pamphlet":

> Go now ye rich men, weep and howl, your miseries that shall come upon you. Your riches are corrupted and your garments are motheaten; your gold and silver is cankered and the rust of them shall be a witness against you and shall eat your flesh as if it were fire. Ye have heaped treasures together for the last day. Behold the hire of the laborers who have reaped down your fields which you kept back by fraud, and the cries of them which have reaped are entered into the ears of the Lord of Sabaoth. Ye have lived in pleasure on earth and been wanton; ye have nourished your hearts as in a day of slaughter; ye have condemned and killed the just.[83]

"Christ built the first organization for the union of men," Keeney declared. "As a result those who lived in their pomp and pride undertook the assassination of the lowly Nazarene." Coal operators and politicians were also condemned because they believed, according to one miner, that they were "created higher and have more forethought and are more qualified than our Lord and Savior Jesus Christ." Another miner compared the coal operators to Pontius Pilate and the Roman authorities because they were always "ready to crucify the cause of right." Describing Jesus as the first labor organizer, another southern West Virginia miner wrote, "Christ says in Divine Inspiration, 'He that is not for me is against me.'" The allegory was simple: to be anti-union was to be anti-Christ.[84] When the coal company at Winnifrede, Kanawha County, posted "No Trespassing" signs on its company church during a particular strike and then fired its company minister for participating in one of the strikers' church services, the local strike bulletin declared, "You can't even talk about God at Winnifrede these days. Who said the coal operators weren't in league with the Devil?"[85]

There could only be one reward for those who sided with the coal operators and stood opposed to unionism. Wrote a southern West Virginia miner about a scab awaiting judgment at the Pearly Gates:

> I ought to get a large reward
> For never owning a union card
> I never grumbled, I never struck
> I never mixed with Union truck.
>
> But I must be going my way to win,
> So open, St. Peter, and let me in.
> St. Peter sat and stroked his staff,
> Despite his office, he had to laugh.
>
> And said with a gleam in his eye,
> "Who is tending this gate, you or I?
> I've heard of you and your gift of gab
> You're what's known on earth as a scab."
>
> Whereat Peter rose in his stature tall
> And pressed a button upon the wall;
> Said he to the imp that answered the bell,
> "Escort this fellow down to hell."[86]

Symbolically, the union took over the church's traditional role of burying the dead. Union miners rather than a minister conducted a union miner's funeral, and a burial service provided by the union constitution, along with a section from the Bible, was read over the grave. Veteran miners often laid out specific instructions for their funerals and burials. On his deathbed, Charles Louie, a charter member of District 17, wrote, "I have been a union man all my life . . . and expect to die a union man. I want union pall bearers, a union undertaker, union men to dig my grave, a union preacher, and a union stone at my head and feet." James Miskell, a UMWA veteran from Hansford, Kanawha County, gave deathbed instructions that he wanted UMWA members for pall bearers, the undertaker to be a "local friend of the UMWA," and the district president to read the burial service.[87]

> Let the flowers be forgotten,
> Sprinkle coal dust on my grave,
> In remembrance of the UMWA.[88]

The miners were not obsessed with elaborateness.

This simplicity did not lessen the dignity of the ceremony, nor its meaning. The character of the funerals demonstrated the maturation of an autonomous working-class culture within the company towns of the southern West Virginia coal fields. Funerals, Genovese has explained, allowed "participants to feel themselves a human community unto themselves . . . respect for the dead signified a respect for the living, respect for the continuity of the human community and recognition of each man's place in it."[89]

A funeral could serve as an occasion for class-conscious rebel-rousing. Referring to the downpour of rain during the funeral service for a UMWA member who had been gunned down by Baldwin-Felts mine guards, the speaker exclaimed: "Even the Heavens weep with grief-stricken relatives and the bereaved friends of this boy." He then launched into a tirade against the men responsible for the murder—the coal operators:

> Sleek, dignified, church-going gentlemen who would rather pay fabulous sums to their hired gunmen to kill men for joining a union than to pay like or lesser amounts to men who delve into the subterranean depths of the earth and produce their wealth for them. At the same time these same men prate of their charities, their donations to philanthropic movements, act as vestrymen and pillars of the churches to which they belong.[90]

The miners' religion provided the day-to-day support that was needed in a harsh, brutal world filled with deadly slate falls and car accidents and murderous gas explosions that killed friends and coworkers in a single flash. Their religion, their theological perspective, provided a protection against despair. It was not a religion of fatalism nor capitulation, but one of comfort and support. It helped them to accept the forces and evils that lay beyond human control and often human comprehension, to accept what had to be endured in the life and work of a coal miner.

Theirs was a religion of spiritual security in a setting in which each miner or his buddy, who was often a relative, was only a spark or slip of the tool from eternity:

> The miner is gone,
>   We'll see him no more,
> God be with the miners
>   Wherever they go.
> And may they be ready
>   Thy call to obey,
> And looking to Jesus

The only true way.

God be with the miner
Protect him from harm,
And shield him from danger
With thy dear strong arm.
Then bless his dear children
Wherever they may be,
And take him at last
Up to heaven with thee.[91]

Knowing that the God of their earthly culture was the God of heavenly comfort, the southern West Virginia miner could face death with assurance and calmness:

Yes sister, I am dying
Soon I'll reach a better shore
Soon I'll gain a home in heaven
Where this coupling will be no more.[92]

Death messages that were found on the recovered bodies of miners who had been trapped by a mine explosion and lived for a brief period before their air gave out read:

We are weakening. Our hearts are beating fast. Good-by everybody.

We're going to heaven. We have plenty of time to make peace with the Lord.

Dear Wife: Still alive, but air is getting bad. Oh, how I love you Mary.

Dear Father: I will be going soon. We are cold. . . . Will meet all in heaven.[93]

This was not fatalism, as Weller claimed, but a religious acceptance of both death and a heavenly reward for the good and faithful. As the Reverend Homer Hicks explained, the miners held "stronger than customary religious beliefs"; that is, the miners possessed an "unqualified acceptance of the immortality of life." Paradoxically, despite and largely because of the exploitation, oppression, and the physical degradation of the company town, the miners developed a culture that allowed them to know and enjoy the fullness of life and to develop the perspective in which, as Erikson explained, "death loses its sting."[94]

Several observers have called attention to the close connection between religious and political extremism. Friedrich Engels, for example, found that early Christianity and the revolutionary workers' movement had "notable points of resemblance," especially in their millennial appeals. Leon Trotsky recognized the relationship as he successfully recruited members for his revolutionary Marxist organization, the South Russian Workers' Union, from the more extreme religious sects. Christopher Johnson has pointed out that Cabe's communist Icarian movement contained a strong Christian impulse, and Eric Hobsbawn has shown that it was more than a coincidence that "Methodism advanced when Radicalism advanced and not when it grew weaker" because there existed a "marked parallelism between the movements of religious, social and political consciousness."[95]

The religion that evolved among southern West Virginia miners had limited political overtones. It helped them to define and sanctify class boundaries, and it gave the miners a sense of mission as they spread the gospel of unionism throughout southern West Virginia. But their religion stopped short of the political ideology that was necessary for an overt attack upon the coal operators.

In its definitions and goals between 1912 and 1919, the union movement remained a political and social movement. But the religion that evolved played an indispensable part in providing the groundwork for the obviously political action that followed World War I. It promoted collective thought and action, gave cohesion and strength to a social class, and permitted the miners to resist the servility and feelings of inferiority that class oppression often breeds in the oppressed.[96] As the miners developed political definitions and goals, their social movement became a political movement that attempted a revolution in the southern West Virginia coal fields.

But the miners never left the religious values and traditions that they had developed. After they had successfully organized all of southern West Virginia in the 1930s, the miners commended their union officials and President Franklin D. Roosevelt, but they also paused to make especially clear that "above all we must not forget to thank and praise God, Who has carried this thing through. Glory to God, he has carried Us through." Virginia West of Mt. Hope, Raleigh County, wrote a ballad that expressed this relation between God and the union. Three lines of it read:

> Now when you meet your boss you don't have to bow,
> He ain't no king—never was nohow.
> Look, Lord, we're independent now.[97]

## NOTES

1. David Selvin, *The Thundering Voice of John L. Lewis* (New York, 1969), 24.
2. The quote is Stuart Kaufman's view of the religious meaning that Samuel

Gompers found in the trade union movement. *Samuel Gompers and the Origins of the American Federation of Labor, 1848–1896* (Westport, Conn., 1972), 21. Other labor historians have found that labor leaders perceived the union as a secular church. Herbert Gutman, for example, explained that UMWA official Richard L. Davis felt that the union "promised redemption from an evil social order. Therefore he gave to his work the zeal and devotion expected of a dedicated missionary." "Protestantism and the American Labor Movement," *American Historical Review* 71 (Oct. 1966): 90–91.

3. Address of Van A. Bittner, *Proceedings of the West Virginia State Federation of Labor, 1929,* 45.

4. Otis Rice, *The Allegheny Frontier* (Lexington, Ky., 1970), chap. 12.

5. For the role and importance of the black preacher and black church in the black community, see W. E. B. Dubois, *Souls of the Black Folk* (New York, 1961), 161, and Charles Hamilton, *The Black Preacher* (New York, 1972).

6. Oscar Handlin, *The Uprooted* (New York, 1951), 117. For another view of this same theme, see Philip Taylor, *The Distant Magnet* (New York, 1971), 217–28.

7. Rice, *Allegheny Frontier*, 22, 267–68, 378. Also see E. E. Northern, "The Religions of the Mountaineer," *West Virginia Review*, Aug. 1933, 317–36, and Edwin Cubby, "The Transformation of the Tug and Guyandot Valleys" (Ph.D. diss., Syracuse University, 1962), 39–40.

8. Fred Mooney, *Struggle in the Coal Fields*, ed. Fred Hess (Morgantown, W.Va., 1967), 49.

9. Carl Degler, *Out of Our Past* (New York, 1970), 339–51; Edmund Brunner, *Industrial Village Churches* (New York, 1930), 127; Liston Pope, *Millhands and Preachers* (New Haven, Conn., 1942), 83–84.

10. Handlin, *Uprooted*, 126–27, and Taylor, *Distant Magnet*, 218.

11. "Religious Work in Mining Towns," advertisement of the West Virginia Coal Operators' Association, *West Virginia Review*, Feb. 1925, inside cover, and A. M. Henderson to Justin Collins, Aug. 7, 1916, Justin Collins Papers, West Virginia University Library, Morgantown, W.Va. For the use of the pulpit to stimulate coal production, see E. E. White's remarks in *UMWJ*, Sept. 1, 1918, 15. That the coal companies built the church and paid the minister was generally known, but for operators' admissions of this, see testimony of Ira Davis, U.S. Congress, Senate Committee on Education and Labor, *Conditions in the Paint Creek District, West Virginia*, 63rd Cong., 1st sess., 3 vols. (Washington, D.C., 1913), 2:1442–43 (hereafter cited as *Paint Creek Hearings*); testimony of R. L. Wildermuth, U.S. Congress, Senate Committee on Interstate Commerce, *Conditions in the Coal Fields of Pennsylvania, West Virginia and Ohio*, 70th Cong., 1st sess., 2 vols. (Washington, D.C., 1928), 2:1490 (hereafter cited as *Conditions in the Coal Fields*); testimony of William Wiley, U.S. Congress, Senate Committee on Education and Labor, *West Virginia Coal Fields: Hearings . . . to Investigate the Recent Acts of Violence in the Coal Fields of West Virginia*, 67th Cong., 1st sess., 2 vols. (Washington, D.C., 1921–22), 1:525 (hereafter cited as *West Virginia Coal Fields*); testimony of Langdon Bell, ibid.,1843.

12. Testimony of Langdon Bell, *Conditions in the Coal Fields*, 2:1843; interview with Sydney Box, Glen White, W.Va., summer 1975; *Paint Creek Hearings*, 2:1465. Fred Barkey has attempted to quantify the denominations of the West Virginia socialists, including the miners. "The Socialist Party in West Virginia from 1898 to 1920" (Ph.D. diss., University of Pittsburgh, 1972), 74–77. I have

avoided such an inclination because, as I stated above, a miner's religion depended upon where he lived and worked, not personal preference.

13. The geographic mobility of the southern West Virginia miners is discussed in chap. 2.

14. Interview with Sydney Box, Glen White, W.Va., summer 1975. Several coal companies attempted to resolve this problem by building "union churches," which would be available to all congregations. Phil Conley, *Life in a West Virginia Coal Field* (Charleston, W.Va., 1923), 42–43, and *Logan* (W.Va.) *Banner*, May 23, 1889.

15. Interviews with the Reverend Shirley Donnally, Oak Hill, W.Va., and the Reverend Homer Hicks, Dunbar, W.Va., summer 1975.

16. W. P. Tams, *The Smokeless Coal Fields of West Virginia* (Morgantown, W.Va., 1963), 62; Mooney, *Struggle*, ed. Hess, 1–5; interview with the Reverend Shirley Donnally, Oak Hill, W.Va., summer 1975.

17. For a discussion of the origins of "We Shall Not Be Moved" as a labor song, see Edith Fowke and Joe Glazer, *Songs of Work and Protest* (New York, 1973), 38–39; George Korson, *Coal Dust on the Fiddle* (Philadelphia, 1943), 286.

18. Jarius Collins to Justin Collins, Oct. 21, 1902, Collins Papers. The miners' resentment of the company ministers can be compared to the black slaves' anger with the "political tasks of their 'white' preachers." Eugene Genovese, *Roll, Jordan, Roll* (New York, 1974), 202.

19. Testimony of William Coolidge, *West Virginia Coal Fields*, 2:936. Also see testimony of Ira Davis, *Paint Creek Hearings*, 3:1443.

20. *Coal Age* 27 (May 28, 1925): 796. Also see Winthrop Lane, *The Denial of Civil Liberties in the Coal Fields* (New York, 1924), 20.

21. Mark Rich, *Some Churches in Coal Mining Communities of West Virginia* (New York, 1951), 25–27, and Carter Goodrich, *The Miners' Freedom* (Boston, 1925), 41.

22. Korson, *Coal Dust*, 45–51; "Negro Miner," letter to the editor, *UMWJ*, June 1, 1916, 9; Rich, *Some Churches*, 25–27.

23. Testimony of Dr. Ira Bradshaw, *West Virginia Coal Fields*, 1:536–37; "Miner," letter to the editor, *UMWJ*, Apr. 15, 1925, 4; A. C. Spaulding, letter to the editor, ibid., Oct. 1, 1925, 17; "Negro Miner," letter to the editor, ibid., June 1, 1916, 9; Sterling Spero and Abram Harris, *The Black Worker* (New York, 1974), 378.

24. Ralph Chaplin, *Wobbly, the Rough and Tumbling Life of an American Radical* (Chicago, 1948), 122; Mooney, *Struggle*, ed. Hess, 43; *UMWJ*, Aug. 15, 1922, 11.

25. *New York Journal*, Apr. 7, 1922. National church leaders throughout the country condemned Sunday for going to West Virginia and making the anti-union sermons; the Reverend Dr. Stephen Wise labeled Sunday's trip a "loathsome instance of the attempted prostitution of the church . . . it is an attempt to use all the power of his eloquence and personality to lead men back to work under conditions just as unjust." *New York Times*, May 15, 1922.

26. Reverend A. J. Jenkins, letter to the editor, *UMWJ*, Aug. 1, 1922, 11, and T. L. Felts to George Bauswine, May 17, 1920, *West Virginia Coal Fields*, 1:215.

27. Mooney, *Struggle*, ed. Hess, 96. Also see the letter of the Reverend Harry Wilson to Norman Thomas, May 13, 1936, Socialist Party Papers, Duke University, Durham, N.C., in which he explained that his disgust with conditions in the company towns caused him to join the Socialist Party of America.

28. Jerome Davis, "Human Rights and Coal," *Journal of Social Forces* 3 (Nov. 1924): 102–6; O. C. Huffman to Justin Collins, Mar. 11, 1907, "A Miner" to Justin Collins, Mar. 11, 1907, James Riley to Justin Collins, Mar. 16, 1907, T. L. Felts to P. J. Riley, May 19, 1907, Collins Papers.

29. Mike O'Leary, letter to the editor, *UMWJ*, July 15, 1922, 9.

30. "Negro Miner," letter to the editor, *UMWJ*, June 1, 1916, 9.

31. *Paint Creek Hearings*, 3:2266; Mooney, *Struggle*, ed. Hess, 43, *Coal Age* 27 (May 28, 1925), 796.

32. James Laing, "The Negro Miner in West Virginia," *Social Forces* 36 (Mar. 1936): 418; interviews with the Reverend Shirley Donnally, Oak Hill, W.Va., and the Reverend Homer Santrock, Black Betsy, W.Va., summer 1975; W.Va. Bureau of Negro Welfare and Statistics, *Annual Report, 1922*, 77–78; David Corbin, "Mine Mules and Coal Tipples," in *Icons of Popular Culture*, ed. Ray Browne (Bowling Green, Ohio, 1978).

33. Korson, *Coal Dust*, 344, and Rich, *Some Churches*, 11.

34. Nettie McGill, *Welfare of Children in the Bituminous Coal Communities of West Virginia*, Children's Bureau, U.S. Department of Labor Publication no. 117 (Washington, D.C., 1923), 59; W.Va. Bureau of Negro Welfare and Statistics, *Annual Report, 1924*, 62; testimony of Ira Davis, *Paint Creek Hearings*, 2:1442–43; Rich, *Some Churches*, 17.

35. W.Va. Bureau of Negro Welfare and Statistics, *Annual Report, 1922*, 77; testimony of William Wiley, *West Virginia Coal Fields*, 1:522, 529; interview with Columbus Avery, Williamson, W.Va., summer 1975. For examples of the disrespect the miners showed the company preachers, see chap. 3.

36. *Paint Creek Hearings*, 2:1437–38.

37. E. P. Thompson, *The Making of the English Working Class* (New York, 1963), 68; Mooney, *Struggle*, ed. Hess, 8–9; West Virginia State Federation of Labor, *Proceedings, 1926*, 178. For examples of the use of biblical passages among District 29 leaders, see *UMWJ*, Apr. 15, 1922, 2, and A. S. Riffle, memo to local unions, Dec. 1, 1919, Straight Numerical File, item number 205194–50–49, Record Group 60, Records of the Department of Justice, National Archives, Washington, D.C.

38. James Maurer, "Has the Church Betrayed Labor?" in *Labor Speaks for Itself on Religion*, ed. Jerome Davis (New York, 1929), 19–28.

39. Dale Fetherling, *Mother Jones, the Miners' Angel* (Carbondale, Ill., 1974), chap. 9, and *Charleston Gazette*, Aug. 2, 1912.

40. Interview with David Cross, Alpoca, W.Va., summer 1975, and Barkey, "Socialist Party," 102–5.

41. Testimony of William Wiley, *West Virginia Coal Fields*, 1:525, and Malcomb Ross, *Machine Age in the Hills* (New York, 1933), 148.

42. Pope, *Millhands*, 140. For the sociological theories of sect emergence to which I referred in developing this assertion, see Bryan Wilson, "An Analysis of Sect Development," *American Sociological Review* 24 (Feb. 1959): 3–16; Clary Boisen, "Economic Distress and Religious Experience," *Psychiatry* 2 (May 1939): 190–95; J. O. Hertzler, "Religious Institutions," *Annals of the American Academy of Political and Social Science* 256 (Mar. 1948): 1–13; John Holt, "Holiness Religion: Cultural Shock and Social Reorganization," *American Sociological Review* 5 (Mar. 1940): 740–47.

43. Erik Erikson, *Childhood and Society* (New York, 1963), 251. For examples of other working-class movements that have assumed a sectarian nature, see

J. F. C. Harrison, *Quest for the New Moral World: Robert Owen and the Owenites in Britain and America* (New York, 1969), 135–39, and Christopher Johnson, "Communism and the Working Class before Marx: The Icarian Experience," *American Historical Review* 76 (June 1971): 642–89.

44. Robert McAfee Brown, *The Spirit of American Protestantism* (London, 1974), 198, and Dow Platt, letter to the editor, *UMWJ*, Nov. 11, 1915, 9.

45. Brown, *American Protestantism*, 198. The ballad is entitled "In the State of McDowell," by Orville Jenks; it is cited in Korson, *Coal Dust*, 304–5.

46. Archie Conway, "A Coal Miner's Goodbye," cited in Korson, *Coal Dust*, 247–48.

47. Ernest Troeltsch, *The Social Teaching of the Christian Churches*, 1 (London, 1930), 44, quoted in Genovese, *Roll, Jordan, Roll*, 166–67.

48. Charles Sydenstricker, letter to the editor, *UMWJ*, Nov. 15, 1909, 5; Lewis Hunt to John L. Lewis, July 24, 1933, District 17 Correspondence Files, UMWA Archives, UMWA Headquarters, Washington, D.C.; Thompson, *Making of the English Working Class*, 386.

49. See Gutman, "Protestantism and the American Labor Movement," and Isaac Brigance, letter to the editor, *UMWJ*, Oct. 22, 1908, 7.

50. Thompson, *Making of the English Working Class*, 388.

51. Genovese, *Roll, Jordan, Roll*, 163.

52. The lay preacher was not new to the southern West Virginia coal fields. Korson (*Coal Dust*, 46–48) related that congregations in mining towns in Great Britain often did not have enough money to pay an ordained minister and therefore they elected one of their own to serve as a voluntary minister.

53. Interviews with the Reverend G. W. "Preacher" Warner, Pineville, W.Va., the Reverend Homer Santrock, Black Betsy, W.Va., and the Reverend Roy Lester, Kopperston, W.Va., summer 1975. See the article on the Reverend Warner in the *Pineville* (W.Va.) *Independent Herald*, Apr. 18, 1974.

54. Interviews with Marion Preece, Delbarton, W.Va., the Reverend Homer Santrock, Black Betsy, W.Va., Elmer Cook, Black Eagle, W.Va., and the Reverend Roy Lester, Kopperston, W.Va., summer 1975. See also Stanley Williams, "Disorganization and Delinquency in Three Coal Communities" (M.A. thesis, West Virginia University, 1954).

55. Lay preachers have generally seemed to have a closer and more successful relationship with working-class movements. See Eric Hobsbawn, *Primitive Rebels* (New York, 1965), vii, and Thompson, *Making of the English Working Class*, 394–97. Similarly, the Roman Catholic Church in France began losing contact with its working-class adherents; priests were able to win back some of the stray flock by putting on work clothes, joining unions, and going to the factories—an effort called the worker-priest movement. Brown, *Spirit of Protestantism*, 252. For the importance of lay preachers to slave society, see Genovese, *Roll, Jordan, Roll*, 258.

56. Interview with Katherine Ellickson, Chevy Chase, Md., fall 1976, and Archie Green, *Only a Miner* (Urbana, Ill., 1972), 255–56.

57. Testimony of Frank Keeney, *West Virginia Coal Fields*, 1:103, and testimony of Roy Combs, ibid., 217.

58. Mooney, *Struggle*, ed. Hess, 111, 113, 117.

59. William Shepherd, "The Big Black Spot," *Collier's Weekly*, Sept. 19, 1931, 12–14, and John Spivak, *A Man in His Time* (New York, 1967), 67.

60. Sydney Mead, *The Lively Experiment* (New York, 1963), 109.

61. Jack Weller, *Yesterday's People* (Lexington, Ky., 1966), 130. For a criticism of general beliefs about fundamentalism, see Charles Hudson, "Structure of a Fundamentalist Christian Belief System," in *Religion and the Solid South*, ed. Samuel Hill (Nashville, Tenn., 1972), 122–23, 136.

62. Mooney, *Struggle*, ed. Hess, 7–9.

63. Isaac Brigance, letter to the editor, *UMWJ*, Oct. 22, 1908, 7.

64. Interviews with Whit Collins, Lanore, W.Va., and the Reverend Roy Lester, Kopperston, W.Va., summer 1975.

65. *UMWJ*, Aug. 15, 1925, 17.

66. Interview with Clyde Scarberry, Keystone, W.Va., summer 1975.

67. West Virginia Mine Workers' Union, *Constitution*, 38, and Mooney, *Struggle*, ed. Hess, 47. The *Constitution* was privately published, ca. 1931. I have a copy in my collection, and a copy is also in the Katherine Ellickson Papers, Wayne State University, Detroit, Mich.

68. *UMWJ*, Apr. 13, 1911, 1.

69. Ibid., Oct. 15, 1927, 16.

70. *Paint Creek Hearings*, 3:1433; West Virginia Mine Workers' Union, *Strike Bulletin*, no. 6, July 23, 1931, Ellickson Papers. For the traditional version of "A Miner's Life," see Korson, *Coal Dust*, 413–15, and *Paint Creek Hearings*, 3:2264.

71. West Virginia Mine Workers' Union, *Constitution*.

72. Ibid. That the miners recognized God's presence in their surroundings is important. Brown (*American Protestantism*, 81, 131) explains: "To know about God is not enough, to know God is not enough either. According to St. Paul, the important thing is 'rather to be known by God' (Gal. 4:9). It is when we are known by God and God makes us aware of the fact, that a new situation is created . . . when people are known by God, they do not first of all write books or work out a philosophy of history or start analyzing their religious experience. They sing, they pray"—and prayer is, according to Brown, any recognition of God.

73. Mary "Mother" Jones, *Autobiography of Mother Jones*, ed. Mary Field Parton (Chicago, 1925), 64.

74. Interview with Katherine Ellickson, Chevy Chase, Md., fall 1976.

75. Emmett Lemons, letter to the editor, *UMWJ*, Feb. 15, 1921, 2; Hubert Kirk, letter to the editor, ibid., Mar. 2, 1913, 2; E. L. Tucker, letter to the editor, ibid., Sept. 1, 1922, 7.

76. Artie Suber, letter to the editor, ibid., Jan. 15, 1922, 7; Charles Sydenstricker, letter to the editor, ibid., Nov. 25, 1909, 6; Emmett Lemons, letter to the editor, ibid., Feb. 15, 1921, 2; Ross, *Machine Age*, 140.

77. Emmett Lemons, letter to the editor, *UMWJ*, May 15, 1921, 15; Charles Sydenstricker, letter to the editor, ibid., Nov. 25, 1909, 6; Jones, *Autobiography*, ed. Parton. 64.

78. Charles Sydenstricker, letter to the editor, *UMWJ*, Nov. 25, 1909, 6; *Charleston* (W.Va.) *Labor Argus*, June 6, 1913; Brown, *American Protestantism*, 190–91.

79. *Montgomery* (W.Va.) *Miners' Herald*, Oct. 3, 1913.

80. Archie Conway, "A Coal Miner's Goodbye," cited in Korson, *Coal Dust*, 247–48.

81. Orville Jenks, "In the State of McDowell," ibid., 304–5. Such perceptions were easily translated into earthly thought and action. Thompson explained that "faith in a better life to come served not only as a consolation to the poor but also as some emotional compensation for present sufferings and grievances: it was

possible not only to imagine the 'reward' of the humble but also to enjoy some revenge upon their oppressors, by imagining their torments to come." *Making of the English Working Class*, 34.

82. The quote is in *Paint Creek Hearings*, 3:2261. F. H. Stewart, letter to the editor, *UMWJ*, Aug. 1, 1921, 15.

83. Maggie Workman, letter to the editor, *UMWJ*, Mar. 15, 1925, 12. Archibald Robertson is quoted in Genovese, *Roll, Jordan, Roll*, 164.

84. Charles Sydenstricker, letter to the editor, *UMWJ*, Nov. 25, 1909, 6; Isaac Brigance, letter to the editor, ibid., Oct. 22, 1908, 7; West Virginia State Federation of Labor, *Proceedings, 1926*, 178.

85. West Virginia Mine Workers' Union, *Strike Bulletin*, no. 5, July 21, 1931, Ellickson Papers.

86. West Virginia Mine Workers' Union, *Strike Bulletin*, no. 4, July 16, 1931, and no. 5, July 21, 1931, Ellickson Papers.

87. *UMWJ*, May 1, 1927, 11; Rice, *Allegheny Frontier*, 22, 267–68, 378. Also see Northern, "Religions of the Mountaineer," 317–36; West Virginia Mine Workers' Union, *Strike Bulletin*, no. 6, July 29, 1931, Ellickson Papers; Genovese, *Roll, Jordan, Roll*, 202; West Virginia Mine Workers' Union, *Constitution*, 38–39; "Death of a West Virginia Miner," *UMWJ*, Apr. 5, 1917, 15.

88. Orville Jenks, "Sprinkle Coal Dust on My Grave," cited in Korson, *Coal Dust*, 65–66.

89. Genovese, *Roll, Jordan, Roll*, 201.

90. *Charleston Gazette*, Aug. 4, 1921; Mooney, *Struggle*, ed. Hess, 88–89; *Huntington* (W.Va.) *Herald-Dispatch*, Aug. 3, 1921.

91. Gladys Smith, "The Hardworking Miner," cited in Korson, *Coal Dust*, 239.

92. The song is "The Dying Brakeman." The version cited here was in a letter from Whit Collins, Lanore, W.Va., to the author, Aug. 14, 1975; it was written by Orville Jenks, McDowell County. For the standard version, see Korson, *Coal Dust*, 246.

93. The death notes are printed in *UMWJ*, June 1, 1927, 17.

94. Interview with the Reverend Homer Hicks, Dunbar, W.Va., summer 1975, and Erikson, *Childhood*, 268.

95. Friedrich Engels, "On the Early History of Christianity," in Karl Marx and Friedrich Engels, *On Religion* (Moscow, 1957), 312–20; Isaac Deutscher, *The Prophet Armed: Trotsky, 1879–1921* (London, 1954), 30–31; Hobsbawn, *Primitive Rebels*, 129–30; Johnson, "Communism and the Working Class," 642–89; Eric Hobsbawn, "Methodism and the Threat of Revolution," *History Today* 7 (Feb. 1957): 124. See also Werner Cohn, "Jehovah's Witnesses as a Protestant Movement," *American Scholar* 24 (1955): 281–99; Donald MacRae, "The Bolshevik Ideology," *Cambridge Journal* 3 (Dec. 1954): 164–77; Seymour Martin Lipset, *Political Man: The Social Bases of Politics* (Garden City, N.Y., 1963), 97–99. Genovese helped to clarify the connection between religion and political movements when he wrote: "Since religion expressed the antagonisms between the life of the individual and that of society and between the life of civil society and that of political society, it cannot escape being profoundly political." *Roll, Jordan, Roll*, 162.

96. Genovese, *Roll, Jordan, Roll*, 597, and Erik Erikson, *Identity: Youth and Crisis* (New York, 1968), especially 59–62.

97. Fred Leykan to John L. Lewis, July 28, 1933, District 17 Correspondence Files, and Virginia West, "We're Independent Now," cited in Korson, *Coal Dust*, 321–23.

CHAPTER VII

# A War for Democracy

> The world must be made safe for democracy.
> Woodrow Wilson

> Kaiserism shall not dominate a people whose forefathers gave their blood that we might stand free.
> Fred Mooney

World War I had a major impact on the miners of southern West Virginia. The coal diggers responded to the nationalism and democratic idealism of the war years and to the tensions that followed—as did the rest of the United States—but for them, the ideology was to encourage not conformity but a highly ideological insistence on their democratic rights. The end of the war was not to prompt a weary cynicism among the miners but a stronger unity in demanding a renewal of their war for democracy against oppressive autocrats who were closer to home than Germany.

The stress on Americanism in the federal government's wartime patriotic propaganda created a pressure for conformity and also made it ideologically acceptable. The demand for conformity certainly helped to ignite three postwar explosions: nativism, antiradicalism, and the open-shop movement.

For the American worker, however, the ideological justification for the war sanctified unity and struggle. The violent mass strikes of 1919 were, as historians claim, undoubtedly attempts to preserve labor's wartime gains as well as a response to postwar inflation and the open-shop movement. But the greatest wave of industrial unrest in American history also involved a militant commitment to a higher ideal—an ideal that the war and the propaganda for the war had brought clearly into focus—freedom. The seemingly conservative president of the American Federation of Labor, Samuel Gompers, found himself supporting some of these struggles, and he later explained that World War I "demonstrated [to me] that the pacifism in which I believed and which I faithfully advocated was a vain hope. I realized that the struggle in defense of right and freedom must ever be maintained at all hazards."[1] The ideological impact of war propaganda on American workers may have been "one of those facts so big," as E. P. Thompson wrote of a similar situation in British working-class history, "that it is easily overlooked, or assumed without question; and yet it indicates a major shift in emphasis in the inarticulate 'subpolitical' attitudes of the masses."[2]

In southern West Virginia the government's effort to sell the war began the political education of the miners, as it gave them a political consciousness that they never before possessed. The wartime rhetoric contained a powerful strain of radical, political ideology. The propaganda that stirred national loyalty also preached the need to conquer autocracy, taught sacrifice and commitment, and justified violence as a legitimate means to achieve righteous goals. In this Americanism, the miners discovered a secular counterpart to the spiritual conviction about justice and brotherhood that they had drawn from their religious faith.

Americanism, as Leon Samson has pointed out, can be seen as a "substitute socialism," for it embodies the principles and goals (equality, classlessness, creation of wealth for all, political liberty, and the pursuit of happiness) of socialism. Here, Samson explained, lay the weakness of socialism in the United States, because its appeal was preempted by a radicalism implicit in the traditional American creed.[3] The southern West Virginia miners were to find the radical thrust of this Americanism that the war stressed.

World War I gave the miners an enormous awareness of their national importance. Coal was essential to the nation's economy and industrial production. When the United States entered the war in 1917, coal furnished 70 percent of the mechanical energy in the country, and the mines of West Virginia furnished 25 percent of that. The nation's dependence upon coal was revealed as coal shortages developed during the war. Coal riots flared in major cities. Cases of pneumonia hit alarming proportions and people froze to death for lack of heat.[4] President Woodrow Wilson bluntly told the nation that a "scarcity of coal . . . is the most serious [danger] which confronts us."[5]

The war effort had increased the nation's need for coal, particularly for the southern West Virginia coal, which quickly became the coal highest in demand. War orders poured into southern West Virginia, especially from the U.S. Navy, which used "almost exclusively" the state's high-quality, smokeless coal. Indeed, as early as 1910, the U.S. Navy had considered the purchase of coal mines in southern West Virginia, and with the outbreak of war, it immediately placed the West Virginia coal first on its "acceptable list."[6]

To help to insure a steady supply of the state's coal, the U.S. War Department instructed the draft boards of southern West Virginia (the first draft boards so ordered) to cease drafting coal miners in August 1918. President Wilson issued a proclamation that urged "essential coal miners" (meaning the miners in West Virginia) to accept deferred classification; he told them that it was their "patriotic duty" to accept exemption and that it was "the patriotic duty of their friends and neighbors to hold them in high regard for doing so." In September the War Department exempted all coal miners from military conscription.[7]

The miners' awareness of their central role in the front line of defense was made more apparent as all segments of American society appealed to them to dig more coal. President Wilson proclaimed that the nation's war program "rested upon the shoulders of the miner," as he and other government officials strived to impress upon the miners their importance to the war effort. Wilson declared that they were the "essential labor for the support of the government and of the liberties of free men everywhere." On another occasion, he told the miners that they were the "chief factor" in America's "second line of defense."[8]

From the European battlefields, General John J. Pershing claimed that the work and support of the "coal miners of America thrills us and helps to make our hearts more strong for the battle. I have always been certain that organized labor would stand steadfastly behind us until victory for democracy is achieved."[9]

From the Catholic diocese in Wheeling, West Virginia, Bishop P. J. Donahue publicly appealed to the state's miners: "Your country calls you as never before. Coal is King. You are his willing subjects. Today he is the greatest power of all." In the ultimate sense of the American "mission," the bishop exclaimed, "I believe that, in the providence of God, our beloved land is now the divinely appointed agent to keep the torch of liberty flaming brightly." He then stressed the miners' responsibility for this great cause—this holy war: "The time is now to help your country. Every additional ton you dig will build a fire to burn despotism to ashes. . . . On your shoulder rests a tremendous responsibility. The pick is as mighty as the sword."[10]

The southern West Virginia coal establishment also saturated the miners with patriotic propaganda to inspire them to dig more coal. As one coal operator explained, the miner should be "aroused to a feeling of his personal responsibility in regard to the prosecution of the war." The coal companies brought in speakers, including soldiers in uniform, to address the coal diggers on the urgency of mining coal.[11]

Company officials almost daily told the miners that upon their work depended the defeat of the "iron hand of autocracy" and the survival of the United States and its blessings of liberty, democracy, equality, and the pursuit of happiness. From the pulpits of the company churches, the ministers urged the miners to dig more coal for God and nation. Any miner at Winding Gulf Collieries who was reported "slacking" in his work received a letter from his coal operator that read: "If you are an American citizen and are in sympathy with our soldiers you should help them in their fight. . . . Every man and woman must do their utmost. Your duty is to produce all the coal you can. You are not working for us any longer, but for your country and your family."[12]

U.S. Coal and Coke sponsored a massive patriotic demonstration at Gary, West Virginia. The rally started with a two-mile parade, consisting of starred and striped banners, fifty floats, and marching schoolchildren singing patriotic songs, through the company town to the company ball park. At the park 5,000 miners and their wives listened to a series of company officials deliver a barrage of patriotic speeches.[13]

Company general manager Edward O'Toole stressed the "Americanism" of all miners and encouraged them to do their part as Americans—to dig coal. Company superintendent L. C. Anderson told the gathering of the importance of the "war for democracy" to world history and of America's mission to bring freedom and democracy to the oppressed: "The effect of this war, both now and in the future, will be far-reaching. The spirit of Americanism and democracy is contagious. . . . Not until the sun of civilization touched the rim of the western continent was the mass of mankind free. Now, as the Israelites in the wilderness looked upon the serpent for deliverance, so the downtrodden of the earth look to America."

A local judge further emphasized the liberating nature of the war, declaring that "when this war is over, our flag will wave not only over a free country, but over a free race. America had no choice but to fight for its rights and for liberty." Then the local minister told the miners that liberty and democracy were ideas and rights worth fighting for. "We have entered this war that all men may be free. I had rather be a corpse than a coward. . . . Victory must be wrested from this power in order that the freedom of the world may be established. Anyone who fails to do their part is a traitor, not only to his country, but to the cause of world democracy as well."

The final speaker was the Catholic priest, who emphasized the need for solidarity during the crisis because victory could only be achieved through collective action: "The establishment of this nation was not accomplished entirely by the native-born American. Kosciusko, Pulaski and Lafayette probably did quite as much to found the new commonwealth as did Warren and Greene and Washington. In the present crisis it is obvious that in union there is strength—'united we stand, divided we fall.'"

The UMWA doused the miners with the spirit of Americanism, too.

Covers of the *UMWJ* were adorned with patriotic pictures and cartoons, such as an American eagle surrounded by stars and stripes. The cover of one issue carried UMWA President Frank Hayes's pledge to General Pershing of the loyal support of American coal miners.[14] The union publication stressed each miner's personal responsibility for the war effort. The cover of one issue read:

This Is Your War!

It is a call to you John Smith and Joe Brown—it is a call to you to hurry that your liberties may live.

It isn't President Wilson's war, it isn't Secretary Baker's war, it isn't the Government's war. . . .

This is your war.

Guard jealously your share in this war of yours! Don't let anyone else do your part. Don't let anyone else rob you of your share in the defense of your freedom.[15]

Emblazoned wording on the cover of another edition of the *UMWJ* read

Dig Coal!

Dig More Coal!

Dig Still More Coal!

The Success of the War Depends on the

Coal You Dig.[16]

An editorial in the *UMWJ* read, "If the coal mining fraternity fails now, it will mean that statesmen, generals and armies will fail."[17]

The miners' union believed that World War I was a fight for American values. "You," the *UMWJ* told the miners, "are fighting the battles that democracy may survive and liberty may be vouchsafed to all of us." The cover of the journal proclaimed: "Let our slogan be 'Win the War for Freedom': The cause of democracy is at stake: Issues vital to Labor are hanging in the balance." Hayes declared that the war was a fight "for the establishment of human freedom, liberty and democracy throughout the world" and a fight "against militarism and autocracy."[18]

A war for democracy was worth the sacrifices of harder work and longer hours. "Ours must be the superior will to conquer," the *UMWJ* proclaimed, "ours the grim courage to endure, to sacrifice and hold unfalteringly until the final victory sounds forever the doom of autocracy." Indeed, no sacrifice was too great. When a mine explosion took the lives of thirteen coal diggers, the *UMWJ* declared that "these loyal boys died in the interests of democracy, they were exerting their manpower in the production of coal with which to help win the war."[19]

The nation's miners responded enthusiastically to the war for democracy. Over 50,000 coal diggers throughout the country disregarded their draft exemptions and enlisted; 3,000 died in combat. It was the miners in the army who dug most of the death-hole trenches, and they who performed the majority of the engineering work on the war front.[20] On the home front the miners also made sacrifices as they raised coal production to all-time highs. Each week southern West Virginia reported new records in the amount of coal mined. In 1917 the state's coal production soared to almost 90 million tons; in 1915 the production was only a little more than 70 million tons. In 1918 mine explosions and accidents took the lives of 404 West Virginia miners. UMWA District 17 President Frank Keeney wrote President Wilson that the West Virginia miners had a higher death rate than the American Expeditionary Force.[21]

From America's entry into the conflict until the armistice, Keeney sent out circulars that urged the miners to ignore minor grievances so that mining could continue. "All the munitions of war and the supplying and shipping of those men [soldiers] depends absolutely upon the production of coal, and the production of coal absolutely depends upon the persistent and steady work of the miner," Keeney told the coal diggers of District 17. "The continued mining of coal," Keeney claimed, "is your full duty as a true UMWA member and an American citizen."[22]

In many areas during the war, for the first time since the rise of the coal establishment, UMWA miners were recognized as important citizens. Keeney served as grand marshal of a three-mile parade that launched the drive for war bonds in Charleston. In 1918 miners of Kanawha Valley were asked to participate in a "monster win-the-war" Labor Day parade in Charleston.[23]

Mother Jones, who in her organizing speeches had been urging the miners to buy war bonds and had been denouncing the Kaiser as a sweatshop owner and exploiter of child labor, herself participated in win-the-war parades. In Charleston, West Virginia, she rode at the head of the parade with the mayor and county sheriff and was followed by a mile-long procession of union miners. In an editorial entitled "What a Difference," the *UMWJ* contrasted the new attitude toward UMWA leaders with "old days of the Paint Creek Strike—not so long ago—. . . when Mother Jones tried to help the coal miners" and "was thrown into jail and guarded by mine-owned militia."[24]

The miners were caught up in the spirit of the war effort. A survey conducted by local school teachers in Kanawha County for the National Women's Defense Work committee indicated that the war effort had stronger support in the coal districts of the county than in the non-coal-producing ones. A teacher on Cabin Creek reported that nearly half the women in her area had contributed to the Red Cross and that a "recent meeting was filled with lots of enthusiasm for our country and our flag." A school teacher at Loudon reported that all the women had signed food pledge cards and that the "prevailing attitude" of her community was "the downfall of Kaiserism." In contrast, a school teacher in Charleston reported that there was "no real pledging for war bonds" and that the people were "not aroused to the situation yet."[25]

When a local newspaper accused the District 17 miners of "frolicking around" instead of digging coal, the miners angrily responded, branding the charges as "absolutely false" and pointing out that they were working, although the coal companies were still violating many state mining laws. Idleness and decreased production were the result of having to work with "old and worn out machinery which continuously breaks down" and of the lack of railroad cars. Instead of enjoying this idleness, the miners claimed that they resented it. One of them wrote the *UMWJ* that "the miners can not support their families under present conditions. Many are forced to give up their Liberty bonds . . . in order to get food for their families."[26]

The miners demonstrated their support for the war through the purchase of Liberty Bonds and War Savings Stamps and contributions to the Red Cross. "Investment of this kind," Keeney advised the District 17 miners, "assures you against loss . . . and at the same time . . . supports your government and renders a patriotic service." The executive board of District 17 practiced what the president preached when it bought a $1,000 Liberty Bond.[27]

The rank-and-file miners responded as enthusiastically in the effort of buying bonds as they had in digging coal. All over southern West Virginia, UMWA members raided their treasuries to buy war bonds and contribute to the Red Cross. The miners at Blaine withdrew their entire treasury of $1,000 to invest in Liberty Bonds. The local union at Thayer, announcing that it had purchased several hundred dollars of war bonds, described itself as a "hot-bed of patriotism." The miners at Ohley contributed $237 for war stamps, intending to "show the world that the United Mine Workers are always in the front in everything the country calls for." A miner from Blair, who reported that his local had gone "over the top" in the purchase of war bonds, related that the miners "are producing more coal in proportion to their numbers than has ever been produced in Blair. The boys here say they are going to help the soldiers burn their way to Berlin and then build a fire on the Kaiser's throne." Another miner wrote the *UMWJ* that "at our last regular meeting of the local, we thoroughly discussed our patriotism and our duty to our country, and decided that there were other ways of helping our country than mining coal." Therefore, they invested an additional fifty dollars in war stamps. "We feel that it is fifty shots at the Kaiser, added to our coal production."[28]

While the miners were making great sacrifices for the war, the coal operators were amassing enormous profits. The outbreak of the war had sparked a runaway coal market, and coal companies across the country reaped huge benefits from their harvests of coal. The greatest profits were accumulated in southern West Virginia, where, according to *Coal Age*, "the demand for high grade West Virginia coal is outstripping production so decidedly." Coal operators in southern West Virginia who had been receiving less than a dollar per ton of coal found that they were obtaining over four dollars for the same amount. The operators in Raleigh County saw the price of their coal skyrocket from $1.00 per ton in 1915 to $4.96 per ton in 1917. In Logan County the price of coal soared from $0.82 per ton in 1915 to $4.39 per ton in 1917.[29]

The federal government acted to stabilize the booming coal industry and to prevent wartime profiteering from the resulting energy crisis. Several months after the United States entered the war, Congress passed the Lever Act, which created the Fuel Administration, under Dr. H. A. Garfield, and gave it control over the price and distribution of coal. Subsequently, Garfield, in conjunction with the coal industry and UMWA, worked out the Washington Agreement, which gave the miners an increase in wages in return for a promise not to strike for the duration of the war.[30]

The southern West Virginia operators initially welcomed the creation of the Fuel Administration. They had feared that with America's entry into the war the federal government might confiscate the much needed supplies of high-grade West Virginia coal. A week after the United States entered the war, Justin Collins had written a fellow southern West Virginia coal operator:

> The Government may decide to take over one or more operations and thus work great harm, not only to the operations so taken over, but to all the others by recognizing the Union, and in other ways bringing about such conditions that it would be hard to ever recover from. Hence, I am very much in favor of trying to arrange some kind of a working agreement with the Government along lines that will be fair.[31]

The operators' support of the regulatory commission quickly soured, however. Although the Fuel Administration did not take over any mines, it forced many coal companies to install weight scales as required by state law (but which neither the state courts nor state government had enforced). It compelled many of the coal companies to pay the wage increases included in the Washington Agreement. Of greatest importance, the labor bureau of the Fuel Administration ordered the coal companies in southern West Virginia to allow the miners to organize.[32]

With the federal government's support and protection and with Keeney's leadership, Districts 17 and 29 grew tremendously during the war years. The membership of District 17 jumped from 7,000 as of January 1, 1917, to over 17,000 within months. By the end of the war membership totaled more than 50,000. The executive board had erased a debt of nearly $17,000 and acquired a treasury of $25,000. Union membership of District 29, owing to what John L. Lewis called a "whirlwind" campaign by Jones, Lawrence Dwyer, and Keeney, soared from 900 to over 6,000. For the first time, the miners in that district obtained the checkoff and the union shop. All of northern West Virginia had been organized, and Keeney and UMWA District 17 were already invading the sacred anti-union bastion of McDowell County. Most important, Keeney was also looking to the nonunion kingdoms of Logan and Mingo counties, which U.S. Senator William S. Kenyon, chairman of the Senate Committee on Education and Labor, depicted as an "industrial autocracy" within the United States.[33]

The coal companies' initial frustration with the Fuel Administration, because of its enforcement of mining laws and recognition of unionism, was compounded by the agency's authority to fix prices. The governor of West Virginia, John J. Cornwell, explained that the state's coal operators "deeply resented the Government's interference in the price of coal, believing that they should be left to sell coal for all they could get and ship it wherever they pleased."[34]

The coal companies unceasingly complained about established prices. They begged and then demanded to be exempted from the wartime controls. The chairman of the Council of National Defense found that the southern West Virginia operators delayed shipping coal out of the state, hoping for an increase in coal prices. Coal companies asked the U.S. Navy to purchase coal from Utah and British Columbia instead of West Virginia. The assistant secretary of the Navy, Franklin D. Roosevelt, replied that the request was "out of the question."[35]

Coal officials in southern West Virginia demanded that the Navy take their companies off its "acceptable list." One wrote that "we do not care anything for this [being first on the acceptable list] at all. . . . At the price you are paying us we doubt whether we will be able to break even on it." Roosevelt responded with a mass of statistics to show that the "quantities allocated to the various companies were very carefully considered. . . . and it is not believed that any great difficulty will be experienced by you in supplying the small tonnage." Roosevelt also noted that this company had not yet furnished its allocation of coal and "requested that it be promptly accomplished." Other coal companies continued to request exemption from federal controls, and Roosevelt continued to reject the pleas, unless the companies provided evidence to show that they were being treated unequally or unfairly. The companies were apparently unable to do that. They did, however, begin to write Secretary of War Newton D. Baker, asking to be excused from supplying the U.S. Army with coal. Baker also refused their requests.[36]

The governor of West Virginia, irritated with the coal companies' opposition to the wartime regulations, blasted them in a speech, declaring, "We can't win this war even now with part of our people working at it and the rest of the people grabbing for labor and . . . profit. So the Government will determine—must determine—who is to consume your coal."[37]

Unable to obtain exemption from the controls, the coal companies simply ignored the federal laws and agreements. The U.S. Department of Justice eventually indicted fifty-two coal companies in southern West Virginia and their officials for "profiteering" in the sale of coal during the war. The Department of Justice claimed that the coal companies had charged "excessive prices" in violation of the Lever Act. Several companies were also indicted under the Sherman Antitrust Law for restraint of trade and foreign commerce.[38]

The miners were aware of the less than patriotic practices of the coal companies. Keeney pointed out to the governor that the coal companies were overcharging for their coal. Rank-and-file miners reported to the Department of Labor that the coal companies were delaying shipments of coal up to thirty days.[39]

The miners were also aware of the companies' refusal to comply with the federally sponsored wartime wage agreements. A number of the coal companies refused to grant the wage increase included in the Washington Agreement. Other companies granted the federally allotted increase, but raised the prices at the company store to offset it. The miners at Sharon, Cabin Creek, wrote the governor that "the prices in their company stores have advanced so much since that agreement was made that we as the laborers of said place . . . have resolved that instead of receiving an advance in wages, we deem it under the prices now in the company store of said place, a reduction." A mass meeting of District 17 miners passed a resolution that declared: "We believe that the increase given by Dr. Garfield was meant for the miners, and not for the . . . company stores of this country."[40]

A major irritation to the miners was that many coal companies, at a time when the federal government was calling for labor-management cooperation in the production of coal, continued to refuse to place a checkweighman on the tipple, although the companies had been ordered to do so by the Fuel Administration. The general manager of one mine declared, "I will die with my boots on before a checkweighman is placed on the tipple."[41]

The miners believed that the coal companies were hypocritical, preaching sacrifice but acting, in their greedy exploitation, as un-American mercenaries. The miners were subjected to the coal companies' "Work or Fight" programs. If the company felt a miner was "slacking" in the amount of coal mined, an elderly miner explained, "The guards took you to the [draft] station, and your deferment was withdrawn." Fifteen mining operations in Kanawha and Raleigh counties devised their version of a draft board, called a "slacker board," a large blackboard that posted the amount of coal each worker mined and the names of the men who did not work. At the end of a certain amount of time, the miners producing the lowest amounts of coal were fired, thus losing their draft exemptions. On the other hand, highly productive miners were given bonuses and "honor medals" and placed on "honor rolls."[42]

Angered over this hypocrisy, a miner complained to the *UMWJ*, "What would the Americans say about the administration if they would send their boys across the water to face the Huns with broken rifles and not enough ammunition. This is just what the operators of West Virginia are doing today. They are sending us into the mines to work with broken and worn-out machinery." The company's refusal to purchase new machinery, the miner noted, "decreases the production of coal that is so badly needed by our country to carry on and win the war." Another miner also pleaded for efficient machinery: "If the operators will cooperate with us in the proper way, we will need no bonuses or urging to induce us to do our duty, for our past record shows we are loyal men."[43]

The complaints of the miners and the gains that the operators saw the UMWA making only served to strengthen the operators' view of the UMWA as a criminal, revolutionary, and traitorous organization. Increasingly, they talked of the UMWA's "treasonable" methods and goals. The speeches of UMWA organizers, the secretary of the New River Association of Coal Operators wrote to the governor, were not "legitimate appeals for a workmen's organization but are rather diatribes of abuse heaped upon the heads of all that are not of a militant, socialistic attitude. . . . If people of this type are permitted to continue their activities they will gradually inflame a sufficient number of ill-balanced minds to a belief that there are wrongs that must be righted." The secretary further claimed that if UMWA organizers, who were of an "anarchistic mold," were successful in organizing southern West Virginia, they "would be in the saddle and would ride the [coal] industry to a downfall," and that would result in "immeasurable disruption of our organization for producing a product that is of vital necessity to our country at this time." A Fayette County coal operator wrote the governor that if the union movement in southern West Virginia was not halted, "under present circumstances of scarcity of labor, the great demand for coal, and the great demand for labor on the part of all coal companies . . . the socialistic and anarchistic element [UMWA leaders] will have full power to ruin and oppress one coal company after another until conditions will be deplorable." Collins, who believed "we should drive the union while the political climate is right," exhorted the governor that unionization would cause "restriction of production" and a "national disaster."[44]

Governor Cornwell responded to the concerns of the coal operators. He secured passage of a bill that authorized the governor to call deputy sheriffs into state service to suppress insurrection and to preserve the peace. A law was enacted that required all "able-bodied" men between the ages of sixteen and sixty to work at least thirty-six hours a week. Idleness (that is, working less than thirty-six hours a week) was punishable by a fine of not less than $100 and sixty days of penal labor. In promoting such legislation, the governor constantly employed labor-baiting tactics, for example, claiming "Bolsheviks" controlled most labor unions.[45]

The governor's attitude toward UMWA organizers was similar to that of the coal operators. A few months after the United States entered the war, the governor asked the U.S. Secretary of Labor, William B. Wilson, to remove UMWA organizers from the state. When Jones embarked upon an organizing campaign in the Winding Gulf coal field, the governor demanded her removal. "I want to advise you," the governor wrote Keeney, "that if there is any trouble resulting from her visit that I shall hold to strict account those who are responsible."[46]

The trouble came when Baldwin-Felts guards attacked and beat UMWA organizers at Clarksburg and Fire Coal, Raleigh County. Angered by the assaults, the miners of Districts 17 and 29 met in a joint convention and declared their intention of "ceasing work until these 'guards' be removed." Stirred by the threat of a massive, wartime, wildcat strike, the governor wrote Keeney, apologizing for the beatings and stating that he was embarrassed. But he refused to force the coal company to remove the guards. Meanwhile, he wrote Secretary of War Baker to request federal troops to prevent the threatened strike. Baker, however, rejected the governor's request, explaining that it was "impractical" and not consistent with "the manner prescribed in the Constitution."[47]

The miners postponed the strike, but the incident enhanced their contempt for a state government that they viewed as an adjunct of the coal companies. "The miners of the state of West Virginia," a coal digger wrote, "are afflicted by the most outrageous and cowardly terms that could be legislated against them by the buying of the control of our courts and legislature."[48]

State wartime legislation gave the coal establishment new devices with which to fight the union. Upon the slightest sign of labor trouble (e.g., the presence of a UMWA organizer), the coal operators demanded that the governor send in the deputy sheriffs, imbued with state authority and power. The coal operators used the thirty-six-hour vagrancy law to suppress the wildcat strikes that occurred over the coal companies' contract violations.[49]

Most important, many coal companies were able to continue their traditional anti-union tactics of discharging, evicting, and blacklisting union miners and prohibiting public meetings in their company towns. The operators were denying the miners at home the very freedoms and democratic rights that they and the federal government were asking the miners to help secure for the people of Europe—the right of free speech, the right of public assembly, and freedom of worship. When the Women's National Defense Work Committee asked a school teacher on Cabin Creek to ascertain the patriotic sentiment in her community, the teacher replied, "I am teaching in a place where I can't have any meetings of any kind." When the miners at Bailey Wood formed a local union and attempted to meet with District officials, a company guard ejected the District representatives with the words, "You are objectionable to this company and you are trespassing on this property and I have instructions to put you off. So you are not going to hold any meeting here, so get." When the superintendent at Madison discovered that his miners had formed a local, he immediately discharged all of them, and the local justice of the peace then jailed four of them for violation of the state's vagrancy law. The vice-president of District 17, William Petry, hurried to Madison and asked the company official for an explanation. The superintendent replied that he would "have no damned union around his mines." Petry told the superintendent that he was violating the order of the Fuel Administration by discharging men for joining a union. The superintendent retorted, "To hell with the Fuel Administration; I'm running these mines."[50]

To the miners, this attitude was Kaiserism in America. After company officials and mine guards broke up a miners' meeting at Thayer, which had met to discuss violations of the Washington Agreement, and then hounded the participants at the meeting, a local miner wrote the *UMWJ* that "the iron hand of autocracy is visible in the Labor world as well as in Germany, so let us put forth every ounce of energy, and every available dollar of finance to abolish the fetters of autocracy."[51] The miners had turned the wartime rhetoric against the oppressor at home.

The ideology that persuaded Americans to fight to make the world safe for democracy in World War I gave the miners a set of political principles upon which they could judge the coal operators. The miners had translated the propaganda against the Kaiser into a broader and higher creed; the war was more than a war against Germany—it became literally a "war for democracy." During a patriotic demonstration at Junior, the miners passed a resolution that read, in part, "We sincerely pledge ourselves to load as much coal as is consistent with our physical endurance—coal that will help win the war for humanity. . . . We stand for freedom and liberty and democracy for mankind over the entire world." The miners at Thayer passed a declaration that stated: "Everyone here is attempting to do their bit in support of this great struggle for democracy." A miner proclaimed to the *UMWJ* that "there is a paramount duty to 'carry on' for the cause of freedom and justice." Another miner declared that "we stand ready to produce every pound that can possibly be produced . . . and thus show that we are the ones on this side that will defeat the Kaiser and insure freedom to the world." Keeney told the District 17 miners that "we must stand shoulder to shoulder in this great battle for universal liberty and democracy that will reach to the four corners of the earth."[52]

Indeed, during the war the meaning of democracy became a major topic of discussion among the southern West Virginia miners. A miner from Kimberly wrote the *UMWJ*, explaining that "the United States and her Allies, after the war, will try to establish a world democracy or as near it as they can." The miner then asked: "What is democracy? I mean pure democracy? . . . How far will it reach? Will it cross all class lines? Will it adjust wages and the prices of necessities so that labor can have a margin for the proverbial day? . . . Isn't democracy the golden rule? I hope that after the war we will have a pure democracy without any dross."[53]

District 17 Secretary-Treasurer Fred Mooney lectured the state Federation of Labor on the meaning of democracy and the potency of its spirit. He exhorted:

> Democracy is not a thing born yesterday to die today. It may meet with obstacles which are hard to overcome; it may travel underground for centuries; its progress may be hidden from view to the

> conventional eye when all at once it breaks forth with renewed vigor as the brook gathers momentum from a cloudburst and with a rush that recognizes no obstacles, shows no favors, leaving in its wake gentle reminders that the barriers which have retarded its progress only make its onslaughts more fierce and severe.[54]

With startling insight into what was happening, Josiah Keeley, the general manager of Cabin Creek Consolidated Coal Company, urged moderation in the use of wartime propaganda. " 'Making the world safe for democracy' is now one of the world's best-known phrases. 'Making democracy a safe thing for the world' is also in the minds of many. . . . It has been impossible to fight Kaiserism abroad without some introspection at home, and it is perhaps natural that the minds of labor turn to their old enemy, capital, and hang on it all the iniquities of Kaiserism." This tendency was particularly easy in the southern West Virginia coal fields, Keeley added, because "from the beginning of the coal industry, the discipline of the coal companies . . . has been the only check on the life of the coal camps, the company being so closely identified with the civil authorities that little distinction was made." Consequently, the company towns, he predicted, will become "fertile grounds for disorder" because "there are enough uninformed individuals who take the idea of an actual emancipation so literally that there is a question whether the more moderate workmen will be able to set any definite limit to what is a necessary authority." Noting the growing democratic ideas and tendencies among the miners, Keeley questioned if the international union officials could continue to control the rank and file. "The French Jacobins, the Roman Plebians, and the Russian Bolsheviki," Keeley explained, "all had some good leaders, but not all of them were able to subdue the spirit which their magic had raised."[55]

Keeley's warning was too late; the wartime rhetoric had added a political dimension to the miners' unified struggle for what was now a secular as well as spiritual liberation. "After conquering the iron hand of autocracy in war-stricken Europe and [then] the labor world," wrote a miner, "[we will] set up a government of democracy that will last as long as the world stands. Now we can not do this with tongue or pen. Each and every one must put forth every effort in our power to help." Mooney was briefer: "Kaiserism shall not dominate a people whose forefathers gave their blood that we might stand free."[56]

## NOTES

1. Samuel Gompers, *Seventy Years of Life and Labour: An Autobiography*, 2 vols. (New York, 1925), 2:524. David Brody has pointed out that "Wilsonian pro-

paganda gave a special appeal to trade unionism" among the nation's steelworkers, and he saw it as a cause of the 1919 steel strike. He did not, however, discuss its political implications as presented in this chapter. *Steelworkers in America: The Nonunion Era* (New York, 1969), chap. 9.

2. E. P. Thompson, *The Making of the English Working Class* (New York, 1963), 73.

3. Leon Samson, *Toward a United Front: A Philosophy for American Workers* (New York, 1972), especially chap. 6.

4. McAlister Coleman, *Men and Coal* (New York, 1943), 90, and James Johnson, "The Wilsonians as War Managers: Coal and the 1917-18 Winter Crisis," *Prologue* (Winter 1977): 203-5.

5. Wilson is quoted in *UMWJ*, Nov. 1, 1918, cover.

6. "Shortage of West Virginia Coal," *Coal Age* 14 (Aug. 8, 1918): 274; W. P. Tams, *The Smokeless Coal Fields of West Virginia* (Morgantown, W.Va., 1963), 69; Secretary of Navy G. L. Meyer to Henry Higginson, Jan. 10, 1910, Document File, item 8426-3, and Assistant Secretary of Navy Franklin D. Roosevelt to John L. Livers, June 26, 1917, Document File, 9454-561, both in Record Group 80, General Records of the Department of Navy, National Archives, Washington, D.C.

7. *Coal Age* 13 (Aug. 1, 1918): 229, and *UMWJ*, Aug. 15, 1918, 4, and Sept. 1, 1918, 9.

8. *UMWJ*, Aug. 15, 1918, 4, Oct. 15, 1918, cover and 2, Nov. 15, 1918, cover. For statements of other high government officials, including the secretary of war, secretary of the navy, and the director general of employment service, see ibid., Sept. 1, 1918, 9.

9. Ibid., Nov. 15, 1918, cover.

10. Ibid., Oct. 15, 1918, 17.

11. W. H. Noone, letter to the editor, *Coal Age* 13 (May 11, 1918): 889-90, 13 (Aug. 8, 1918): 289, 13 (Aug. 22, 1918): 361, 13 (Sept. 5, 1918); *UMWJ*, Apr. 11, 1917, 6; interview with Walter Hale, Elkhorn, W.Va., summer 1975; C. A. Cabell to John J. Cornwell, June 26, 1917, John J. Cornwell Papers, Department of Archives and History, Charleston, W.Va.

12. *UMWJ*, Sept. 1, 1918, 15, and Justin Collins to Charles Aker, July 9, 1918. Justin Collins Papers, West Virginia University Library, Morgantown, W.Va. Also see George Wolfe to Collins, July 6, 1918, Collins Papers.

13. The description of the parade and rally and the quotations of the speakers are in Frank Kneeland, "Patriotic Demonstration at Gary," *Coal Age* 11 (May 12, 1917): 826-27.

14. *UMWJ*, Nov. 1, 1918, cover, 5. Also see ibid., Aug. 15, 1918, cover.

15. Ibid., July 11, 1918, cover.

16. Ibid., Aug. 1, 1918, cover.

17. Ibid., Oct. 15, 1918, cover.

18. Ibid., July 25, 1918, cover, Dec. 1, 1918, 3, Oct. 15, 1918, cover.

19. Ibid., May 23, 1918, cover, May 30, 1918, 10.

20. Ibid., July 25, 1918, cover; *Coal Age* 13 (Mar. 30, 1918): 596; Helen Wright, *Coal's Worst Year* (Boston, 1925), 130.

21. James H. Thompson, *Significant Trends in the West Virginia Coal Industry, 1900-1957*, West Virginia University Business and Economic Studies, Bulletin no. 58 (Morgantown, W.Va., 1958), 6; *Coal Age* 13 (Aug. 8, 1918): 289; Charles Fowler, *Collective Bargaining in the Bituminous Coal Industry* (New York, 1927), 106-8; Keeney to Wilson, Nov. 3, 1919, Straight Numerical File,

item 205194-50-4, Record Group 60, General Records of the Department of Justice, National Archives.

22. Winthrop Lane, *Civil War in West Virginia* (New York, 1921), 87-88; Frank Keeney, "Circular to All the Local Unions throughout District 17," *UMWJ*, Oct. 15, 1918, 13; West Virginia State Federation of Labor, *Proceedings, 1918*, 9. Also see C. Frank Keeney, "Clean Coal: The Keynote for an Early Victory," Kanawha Valley Central Labor Union, *Win-the-War Labor Day Celebration* (Charleston, 1918).

23. *Charleston Gazette*, Apr. 6 and 7, 1918, and *UMWJ*, July 25, 1918, 5.

24. *UMWJ*, Apr. 26, 1917, 10, July 12, 1917, 5, and *Coal Age* 14 (Nov. 7, 1918): 869.

25. Calvin Nichols to Department of Educational Propaganda, Feb. 21, 1918, Helen Mason to Women's Defense Work, n.d., Gusto Shriver and Hattie Hoffman to Department of Educational Propaganda, Feb. 25, 1918, and also Morgan Bailey to Women's Defense Work, Feb. 16, 1918, Reports of School Survey, Women's Defense Work File, item 135-A3, Record Group 62, Council of National Defense, National Archives.

26. *UMWJ*, Jan. 3, 1918, 13.

27. Ibid., May 16, 1918, 27, Nov. 1, 1918, 12.

28. Ibid., Dec. 27, 1917, 5, Apr. 15, 1918, 11, June 20, 1918, 17; Thomas Ball, letter to the editor, ibid., Dec. 1, 1918, 12; Sam Bradley, letter to the editor, ibid., Aug. 15, 1918, 15. For the responses of other local unions in southern West Virginia, see ibid., Dec. 27, 1917, 5, Apr. 25, 1918, 11, June 20, 1918, 17, Aug. 1, 1918, 16, Oct. 1, 1918, 13.

29. Thompson, *Significant Trends*, 54-55; *Coal Age* 13 (Oct. 10, 1918): 695; Tams, *Smokeless Coal Fields*, 69. The U.S. Treasury Department reported that in 1917 one coal company with $10,000 in capital made a profit of 504 percent. Another coal company with a capital of $2,000,000 obtained a profit increase of nearly 20 percent, a jump from $171,000 to $562,000. *UMWJ*, Aug. 1, 1918, 6.

30. *Coal Age* 13 (Oct. 10, 1918): 695; Tams, *Smokeless Coal Fields*, 69; Arthur Suffern, *The Coal Miners' Struggle for Industrial Status* (New York, 1962), 94-96; J. Walter Barnes, "Federal Fuel Administration," in *West Virginia Legislative Handbook and Manual and Official Register, 1918*, ed. John Harris (Charleston, W.Va., 1918), 882-86; Johnson, "Wilsonians as War Managers," 196-202. The Council of National Defense had tried for voluntary control of coal prices before Congress gave the Fuel Administration its compulsory power.

31. Collins to I. T. Mann, Apr. 9, 1917, and also Wolfe to Collins, May 15, 1917, Collins Papers; Johnson, "Wilsonians as War Managers," 198.

32. *UMWJ*, Jan. 25, 1918, 15, Apr. 4, 1918, 7, Aug. 15, 1918, 5, Nov. 1, 1918, 12, Nov. 15, 1918, 12, Aug. 1, 1919, 17.

33. Ibid., Mar. 29, 1917, 4, 6, 10, Apr. 12, 1917, 8, 16, Apr. 29, 1917, 1, June 28, 1917, 10-14, Mar. 4, 1918, 7, 14, Aug. 1, 1918, 10-11, Dec. 1, 1918, 4, Aug. 15, 1919, 3, Sept. 1, 1919, 12-13, Nov. 15, 1919, 8, and West Virginia State Federation of Labor, *Proceedings, 1918*, 9. Kenyon is quoted in Winthrop Lane, *The Denial of Civil Liberties in the Coal Fields* (New York, 1924), 16.

34. Cornwell is quoted in *Coal Age* 13 (Oct. 10, 1918): 695.

35. Secretary of Navy Josephus Daniels to New River Coal Company, July 24, 1917, General Correspondence File, item 9454-615-1, Daniel Willard, chairman, Council of National Defense, to Secretary of Navy Daniels, Apr. 13, 1917, Willard to Roosevelt, Apr. 14, 1917, General Correspondence File, item

9454–541; Roosevelt to John P. Liver, vice-president and general manager, Charlottesville and Albemarle Railroad Coal Company, June 2, 1917, Document File, item 9454–561, Record Group 80, General Records of the Department of Navy.

36. G. C. Hedrick, secretary-treasurer, West Virginia Coal Mining Company, to Daniels, June 26, 1919, Roosevelt to West Virginia Mining Company, General Correspondence File, item 9454–1373; B. C. Hedrick to Daniels, July 23, 1919, Roosevelt to West Virginia Coal Mining Company, July 30, 1919, General Correspondence File, item 9454–1416, Record Group 80, General Records of the Department of Navy. W. H. Adams, Crozer Pocahontas Coal Company, to Secretary of War Newton D. Baker, Oct. 10, 1919, H. L. Rodgers, Quartermaster General, to Assistant Chief of Staff, Oct. 27, 1919, F. C. Boggs, chief, Purchase Branch, to Quartermaster General, Oct. 16, 1919, P. S. and T., Chief of Staff, item 463.3, Record Group 165, Records of the War Department General and Special Staffs, National Archives.

37. Cornwell is quoted in *Coal Age* 13 (Oct. 10, 1918): 695.

38. *Coal Age* 12 (Mar. 10, 1917): 439–40, 12 (Aug. 26, 1920): 462, 12 (July 29, 1920): 253.

39. Keeney to Cornwell, Oct. 27, 1919, Cornwell Papers, and W. E. Zirkle to Secretary of Labor William B. Wilson, Dec. 18, 1917, Correspondence File, item 16–514, Record Group 174, General Records of Department of Labor, National Archives.

40. J. C. Brewer to William B. Wilson, Dec. 9, 1919, W. H. Rider to Wilson, June 10, 1919, Hugh L. Kerwin to Joe Addis, July 12, 1919, Kerwin to E. C. Marshall, July 12, 1919, H. D. Hall to Wilson, Sept. 12, 1919, T. B. Knight to Wilson, Aug. 12, 1919, Magee McClung to Wilson, Sept. 11, 1919, General Correspondence File, item 16–514–A, Record Group 174, Records of the Department of Labor; *UMWJ*, Apr. 4, 1918, 7, Aug. 15, 1918, 5, Aug. 15, 1919, 5; testimony of J. H. Reed, U.S. Congress, Senate Committee on Education and Labor, *West Virginia Coal Fields: Hearings to Investigate the Recent Acts of Violence in the Coal Fields of West Virginia*, 67th Cong., 1st sess., 2 vols. (Washington, D.C., 1921–22), 1:480–81; newspaper clipping, dated Apr. 15, 1918, District 17 Files, UMWA Archives, UMWA Headquarters, Washington, D.C.; William Burns et al., Petition, Sharon local to John J. Cornwell, Apr. 10, 1918, Cornwell Papers; District 17 Resolution, *UMWJ*, Nov. 29, 1917, 17. See also Sam Boykin, letter to the editor, ibid., Feb. 15, 1920, 8.

41. Keeney to Cornwell, July 13, 1917, Cornwell Papers.

42. *Coal Age* 13 (Aug. 8, 1918): 289, 13 (Aug. 22, 1918): 361, 13 (Sept. 5, 1918): 433; *UMWJ*, Apr. 11, 1917, 6; interview with Walter Hale, Elkhorn, W.Va., summer 1975; C. A. Cabell to Cornwell, June 26, 1917, Cornwell Papers.

43. S. D. Buckley, letter to the editor, *UMWJ*, Aug. 8, 1918, 10.

44. William Warner to Cornwell, Aug. 19, Aug. 30, 1917, Warner to J. W. Weir, Aug. 20, 1917, N. S. Blake to Cornwell, June 4, 1917, Collins to Cornwell, July 17, 1917, Cornwell Papers.

45. A. R. Montgomery to Cornwell, July 17, 1917, Cornwell Papers; West Virginia Report, chap. 12, Women's Defense Work File, item 15–G3, Record Group 62, Records of the Council of National Defense, and Evelyn Harris and Frank Krebs, *From Humble Beginnings: West Virginia State Federation of Labor, 1903–1957* (Charleston, W.Va., 1960), 89–90.

46. Cornwell to Wilson, June 16, 1917, and Cornwell to Keeney, July 27, 1917, Cornwell Papers.

47. Keeney and J. R. Gilmore to Cornwell, Aug. 27, 1917, Cornwell Papers. The papers also contain affidavits of the organizers beaten by the guards. Cornwell to Keeney, Aug. 28, Sept. 14, Sept. 20, 1917, Cornwell Papers. For Cornwell's pleas to the coal operators, see Cornwell to Ernest Chilson, Sept. 18, 1917, and to E. Drennen, Sept. 17, 1917, Cornwell Papers. Cornwell to Baker, Aug. 31, 1917, Baker to Cornwell, Sept. 8, 1917, Secretary's Office File, item 370.6, West Virginia Documents, Record Group 407, Adjutant General's Office, 1917–1925, National Archives.

48. *UMWJ*, Oct. 1, 1919, 14. As the miners could expect no help or support from the wartime state government and legislation, neither could they expect it from wartime state agencies. The West Virginia State Council of Defense, for example, although endowed with "extraordinary large powers," was dominated by "conservative" state officials and businessmen. Jessie Sullivan, secretary of the Winding Gulf Coal Operators' Association, held the most important and influential position on the state council. A member of the National Council of Defense reported that the West Virginia Council of Defense, "although containing a representative of labor, is conservative" and that the council would thus favor "continuance of the pre-war status-quo." He further claimed that most members intended to use their positions on the state council merely "to finance their own departments." Report of Sullivan, Nov. 30, 1917, and report of E. D. Smith, Dec. 6, 1917, item 14–G1, Membership List of the West Virginia Council of Defense, item 15–G1, Record Group 62, Records of the Council of National Defense.

49. Secretary to the governor to Mr. Tutweiler, Aug. 18, 1917, A. R. Montgomery to Cornwell, July 17, 1917, Cornwell to William Warner, Sept. 22, 1917, Cornwell Papers; *UMWJ*, Mar. 7, 1918, 12, Nov. 20, 1917, 15. The coal operators also called upon the federal government, especially the Council of National Defense, to keep UMWA organizers out of southern West Virginia. Montgomery to Cornwell, July 17, 1917, Cornwell Papers, and George Wolfe to Collins, Apr. 30, 1917, Collins Papers.

50. Anna Wiseman to Women's Defense Work, n.d., School Survey Reports, Women's Defense Work, File 135–A3, Record Group 62, Records of the Council of National Defense; *UMWJ*, Aug. 30, 1917, 11, June 13, 1913, 29. For other instances of coal companies prohibiting public meetings, see Keeney to Cornwell, July 13, 1917, Cornwell Papers, and *UMWJ*, Nov. 29, 1917.

51. Thayer Local, Resolution, *UMWJ*, Apr. 4, 1918, 14.

52. *UMWJ*, Aug. 8, 1918, 10, May 23, 1918, 15, July 25, 1918, cover, Oct. 1, 1918, 12. Also see Keeney, "Clean Coal."

53. C. M. Walker, letter to the editor, *UMWJ*, May 3, 1917, 11.

54. Mooney is quoted in West Virginia State Federation of Labor, *Proceedings, 1918*, 56.

55. Josiah Keeley, "After the War," *Coal Age* 13 (Oct. 10, 1918): 868.

56. Reporter for Thayer local, letter to the editor, *UMWJ*, Apr. 4, 1918. Mooney is quoted in W.Va. State Federation of Labor, *Proceedings, 1918*, 56–57.

CHAPTER VIII

# "I'm Gonna Fight for My Union"

Political power grows out of a barrel of a gun.
Mao Tse-tung

The only way you can get your rights in this state is with a high-powered rifle.
Frank Keeney

In late August 1921, a sixteen-year-old miner, recently fired from a job, was walking to his parents' house in Mingo County. He had just started climbing the huge, steep ridge, called Blair Mountain, when he was suddenly grabbed from behind, an arm wrapped around his chest, and a hand placed over his mouth. A voice whispered: "Any noise and you are dead." Looking around, the young coal digger saw hundreds of miners, dressed in blue jeans and wearing red handkerchiefs around their necks, hiding in trenches and behind trees, armed with rifles, shotguns, and machine guns.[1] The young coal digger had walked into the greatest domestic armed conflict in American labor history, the armed march on Logan. This conflict was the culmination of three years of insurrectionary fury in the southern West Virginia coal fields that followed World War I.[2]

During the years 1919–21, the miners expressed their antagonisms toward the coal companies in World War I rhetoric: "We came out to stay until the job is finished," a miner wrote during the Mingo County strike, "for the iron hand of oppression has ruled us long enough." But they were giving a new twist and a new meaning to that rhetoric. The miner finished his letter: "We, here, are followers of Patrick Henry, whose immortal words, 'Give me Liberty or give me death' will go ringing through the history of the ages."[3]

The miner was not indulging in hyperbole; he and thousands of other southern West Virginia coal diggers felt robbed of their American birthright and their constitutional rights, and this conviction was deeply felt. "The constitution and bill of rights have been repealed, free speech and free assembly absolutely denied, the elementary laws of justice contemptuously kicked into discord," exclaimed District 17 President Frank Keeney in a telegram to West Virginia Governor John J. Cornwell in

November 1919. A Raleigh County miner lamented: "For a people bred in a Democracy, and been taught the sacred traditions of our liberty loving fathers, it is very humiliating to be robbed of our rights and be treated like chattel."[4] Thus began three years of bitter and bloody warfare to regain those rights.

Between 1919 and 1921, the southern West Virginia coal fields exploded in wildcat strikes. In one coal field alone, sixty-three work stoppages occurred within eleven months. At one time seventeen wildcat strikes were in progress simultaneously. Keeney urged the union miners to honor their contracts, but he never suspended any person or local for a work stoppage, because he believed that the local miners knew what their problems were and what had to be done.[5] They conducted wildcat strikes to secure the dismissal of nonunion personnel and company spies and to obtain wage scales, checkweighmen, and the closed shop.[6]

In addition to wildcat strikes, nonunion miners all over southern West Virginia were striking for union recognition. Where violence was needed to gain or protect their union, the miners showed they were prepared to fight. One hundred miners met at Roderfield, McDowell County, to form a local. The miners posted sentries armed with high-powered rifles to guard against intrusion. When the county sheriff and a squad of Baldwin-Felts guards (who were also deputy sheriffs) arrived to break up the meeting and the local, a gun battle ensued, leaving four dead and four wounded. Hearing of the shooting the next day, R. B. Page, a black miner in McDowell County, quickly gathered together seventy-five men and started for Roderfield to help his union brothers. Page's plans were aborted when he and his followers were stopped and most of them were arrested by a police force of over one hundred deputies, but the intent to fight was there.[7] Following the organization of a union in a town, two organizers were accosted by fifteen mine guards with high-powered rifles who were prepared to perform their usual treatment of union organizers. Suddenly, according to the organizers, twenty-five miners also carrying rifles appeared "to see us safe and the premeditated murder was not carried out."[8]

All of these encounters required planning. First, the miners had to acquire guns. With little money, a company store and post office controlled by company officials, and the unreasonable searches and seizures conducted by the mine guards, this was no easy task. The miners acquired guns in several ways. As Keeney explained, they took them "from people who already had them"—the operators and the mine guards—or pooled their funds and bought them in groups. They either went to an independent, pro-union merchant in a nearby town or ordered them through the mails, giving false addresses. A miner at Dawes Creek obtained a cache of rifles, 1,000 rounds of ammunition, and 300 dum-dum bullets that way. The miners also had a procedure for selecting the men who would take part in either a hit-and-run organizing effort or a skirmish—they were drafted by an unknown central authority.[9]

Against this backdrop of almost continuous skirmishing, labor struggles in postwar southern West Virginia erupted into mass conflict. Such was the case in the summer and fall of 1919 in the New River field, located in Raleigh and Fayette counties. In September 1919, a UMWA national convention called for a nationwide coal strike to begin on November 1 to obtain wage increases proportional to the postwar increases in prices. With the nation already suffering from a lack of coal, President Woodrow Wilson and Attorney General A. Mitchell Palmer feared a severe energy crisis and industrial paralysis if the miners struck. Consequently, both men appealed to union officials to postpone the strike, but the UMWA officials ignored the pleas. The southern West Virginia miners pledged to support it "one-hundred per cent." Miners at Gypsy, West Virginia, wrote UMWA President John L. Lewis: "We . . . will see that no damn scabs work in our mines, and furthermore we will all come out, and stay out, until we get what we are after."[10]

The West Virginia state government was alarmed by this show of support. Governor Cornwell asked Keeney not to honor the international union's strike call. When Keeney refused, Cornwell ordered federal troops into the state's coal fields, under the pretext of preserving law and order, two days before the walkout was scheduled to begin.[11]

The miners were not intimidated. On the eve of the strike, Keeney telegrammed Lewis: "The miners of District 17 will respond to the strike order notwithstanding the fact that federal troops are already in the vicinity of Charleston."[12] The next day miners in southern West Virginia, as throughout the nation, walked out. State authorities immediately began to arrest local union officials under the state's wartime "Work or Fight Law." Attorney General Palmer obtained a special injunction that prohibited Keeney from talking about the strike or even sending out letters concerning it.[13] Although the rank-and-file miners remained loyal to the national strike, the international union capitulated.

The national coal strike was aborted when Palmer obtained a federal injunction that ordered the miners back to work under the Washington Agreement. President Wilson followed the injunction with a proclamation that granted the miners a minimal wage increase, but that otherwise called for a resumption of work based upon the prewar status quo. It also provided for the appointment of a presidential commission to investigate the coal industry and to adjust future grievances. But for a few minor incidents, the 1919 national strike had been settled—except in southern West Virginia. In the New River coal field, which had been organized during World War I, the coal operators attempted to use the strike and the federal court injunction to bust the union movement. They blatantly disregarded the presidential proclamation and declared their mines nonunion.[14]

The miners telegrammed Lewis and Palmer about the violations. Both men advised the miners to await the appointment of the presidential commission that would "pass upon the matter and render a definite decision."[15]

The 5,500 miners in the New River field, encompassing 106 locals, were irate with the government's position. During the war, they had worked extra hours and ignored company violations to produce the coal that the country needed, and now, as the nation faced a coal shortage, the miners were prepared to resume work with only a minimal wage increase. They were not, however, about to return to work without their union. "In making our contract we sacrificed nearly everything to get the checkoff and when we lose that we lose everything," the president of the Dun Loop local wrote Palmer. "We cannot live in peace and security without our union."[16]

Despite court injunctions, presidential requests, and instructions from the international union, the miners of the New River field were prepared to fight for their union. They all walked out. "We are loyal American citizens," a miner on Keeney's Creek wrote Mitchell, and "we are on strike so that we might git justice." This time it was the miners who turned violent. The operators of the New River field moved to break the strike and, consequently, the union movement, in their traditional method of discharging, evicting, and blacklisting miners and importing strikebreakers.[17] The miners refused to settle for a war of attrition; they moved to stop production.

The miners acted quickly and harshly against the coal companies that tried to maintain operations. At Glen White, nearly fifty miners, wearing red handkerchiefs on their arms for identification, met at the company store and then proceeded to the mines, where they drove out the mine guards and strikebreakers and then destroyed the company's mining equipment.[18]

Protracted and bitter guerrilla warfare broke out at Willis Branch. When the coal company tried to continue production, the miners initially broke into buildings, smashed equipment, and cut down the company's power lines. The company had the damaged equipment repaired, brought in more Baldwin-Felts guards, and resumed operations. In May dozens of miners with high-powered rifles banded together, marched into

Willis Branch, drove out the guards, saturated the tipple with gasoline, and gutted it, forcing another temporary closing of the plant.[19]

The coal company responded by importing still more Baldwin-Felts guards and reopening its mines. In August the guards began to attack the local independent merchants who had been supplying the striking miners with food and guns. The miners now declared open war on the guards and the company. They shot guards on sight and ravaged the company, dynamiting the hoist house, power house, company store, and the houses of the company officials. The Willis Branch Coal Company finally closed its mines, but the strike and fighting continued for over a year.[20] The miners never lost faith. In December 1920 a striking miner in the New River field wrote the *UMWJ* that "we are still holding out for right and justice. . . . As our new year draws near we renew our faith and hope that the day is close at hand when every man who works in and around the mines will join our organization that we may become one in body and purpose."[21]

Of all the larger conflicts in southern West Virginia in these years, none matched the extraordinary events that transpired in Mingo, Logan, and McDowell counties, which, it will be recalled, Samuel Gompers had cited as "the last remains of industrial autocracy" in the United States.[22] At the District 17 convention in 1919, Keeney announced that the miners of these counties "have repeatedly appealed . . . for assistance in organizing" and for union recognition, and because of this, "mine guards have beaten up men." "It is plain," Keeney declared, "that the miners of Guyan and Norfolk and Western [coal fields] have no security whatever in their civil and constitutional rights so long as the criminal and unlawful mine guard system is maintained." The district president then asked for the support of the rank-and-file miners in sending fifty organizers into Mingo County to help the unionized miners and "to end the reign of terror."[23]

The rank-and-file miners of Districts 17 and 29 pledged their full support. They met in joint convention and agreed that "both districts would jointly go to the assistance of the men . . . in the Guyan Field, in order that they might exercise their lawful rights and become members of our organization." "Organize Logan and Mingo Counties we will," declared Keeney, "and no one shall stop us. If our organizers come back in pine boxes, neither heaven nor hell will stop the miners."[24]

In November 1919 rumors reached the Kanawha County miners that the union organizers whom Keeney had sent into Logan County were being beaten and murdered, as were the miners in the county who dared join the UMWA. Instantly, between 3,000 and 5,000 coal diggers with guns assembled just outside of Charleston, intending to march into Logan County to eliminate the mine-guard system and, according to Keeney, "to rehabilitate the constitution and bill of rights of the state."[25]

The first armed march on Logan was halted and bloodshed was avoided when Keeney and Governor Cornwell made hasty trips to Lens Creek and pleaded with the miners to return home. The miners agreed to disperse, but not before they had told the governor: "There is a group of men in this audience who have been overseas fighting to save the world for democracy, but we found, we find the conditions here more hellish than they ever were over there."[26]

The instigators and exact origin of this first armed march have never been revealed. The common assumption is that the gathering was simply a spontaneous development without leaders and organization and that Keeney's pleas to the miners to stop the march revealed a rift between the thinking of the rank and file and the district officials.[27] John Spivak, who became a nationally prominent journalist, was present at the first armed march. Later, he reflected:

> I personally did not believe this had been a "spontaneous" gathering of miners. . . . I just could not believe that several thousand miners from dozens of communities had spontaneously taken their guns and marched to a specific spot to join similarly armed brethren and that the first inkling the District officials had that this was going on was when a miner appeared to report it. . . . I strongly suspected that word for this gathering had gone out from Keeney and Mooney.[28]

By the spring of 1920, Keeney believed that the UMWA had enough union locals in Mingo County to justify recognition by all the coal operators in the county. He asked them to sign union contracts with their miners and sent additional UMWA organizers into the area.[29] The coal operators rejected Keeney's proposals, declined to meet with UMWA officials in conference, and refused to sign union contracts. Instead, they moved to break the union movement; they offered wage increases to miners who refused to join the union, locked out the miners who did, and brought in Baldwin-Felts guards to evict union miners from their company houses.[30]

Keeney pledged the full support of District 17. Early in the Mingo strike, he had warned, "The campaign to organize southern West Virginia is well underway. . . . In their present temper the [Mingo] miners are not to be fooled with. They will oppose to the last the use of private armies enlisted by the coal companies." Ignoring Keeney's warnings, the Mingo County operators proceeded to bring in more mine guards to intimidate and evict the striking miners. Union miners were forced out of the company houses, often at the "point of a Winchester rifle," and watched their furniture dumped into the streets and, at times, the roofs torn from their houses.[31]

Fred Mooney, obviously expecting no help from the state government, telegrammed Palmer, complaining of the humiliation, brutality, and "violations of due process of law" and asking "cannot some action be taken by your department?" An assistant attorney general replied that the department could not do anything. Subsequently, Mooney tried unsuccessfully to arrange a deal for the evictions to be executed by the sheriff of Mingo County instead of the Baldwin-Felts guards.[32]

On May 19, 1920, eleven Baldwin-Felts guards arrived in the town of Matewan and evicted the union miners who worked for the Red Jackett Coal Company. When the guards tried to board a train, they were blocked by the mayor and the chief of police, Sid Hatfield, a former miner and a member of the UMWA. Following an exchange of words, a gunfight broke out in which three townspeople, including the mayor, and seven of the eleven Baldwin-Felts guards were killed, including Albert and Lee Felts, brothers of the head of the agency, Tom Felts.

Overnight, Matewan became known throughout the nation, as did Mingo County, which is now called "bloody Mingo," and Hatfield, credited with the killings of the Felts brothers, became a folk hero to the southern West Virginia miners. When news of the event reached District 17 headquarters, all the people in the office "seemed pleased." One person began shaking hands with himself and dancing around the floor, shouting that this was the best news he had heard in a long time. District 17 miners asked that the families of the men killed by the Baldwin-Felts guards during the fight receive union compensation, and District 17 officials made plans to construct a monument in honor of the mayor who was killed by Albert Felts.[33]

Keeney, however, was angry. On the day of the tragedy, he telegraphed the attorney general:

> On May 8th Fred Mooney Secretary Treasurer of District Number 17 United Mine Workers of America telegraphed to you from Huntington West Va. Calling attention of your department to unlawful procedures occurring in Matewan Mingo County West Va. You were requested to act immediately in an effort to stop the unlawful evictions of miners some of which was done at the point of Winchesters. This telegram was unanswered. As a result of the lack of action upon your part twelve men are dead including the Mayor of Matewan who attempted to defend the rights of the citizens in that community. We want to know why your Department did not take any action when there was time to avoid the shocking tragedy. Furthermore we request that your Department take immediate steps to

wipe out the rule of gunmen which is prevalent in this state. This must be done and done immediately.[34]

Following the Matewan Massacre, the push for unionization of Mingo became stronger. By July 1, 1920, over 90 percent of the miners in the county had taken the union oath and had joined one of thirty-four locals. Feeling that the gun battle had helped to spark the movement, the *UMWJ* commented: "Once these outlaws [Baldwin-Felts guards] were out of the way there was a great rush for membership in the union." Keeney was thrilled and optimistic: "The campaign to organize the miners is well under way and we expect to have it completed before the snow flies."[35]

Excited about the tremendous enthusiasm for the union among the Mingo County miners, the international union promised Keeney its support and immediately began sending money to meet the enormous expense of the strike. After receiving one large grant of money, Keeney wrote UMWA Secretary-Treasurer William Green, "We have Mingo County nearly completed and are breaking into McDowell. We do not intend to stop until every miner in the state is in the United Mine Workers' Organization. I do not propose to be blocked, bluffed, or brow-beaten in this campaign until every miner is in the organization."[36]

The Mingo County miners proved worthy of the support of Keeney, the international union, and the miners of Districts 17 and 29. Eight months into the strike, a striking miner living in the Lick Creek tent colony wrote the *UMWJ*, "There are about 300 of us here in Lick Creek, living in tents. I have been living in a tent ever since June 1, and if it takes that to make the company come to their milk, I am willing to stay there five years, and the men all feel the same way. . . . we will never go to work for them until they do sign up." A month later, a miner in Sprigg declared that the "members in our local . . . are standing firm. They are standing like a stonewall. If it takes ten years to win this strike, we will be right here." Another miner simply wrote, "If my brothers will stick to us, we will stick to you."[37]

The Mingo County operators were not about to surrender to the institution that they hated and that they claimed would cause the downfall of the United States. They brought in trainloads of strikebreakers from the South and New York and Chicago to run the mines and imported more and more guards to protect their plants and the strikebreakers.[38] Like their union brethren in Raleigh and Fayette counties, the Mingo County miners would not settle for a war of attrition; they were prepared to fight for their union and they, too, took the offensive to stop production. The Mingo County strike had now become the Mingo County war.

The miners brought in and distributed guns. They attacked, beat, and sometimes killed mine guards and strikebreakers at, among others, the following mines: Matewan (October 6, October 21, November 29, 1920); Thacker (October 14, 1920); Vulcan (October 21, 1920); War Eagle (October 21, 1920); Rawl (October 21, November 7, 1920); Gates (October 23, 1920); Borderland (November 7, November 27, November 31, 1920); Chatteroy (November 22, November 27, 1920); Nolan (November 23, 1920); Kermit (November 23, 1920); Rose Siding (November 7, 1920); Merrimac (May 20, 1921).[39]

With so many Baldwin-Felts guards—and later state police—in the area, these attacks could not be haphazard. On several occasions, the striking miners placed material across the tracks of the mine cars. When the driver of the car stopped to remove the obstacle, the miners ambushed the strikebreakers. The striking miners also blocked and shot up trains that were importing potential strikebreakers into the county, sending the passengers fleeing back to New York.[40]

The miners also stopped production by destroying coal company property. The tipple at Rose Siding was blown up (November 7, 1920), the railroad tipple at Thacker was dynamited (November 10, 1920), the drum house at Ajax was dynamited and the head house gutted (November 10, 1920), and the head house at War Eagle was dynamited (May 19, 1921). Fires of incendiary origin destroyed the head house of Stone Mountain Coal Company (May 22, 1921) and the company store at Lynn (June 1, 1921). The striking miners also attacked the superintendents, foremen, plant managers, and other company officials who attempted to keep these mines running.[41]

The striking miners of Mingo County aggressively, at times ruthlessly, attacked anything or anyone standing between them and their union. A man distributing anti-union material was castrated and left bleeding to death on railroad tracks. Casually, a Mingo County miner wrote of the new means of responding to the activities of the mine guards: "A thug was killed last week on the Mingo County line. He had just driven two union men out of the county. . . . Our people want no trouble, but the thugs can sure get it if nothing else will do them."[42]

Neither did the striking miners tolerate the strikebreaking activities of the state police who were sent into Mingo County to keep the mines from closing. They shot and killed at least three policemen at Vulcan and Nolan. "We are going to fight to the last ditch," declared Keeney. "There will be no abatement of effort and neither time nor money will be spared to protect the lawful and statutory rights of the mine workers of Mingo County. Those who are wondering whether they will be organized may rest assured that this means the organization of the coal miners in Mingo."[43]

Gun battles between the striking miners and the mine guards and state police became frequent occurrences: Nolan (June 1920), Freeborn (July 1920, September 1920), and Merrimac and Chatteroy (November 1920), to name only four. Several of the battles lasted for hours and resulted in numerous deaths.[44]

The destruction of company property and the guerrilla warfare involved more than the miners' hatred of the mine guards and their vindictiveness toward the coal companies. There was a purpose to their violence: to close the nonunion mines. The coal company at Mohawk, for example, attempted to continue operations with imported strikebreakers. The UMWA sent an organizer into the town to talk to the miners; he was run out of town by the company's mine guards. A second organizer was sent in, and he was accompanied out of town by the mine superintendent, a mine guard, a deputy sheriff, and three state policemen. The strikers tried a third time to organize the miners and to persuade the company to cease operations. Several hundred of them gathered in the hills surrounding Mohawk and shot into the town until the company officials agreed to close the mines.[45]

An intelligence officer for the U.S. Army reported:

> To prevent miners from working in these so-called "scab mines" every character of intimidation has been practiced, with frequent instances of assault and destruction of property. Operations have been completely closed down, not even permitting pumps to be worked or necessary repairs to be made, occasioning most serious loss and destruction.

The report further pointed out that the miners

> have a well defined scheme for the employment of armed forces and the use of high explosives in the furtherance of their campaign of unionization, and the organization [UMWA] sanctions, advises, abets and protects their members and those who assist them in the use of such.[46]

At this point, the striking Mingo County miners had gained control of most of the county. The state police and mine guards did not try to reopen the closed mines. The strikers posted sentries who patroled the streets and company towns, preventing lawlessness and scabbing. Telephone repairmen were forced to ask the strikers for permission to fix telephone lines that had been shot down during one of the gun battles.[47]

The miners' hold on the county was broken when the governor placed the area under martial law and had the federal government send troops into the county under the pretext of restoring peace. The soldiers temporarily restored order, and some of the mines resumed operations, under the protection of the federal troops. The soldiers, under the direction of the coal operators, guarded strikebreakers and prevented union miners from attempting to persuade imported strikebreakers to join the strike.[48] Following a visit to southern West Virginia during the election campaign in 1920, Secretary of the Navy Josephus Daniels recorded in his diary:

> The saddest day I had [on my campaign trip] was in speaking at a number of points from Bluefield to Huntington, West Virginia. There had been a strike by mine workers and in the cold fall days, with a sprinkle of early snow, the miners and their wives and children had been evicted from company houses and were suffering. Worst of all, men in the Army uniform were being used by the mine-owners under the pretense of "preserving order." . . . When I reported how men in the Army were being used by mine operators, Newton Baker [secretary of war] was as indignant as I.[49]

The miners were infuriated. They began attacking the federal soldiers, killing several of them. Keeney threatened a general strike of the state's 100,000 coal diggers, unless the soldiers were removed. In September 1920 Governor Cornwell withdrew the troops. The miners now acted to close the mines reopened during the military occupation. Three coal tipples were blown up during the following week.[50]

Three days of unabated violence flared at the White Star Mining Company at Merrimac. Following the removal of the soldiers, the miners shot up the town to scare away the strikebreakers. The company brought in more strikebreakers and continued operations. The miners blew up the company's power plant. The company had the power plant repaired and resumed operations. On May 12, the miners launched a full-scale attack on the town. Scores of miners gathered in the hills surrounding Merrimac, cut down the telegraph and telephone lines, and then opened fire on the town, its buildings, its mines, and its inhabitants—the company officials and strikebreakers. The shooting was controlled by blasts from a cow horn, one blast to start shooting and three blasts to stop.[51]

Mine guards, state police, and deputy sheriffs from the surrounding communities quickly rushed to Merrimac to repel the attack. By this time, the army of attacking miners had also grown, and the attack was carried up the Tug River to include the towns of Blackberry City, Alden, Sprigg, New Howard, Rawl, and across the river into McCarr, Kentucky. The battle raged on this ten-mile front for three days. At one point, shooting continued for forty-eight hours without a stop. When the battle was over, at least twenty men had been killed. Keeney told the U.S. Senate investigating committee that the "dead [were] brought out of the woods for eight days."[52]

The "three days battle," as it was then called, was over, but the Mingo County war was not. A military intelligence officer who had been rushed to the area telegraphed the adjutant general of the U.S. Army that

> suspension of firing regarded . . . as but temporary and further outbreaks momentarily expected. . . . Lawless forces unrestrained by presence of present force of Kentucky militia and West Virginia state constabulary. Deadly warfare will instantly follow any real or fancied violation of alleged armistice. . . .
>
> Under orders men are mobilizing in the battle zone. Entire border through coal field is smouldering volcano with an eruption all the more imminent because of expected demonstration Thursday which is the first anniversary of battle of Matewan.[53]

The assistant chief of staff of Military Intelligence, also investigating the warfare, reported that "public hysteria" had gripped the people of Mingo County, especially the coal operators.[54] One coal operator wrote the secretary of war: "In the name of God and humanity please hurry Federal aid to Mingo. Our citizens are being shot down like rats. Men and women are being fired upon by striking miners and are held in cellars without aid from State Constabulary who are powerless. Our mining property is being fired upon all day, five deaths known to be the result and hundreds will be killed unless aid comes quick."[55] Another local coal operator wired U.S. Attorney General Harry Daugherty: "Only those who live in this section can realize the gravity of the situation arising from the Civil War that has raged here practically unrestrained."[56]

The coal operators were not alone in their fear of the insurgent miners. A group of businessmen in Williamson, including a newspaper publisher, a postmaster, and several merchants, druggists, dentists, real estate dealers, attorneys, and doctors requested "that federal troops be sent here in numbers large enough to police Mingo . . . County without an hour's delay. For three days firing on ten mining towns in this county has been incessant. Families driven from their homes, other penned in cellars forty-eight hours without food or water not yet rescued. . . . State constabulary and local officers unable to cope with situation. Condition is one of Civil War." Hysteria also had gripped the congressional representatives of West Virginia. U.S. Senator Howard Sutherland wrote the secretary of war, wanting a national police force stationed at Nitro, Kanawha County.[57]

Hysteria also possessed the statehouse at Charleston. The newly elected governor, Ephraim F. Morgan, had been begging for federal troops since assuming office in March 1920. With the three days battle his requests became frantic appeals. On the first day of the battle, Morgan telegrammed the secretary of war, John Weeks: "Situation absolutely beyond control of State and county authorities. Nothing short of one hundred per cent martial law will accomplish result." Two days later, Morgan again wired Weeks: "No security possible except presence of federal troops which should be sent without an instant of delay."[58]

Unknown to Governor Morgan, President Warren G. Harding had secretly signed a proclamation of martial law, but had not issued it publicly. Although the president had declared martial law, he gave orders that federal troops were not to enter the area without his authorization. According to the telephone instructions of General Henry Jervey: "The President has taken the matter into his own hands and troops will not be sent until orders are received from the War Department."[59]

Why did the White House and the War Department hesitate to send federal troops to Mingo County? Both the president and the secretary of war knew of the violence and the seriousness of the situation, as military intelligence officers kept them informed daily. Possibly Harding believed, as historians have claimed, that West Virginia governors perceived federal troops as federal police, whose purpose was to quell domestic disorder, so the state did not have to develop its own state police force to handle its own affairs.[60] It is also possible that the White House and the War Department were concerned about the effectiveness of federal troops in the area. One military advisor had pointed out that "federal troops have been sent twice into Mingo County within the past nine months with no effect."[61] Quite possibly, Harding and the War Department were also concerned with the past abuses of martial law in West Virginia and were determined to prevent further ones.

When the federal troops did not come, Governor Morgan proceeded on his own. He declared martial law in Mingo County, doubled the state police force, sent an additional seventy-five state troopers to the county, and called for the establishment of a special police force, made up of 250 citizens of Williamson, under the command of a state trooper, Captain J. R. Brockus. The members of the special police force, chosen from lists provided by the coal companies, were given a rifle, ammunition, official arm bands, and trained in military tactics. Morgan also speedily reorganized the West Virginia national guard, which accepted all able-bodied men—except union men. "It is a rule of the national guard," declared Major T. C. Davis, "not to take in union men." Fifteen companies were formed, and fourteen of them were immediately sent to the southern West Virginia coal fields to enforce the governor's proclamation of martial law.[62]

The state national guard, the state police, and the special police empowered with state authority were now to enforce the governor's proclamation of martial law and assume the responsibilities of the mine guards, who had long since lost control of the strike. Declaring that "it's these [union] agitators that make all the trouble," Major Davis banned all meetings in Mingo County, which nullified the right of an organizer to address a gathering. Further, he decreed that the gathering of two or more union men constituted a union meeting and that he would arrest the men. Miners were forced to walk single file through town, were arrested without warrants, and were forbidden to meet in their union headquarters in Williamson.[63] A miner arrested for having a copy of the *UMWJ* in his possession wrote the union journal that "we are not allowed to talk under Governor Morgan's martial law, but the boys are brave. We have stood firm and held on for a year and a month. This arrest will show you what we are up against in West Virginia."[64] "The big advantage of this martial law," explained Major Davis, "is that if there's an agitator around you can stick him in jail and keep him there."[65]

Major Davis, the state militia, and state police also began to intimidate striking miners. Miners watched the state authorities invade tents, smash furniture, and cut up clothes (including a miner's World War I uniform). Miners saw the state police and state militia humiliate their wives, stop them from going into town to get food for the family, curse and abuse them, and even pour kerosene in their babies' milk![66]

Apparently not satisfied with these individual acts of intimidation, on June 14, 1921, the special police and seventy state police, led by Captain Brockus, raided the Lick Creek tent colony. The state authorities rounded all the men and women in the colony into a herd. A member of the special police, also a deputy state mine inspector, shouted to Brockus: "All the God damn ______ ______ ought to be burned; the women ought to be piled up and the tents put on top of them and burned." This was not done, although the state authorities did arrest forty of the miners and marched them, hands tied, to jail in Williamson. But before the arrests, they sliced the tents with their knives and demolished the furniture. After they destroyed the tent colony, the state authorities then pulled a miner, Alex Breedlove, out of the group. A state policeman shouted to Breedlove: "Hold up your hands, God damn you, and if you have got anything to say, say it fast." Breedlove mumbled, "Lord, have mercy," at which point the state policeman killed him.[67] The special police and state police had assumed the guards' function of murdering striking miners.

Three weeks later the state police and state militia, this time led by Major Davis, raided the UMWA headquarters in Williamson. The twelve union officials in the headquarters, preparing to distribute food to striking miners, were charged with "unlawful assemblage," arrested, jailed without bond, and transferred to Welsh, McDowell County, where they were confined for several days and refused bail and habeas corpus. The men were released from jail only after they had promised to leave the state (in violation of the state constitution).[68] "Your organization," Phil Murray, vice-president of the international union, told a UMWA convention, "is face to face with the combined opposition of the state of West Virginia, state constabulary, state militia, deputies, sheriff deputies, coal operators, and every other agency that can be mustered out to defeat the aims of the strikers."[69]

As the stories of the atrocities and strikebreaking efforts of the state police and state government poured out of Mingo County, the southern West Virginia miners became enraged and aware of the link between the coal establishment and the state government. The arrest of the UMWA officials in Williamson, wrote one miner, "is another illustration of the high-handed methods employed by the employers aided by the state officials. . . . It is time that some people learned that working men have some rights under the Constitution . . . and these rights we are going to obtain at all costs."[70] A Raleigh County miner explained to President Harding that "the Governor of West Virginia has declared Martial Law . . . : [and] this Martial Law has the jails of Mingo and McDowell County filled with innocent miners which were locked out by the coal operators of that section for trying to exercise their constitutional rights for not being willing to submit to feudalistic conditions being imposed upon them."[71] Approximately 1,000 people met at Mount Clare and passed a resolution that declared: "We wish to let the brothers of Mingo County know that, we are with them to a man, and ready to assist them in any way. The Lick Creek tent colony outrage is the last straw. Indignation is running very high in West Virginia." The resolution finished by warning Governor Morgan that unless the state police and mine guards were removed, "the fair name of our state may be stained with another Ludlow, Stanaford, or Holly Grove massacre."[72]

The miners throughout southern West Virginia let it be known that they were not going to allow the state authorities to break the strike. Meeting in convention, the miners of District 29 announced that they were "ready and willing to make any sacrifice" for the success of the Mingo County strike. "By the eternal gods," declared Keeney, "before I sacrifice these men, I will go and fight myself."[73] On July 10, Keeney wrote Lewis, explaining that the "military, made up of company clerks, bookkeepers, superintendents and non-union men, and the Law and Order gang [the state police] is ever threatening the lives of our men." "We are putting the machinery of District seventeen and the central bodies into action," Keeney continued, "to protect against the total destruction of the Constitutions of state and Nation."[74]

During the last week in July, Keeney announced that he was going to "invade" Mingo County by sending large numbers of organizers into the county for the purpose of forcing the state police to arrest them and fill the jails of all of the southern counties of the state. Keeney abruptly called off this plan on August 1, when he heard that Hatfield and Ed Chambers had been murdered by Baldwin-Felts detectives on their way to trial in Welsh, McDowell County.[75]

The killing of Hatfield probably was, as contemporaries and historians claim, the spark that prompted the armed march on Logan. The real significance of the event to the miners, however, is often lost; it involved much more than the murder of a regional folk hero; the murder was a violent demonstration of the destruction of the last refuge of justice that the miners had in what Murray had called "this greed cursed state."[76]

The miners had long abandoned local politics as a possible source of protection and a means of acquiring their rights, for the coal companies' complete and total dominance of local politics had closed this traditional political channel to them. In Logan County loomed the omnipresent figure of the county sheriff, Don Chafin, always reminding the miners of how far local authorities and the coal establishment would go in their efforts to keep out the union. District 17 Vice-President William Petry explained that "Chafin, being the political boss, controls other county officials. He secures the election of judges, prosecuting attorneys, members of the county court, justices of the peace, constables, etc. They are all his willing tools. . . . He has a complete ring of official and non-official henchmen who do the bidding of the coal operators."[77]

By 1919–21, state politics had become as meaningless to the miners as had local politics. Even before the Paint Creek–Cabin Creek strike, the miners had questioned a state government that not only allowed the mine-guard system to exist, but that clothed it with state authority.[78] During and after that strike, the miners' questions turned into contempt as subsequent governors either supported the system or rhetorically claimed that they lacked the power to abolish the murderous system that denied the miners their rights as Americans—including the right to vote.[79]

That contempt deepened during World War I as Governor Cornwell and the state legislature passed more and more labor-restricting legislation, and it lingered in the postwar period. "For sometime past," Keeney wrote to Cornwell before the outbreak of violence in 1919, "many members of United Mine Workers District 17 have confided to me their beliefs and suspicions that you are not dealing openly and above board with them. . . . They have repeatedly expressed their belief that you were not sincere in your utterances concerning the desire to see that the miners receive their constitutional rights." Keeney's personal distrust of the governor was so overwhelming that when Cornwell decided to appoint a governor's commission to investigate conditions in Logan and Mingo counties, Keeney refused to cooperate. He publicly announced that he had "broken off diplomatic relations" with the governor.[80]

As the miners' rebellious behavior exposed the state government's antiunion proclivities and caused the miners to understand—and feel—the connection between the coal establishment and the state government, contempt and distrust turned into rage. During the Mingo County strike, Cornwell publicly stated that there were no gunmen or mine guards employed by the coal companies of that county. Everybody in the nation knew differently; the Mingo County coal operators had admitted as much.[81]

Little wonder that the West Virginia state government was scorned by organized labor. For example, Gompers declared that a "political autocracy" governed West Virginia. He continued:

> The state government of West Virginia is the most incompetent and least responsive to the common welfare of any in the United States. . . .
>
> West Virginia must be brought into line with modern thought. West Virginia must restore government by democratic methods. West Virginia must forbid the existence of privately owned armies. West Virginia must guarantee to the miner the same freedom, the same rights, the same equality of opportunity that she has given so generously to mine operators.[82]

Newspapers and magazines throughout the nation were vocal and prolific in their denunciations of the West Virginia state government. The *Memphis Press* wrote about the "great feudal baronry known as West Virginia." The *Nation* explained that the government of West Virginia was "historically an operators' government." In an editorial, "Is Anybody Governing West Virginia?" the *New York World* pointed out that the mine guards "suppressed free speech and denied the rights of citizenship to any man who sympathized with the union cause." The newspaper continued: "No Governor of West Virginia . . . could be ignorant of the facts, yet there has never been a serious attempt on the part of a State executive to restore order, liberty and popular government." The *Baltimore Sun* was less subtle: "Through the power of their money these [West Virginia] operators have brazenly bought up the politics of the state. Through their control of successive state administrations they have controlled the law officers and the courts . . . [they] have ruthlessly brushed aside every constitutional guarantee claimed by the miners."[83]

The most devastating rebuke of the political system in West Virginia came from the governor of Indiana. A grand jury in Mingo County had indicted UMWA organizer David Robb for murder. Robb had already returned to his home state of Indiana when the indictment was issued. The governor of Indiana refused the governor of West Virginia's request for extradition, explaining that a union organizer could not get a fair trial in West Virginia. The *UMWJ* commented that the decision of the governor of Indiana "stands out as a strong indictment of the deplorable conditions in West Virginia, where the state government has completely broken down and been supplanted by thugs and gunmen in the employ of the anti-union coal operators."[84]

No one needed to tell the southern West Virginia miners about the inclinations of their state government. During the Mingo County strike, the miners reflected upon the relationship between the state government and the coal establishment, and they did not like what they perceived. "I am fifty years old," explained one miner, "but I cannot remember when the state militia, private guards or the new state constabulary were ever called out except for the purpose of strike-breaking." Mooney wrote Cornwell that "your mental processes run along private rather than public interest." The miners' suspicions were confirmed in the spring of 1921. After Cornwell left office, he became a member of the board of directors of the Baltimore and Ohio Railroad—one of the leading coal producers in West Virginia. In May 1921, as a coal operator, Cornwell warned a meeting of the National Coal Association in the Waldorf-Astoria Hotel of "the desperate efforts . . . of the United Mine Workers to organize the coal fields of this country."[85]

The miners did not need any time to become suspicious of Cornwell's successor, Morgan, as it was quickly revealed that the coal companies had paid $750,000 to secure his election.[86] Morgan disappointed neither the coal companies nor the miners. In addition to attempting to break the Mingo County strike with state police and state militia, from his executive mansion in Charleston Morgan praised nonunion labor as more efficient and productive than union labor and blamed the Mingo County war on three problems: moonshining, pistol-toting, and the automobile. A District 17 official wrote that Morgan and other state officials "display their design in accordance with the directions of those who secured their elections, i.e., Wall Street."[87] Several UMWA locals demanded the recall of Morgan, and others advocated a statewide movement to impeach him.[88]

More and more miners, however, began to accept the facts of politics in West Virginia—that the state government and the coal establishment were inextricably bound together. The miners now regarded the state police of West Virginia as the "things and tools" of the coal operators. And this view may explain why the miners were as willing to gun down a state trooper as they were a Baldwin-Felts guard. State politics had become as meaningless as local politics. "Morgan was nominated, financed, and elected by the operators," Petry explained. "He represents them first, last and all the time. It is useless to look to him for any change in the system."[89]

The miners also knew it was useless to look to their congressmen for any change in the system. The congressman from southern West Virginia, Wells Goodykoontz, was too busy describing McDowell County as a "land of milk and honey" for miners and requesting that the Department of Justice send secret agents to that county to drive out the Ohio agitators responsible for the industrial unrest there. J. Edgar Hoover obliged him.[90]

Neither did the miners expect protection or change from the state judges. Judgeships, like local politics and the state government, were under the control of the coal companies. Mooney remembered that Judge Robert Bland of the criminal court of Logan County "made no move whatsoever without consulting Sheriff Chafin." Judge James Damron of the Circuit Court of Mingo County, while he served on the bench, was hired by Harry Olmsted—a Mingo County coal operator—and Tom Felts to serve as the counsel for the prosecution of Hatfield. When a state policeman was indicted for paying money to Mingo County deputy sheriffs to persuade union miners to leave the county, a local judge quickly quashed the indictment, claiming that such a form of bribery did not constitute an offense against the state of West Virginia.[91]

The anti-union decisions and injunctions of federal Judge A. G. Dayton, who had thrown several UMWA organizers into jail, led to several nationwide movements to impeach him. Dayton was never impeached, but the judiciary committee of the U.S. House of Representatives investigated him and reported that "the judge had issued . . . very drastic and comprehensive orders and that his conduct on the bench at times had been that of one who had prejudged the case before him."[92] It was Judge Dayton who delivered the decision in the case of *Hitchman Coal and Coke Company vs. Mitchell et al.*, which established the legal basis for the yellow-dog contract. Following the U.S. Supreme Court decision that upheld Dayton's ruling, *Survey* editorialized that "the same West Virginia in which eleven men were killed in a recent street battle between coal operators' detectives and coal miners [Matewan Massacre] has also contributed to the nation certain legal principles which promise much bitterness for the future."[93]

With the Mingo County strike, the courts initiated what one journal called the "state's judicial terrorism against unionism." Injunctions and restraining orders against the UMWA poured out of the courthouses. At the request of Borderland Coal Corporation, a judge declared the check-off illegal in West Virginia. In the U.S. Supreme Court's decision in *Swift vs. United States*, Justice Oliver Wendell Holmes declared that courts could not issue "sweeping injunctions." The ruling made no difference. At the request of Pond Creek Coal Company, a West Virginia judge prohibited the miners "from advertising, representing, stating by work, by posted notices, or by placards displayed at any point in the State of West Virginia or elsewhere, that a strike exists in the Pond Creek Field, or at plaintiff's mine, and from warning, or notifying persons to remain away from said Pond Creek Field or from plaintiff's mines."[94]

The coal operators were not content with only token legal victories; they filed suit asking that the UMWA be declared an illegal institution and that it be prohibited from organizing in Logan and Mingo counties. At a UMWA convention, Lewis declared that if the court ruled in favor of that suit, he would surely end up in jail. Gompers exclaimed, "John, in defense of that principle, I would like to visit you in jail." Lewis replied, "I have every expectation that Mr. Gompers will be accorded that privilege."[95]

The court apparently did not rule on that suit, but the "state's judicial terrorism against unionism" still earned it the wrath of labor. "It is almost traditional that the Constitution means nothing to the mine owners of West Virginia," Gompers declared, "and there have been many indications that it means little less to some of the courts of West Virginia." The West Virginia State Federation of Labor denounced the "using of the courts and the machinery of law as a means of persecuting the members of the labor movement." The southern West Virginia miners felt the same way: "The greatest and most powerful enemies of organized labor in the United States today," Mooney later wrote, "will be found sitting on the bench wearing judicial ermine."[96]

Miners throughout all of southern West Virginia were asking, "Is there any law by which mine workers can secure their rights before this state drops into conflict?" Most of them already knew the answer. When Governor Cornwell asked the miners on the first armed march to return home, he remarked, "Boys, do you know that everyone of you is acting in violation of every law against bearing arms and that you are taking the law into your own hands?" The miners replied, "There is no law in West Virginia except that decreed by the coal operators, and you know it, Governor."[97]

Keeney had forced Cornwell to withdraw the federal troops by threatening him, not with the ballot box, but with a general strike of 100,000 coal diggers. When Keeney first announced that he would try to unionize Logan County, Cornwell and District 17 officials agreed to wait thirty days. Keeney, however, ignored the agreement and promptly sent fifty organizers into the county. Concerned over this act, Mooney asked the District 17 executive board, "Are we going to take this action in the face of our understanding with the Governor?" He was quickly informed that "this organization has reached such proportions that it is no longer compelled to await the whims of every politician in the state." The mode of political action among the miners was now the union.[98]

The political power that the miners found in the union came into full force when the coal establishment attempted to close off the last remaining refuge of justice and fairness available to the miners—the jury system. Since the rise of the coal establishment and its domination of local and state political machinery, the coal diggers had found some protection through trials by juries of their peers. During the 1902 strike, a superintendent wrote to his coal operator that "juries have ignored the law and given decisions in favor of the miners regardless of law and evidence." Miners injured on the job found relief through juries that often held the coal company responsible for mine accidents. In a single session of the circuit court in McDowell County, six damage suits were filed against coal companies, and in each case, the jury ruled in favor of the miner. In one case, a miner was awarded $10,000 for injuries received.[99] What the state mine inspectors, usually coal company officials, failed to punish the coal companies for, juries did. Workmen's compensation closed this loophole to the miners,[100] but to the coal establishment greater loopholes remained: In criminal cases, miners were still tried by juries of their peers.

In January 1921, Hatfield and fifteen other men were tried in Matewan for their participation in the Matewan Massacre, charged with the murder of Albert Felts. Determined to get the men responsible for his brother's death, Tom Felts had hired "seven of the best lawyers in the South" to prosecute the case. Once again, the miners found safety in the jury system, as all sixteen men were acquitted. Felts was irate. Although the courtroom had been cleared since the opening days of the trial, he charged that the jury had been intimidated by the "waving of the red flag." He further declared, "I shall spend my last cent and the last atom of my energy in trying to obtain Hatfield's punishment."[101]

While that trial was in process, state representative Joseph M. Sanders of Bluefield—the home office of the Baldwin-Felts Agency—introduced, and the state legislature passed, the "jury bill." This law allowed a judge in a criminal case to summon a jury from another county. The miners were alarmed, for they knew that the law was specifically aimed at getting Hatfield, but they were also concerned with the larger ramifications of it. As a miner in Fayette County explained in the *UMWJ*, the jury bill "was enacted into law for the sole purpose of convicting certain working men of Mingo County, whom the coal interests of that vicinity were particularly anxious of getting out of their way so that they could carry out their sinister designs against our fellow workers without hindrance."[102]

Following the passage of the jury bill, the officials of District 17 wrote and presented to the West Virginia legislature a *Manifesto of Labor in West Virginia on Constitutional Liberty and the Bill of Rights as Annunciated by our Fathers and Written into the Organic Laws of the State and Nation.*[103] The *Manifesto* began by explaining that "to bring about justice, one man to another, the law had its birth." But such justice was denied to the miners of West Virginia because "the majority of both branches of the present legislature are representative of entrenched privilege, both within and without the state, and that too many of them are in sympathy with the absentee landholders, who are only interested in the state to the extent of exploiting its natural resources." The *Manifesto* continued, "We are led to believe that the slogan of the present legislature is: Plenty of money for jails, money for grooming and equipping a state constabulary, money for anything except education and safety appliances for the workers in industry."

The *Manifesto* then focused in on its real target—the "Jury Bill, that abridges the right of a defendant to be tried by a jury of his peers in his own vicinity." It recognized the immediate circumstances that prompted the legislation to "frame up the men who were recently acquitted at Williamson in connection with the Matewan trouble." But of more concern was that the bill "will permit judges who are unfair to persecute citizens, and let them rot in jails because they are unable to secure trials in manner provided by the Bill of Rights and the Constitution. . . . [It] seeks to deprive the wage earners of those fundamental liberties for which our forefathers shed their blood from Bunker Hill to Yorktown."

The *Manifesto* called upon the workers throughout the state "to resist any further encroachment upon our civil liberties and constitutional rights with intelligent co-ordinated use of their economic power in every instance where the constitution and bill of rights are infringed." It concluded by warning the legislature that "should the dark clouds gather on the horizon and bring discontent among the people of the state in the future, we can only meet the conditions that develop sanely and soundly, with the assurance that the wage earners of the entire state were not to blame."

In late summer of 1921 a new trial of the participants in the Matewan Massacre began in Williamson, with a jury summoned from Pocahontas County. This time the charge was the murder of the other six Baldwin-Felts guards who had been killed in Matewan.[104] Simultaneously, Hatfield and Chambers were also being tried in Welsh, McDowell County, for the alleged shooting up of Mohawk, a small coal camp in McDowell County, which occurred about two weeks after the Matewan Massacre. On August 1, 1921, while Hatfield and Chambers, unarmed and with their wives, were climbing the steps to the county courthouse, Baldwin-Felts guards rushed the two men and murdered them. The next day a placard was placed in the window of the District 17 headquarters: "Shall the government live of the people, for the people, and by the people of West Virginia, or be destroyed by the Baldwin-Felts detective agency, which substitutes itself for constituted authority."[105] The two trials, as well as the murders, were vivid proof to Keeney and the miners that the only justice, the only means of protecting and procuring their rights was with guns—"political power grows out of a barrel of a gun."

On August 7, six days after the murder of Hatfield and Chambers, 5,000 coal diggers met in Charleston, ostensibly to present a list of demands to the governor. During the ten-hour meeting, they listened to speeches from Mother Jones, Bill Blizzard, and then Keeney, who reportedly told them, "You have no recourse except to fight. The only way you can get your rights is with a high-powered riflle, and the man who does not have this equipment is not a good union man."[106]

Keeney and the southern West Virginia miners were determined to organize the entire area—at any cost. "By the eternal Gods, before I sacrifice these [Mingo County] men," Keeney had told the West Virginia State Federation of Labor, "I will go there and fight myself." He was now prepared to go. On August 14, Keeney told miners in Fairmont that there were 500 men under arms in the southern part of the state and that others were arming, intending to go to Mingo County to organize that coal field. "It is my opinion," Keeney declared, "that if we meet any resistance . . . the Matewan affair will look like a sunbonnet parade compared with what will happen."[107]

During the following week, UMWA locals throughout Kanawha, Fayette, Raleigh, and Boone counties held meetings to protest martial law in Mingo County and the governor's refusal to lift it. At a meeting in Laurel City, a letter from Keeney was read: "Every drop of blood and every dollar of the Union will be spent in the attempt to lift martial law in Mingo County." At another local meeting, UMWA organizer Savoy Holt assailed Morgan's refusal to lift martial law in Mingo County and declared, "Now it is up to us." Holt explained that District 17 was taking actions to unionize the county, regardless of the existence of martial law. A place would be designated, and at an appointed time the miners were to meet to put these plans into effect. "If you are men, you will be there, prepared as instructed."[108]

The place was Lens Creek, about ten miles south of Charleston, and the date was August 20. For three days, thousands of miners descended upon that place forming a "citizens' army." Their announced intentions were to march through Logan County, hanging the county sheriff (Chafin) and blowing up the county courthouse on the way, and then to move on Mingo County, where they would overthrow martial law and liberate their union brothers in the county jails. In the process, they would abolish the mine-guard system and unionize the remainder of southern West Virginia.[109] The marchers were going to fight for their union. A participant later wrote:

> I'm going to Hart's Creek Mountain
> Going to ol' Blair Mountain Hill
> I'm gonna fight for my union
> I know it's Mother Jones's Will.[110]

Actually, the armed march was not "Mother Jones's will." Shortly after the miners began assembling at Lens Creek, she had asked them to discontinue the march. According to a participant in the march: "She didn't think we could make it into Logan." Keeney, however, followed Jones to the platform and urged the miners to march—and they did.[111]

On August 26, the miners arrived at a twenty-five-mile mountain ridge that surrounds Logan and Mingo counties. Here they met an equally strong, determined, and well-entrenched army composed of the deputy sheriffs of these two counties, state police, state militia, and Baldwin-Felts guards. The army was commanded by Chafin.[112]

The full story of the armed march on Logan, particularly its organization and leadership, is not yet known. The miners who participated in the event swore themselves to secrecy, a secrecy that the marchers honor today. Furthermore, the marchers used sentries, patrols, codes, and passwords to guard these secrets from spies and reporters. The secrecy was so tight that agents for the Department of Justice and Bureau of Investigation, as well as reporters, though disguised as miners, were unable to attend the most important meetings.[113]

Available evidence, however, does provide a good picture of the events surrounding the march, eliminates many of the myths and misunderstandings about it, and answers some of the most important questions. For example, how many miners took part in the march? Estimates have varied from as low as 7,000 to as many as 20,000.[114] These discrepancies possibly stem from the time when the count was made. About 4,000 miners constituted the original army that gathered at Lens Creek, but more miners joined the march after it was under way. The miners in Boone County joined the march as it came through their county. Groups of miners from Raleigh, Fayette, Wyoming, and McDowell counties also united with the original army after the march had begun. Rogers Mitchell led a band of miners from Dorothy, Fayette County, that connected with the marchers just before the ascent up Blair Mountain. All of the miners working at Ed Wight, Raleigh County, laid down their tools, pilfered the company store for guns, and joined the march already in progress.[115]

Still other miners arrived after the battle had begun. Columbus Avery reflected that the miners in McDowell County "were madder than hell" about the Mingo strike, so that when they heard of the fight along Blair Mountain, they organized into squads, crossed the mountain ridge, and joined the miners the next day. Still more miners, from McDowell, Mingo, and even Logan counties, "snucked" out at night, took part in the fight, and returned home before daybreak.[116] Ten days after the miners had assembled at Lens Creek, Governor Morgan reported that the "number of insurrectionaries are constantly growing." Although an army officer sent to the battle observed that it is "humanly impossible" to say how many miners participated,[117] an estimate of between 15,000 and 20,000 is probably safe.

The armed march assumes an even larger size when the number of persons who actively aided the march, though not directly participating in the fighting, is considered. A reporter for *Nation*, who estimated the size of the miners' army at between 10,000 and 14,000, claimed that "perhaps two to three times that number of people actively abetted . . . the march." Harry Hypes, for example, strongly opposed violence and refused to participate in the fighting, but for three days he transported miners, food, and provisions in his work wagon. All along the march, women set out food for the marchers, while independent merchants and other miners loaned them guns.[118]

That miners from Boone, Raleigh, Fayette, Mingo, McDowell, and Logan counties also participated in the march destroys the assumption that this was an armed march of Kanawha County miners upon Logan; this was an uprising of the southern West Virginia miners against the coal establishment. Exploitation, oppression, and injustice had created a common identity and solidarity among the miners, and their geographic mobility had turned the hundreds of seemingly isolated company towns into a single gigantic community. "In each community each individual knew everybody else," Mooney recalled. "This knowledge of one's neighbors and their whereabouts over hundreds of square miles was prevalent in all communities [throughout southern West Virginia] . . . in the years from 1917–1925." Thus, many of the Kanawha County miners in the march had once worked in Mingo and Logan counties, just as many of the Mingo strikers had once mined coal in Kanawha County.[119]

Nor was the march a march of "guntotin'," "moonshining," and "feuding" mountaineer miners.[120] A sizable proportion of these marchers were from out of state, approximately 2,000 of them were World War I veterans,[121] and all of them were from industrialized backgrounds. Indeed, the older patriarchs of the clans of the hills had no identity and little sympathy for labor unions and industrial protest. For example, unable or unwilling to understand the march, "Uncle" Talbert Hatfield told a *New York Times* reporter that the "corn liker in the hills" was responsible for the outbreak of violence. "It jest happens to be over the unions now," Hatfield explained, "but if'n they gets that settled they'll think o' somethin' else to crow about."[122]

That this particular explanation about the armed march has been allowed to persist may be due to historians' unwillingness to take class violence in the United States seriously. Not only is there no existing credible evidence to indicate that liquor was a cause of the march, but evidence shows that the miners were careful to keep it out. A minister from Marmet who was allowed to inspect the marchers' campsites related that the patrols that guarded against infiltrators also kept out "bootleggers." Sentries stopped and searched all cars, including those of union officials and even the minister, to make certain they were not transporting liquor. Furthermore, on their way to Logan, the marchers came across twelve moonshine stills—and destroyed all of them. An army intelligence officer also reported, "I have seen no drunkenness. Union officials have given strict orders that there will be no drinking in their ranks and have gone as far as holding known bootleggers and destroying their stills."[123]

This army of miners was organized and prepared to fight. The marchers had their own doctors, nurses (who wore UMWA, instead of Red Cross, headbands), and hospital facilities. They had sanitary facilities. The marchers were fed three meals a day in mess halls and commissaries. In addition to robbing the local company stores for food, the marchers bought every loaf of bread (1,200 dozen) in Charleston and transported the loaves to their campsites.[124]

To guard against infiltrators and spies, the marchers used patrol systems and issued passes. Orders were given on papers that carried the union seal and had to be signed by a union official. The marchers used passwords and codes. To attend a meeting during the march, a miner had to give the password and his local union number to the posted sentries. If the sentry was suspicious, a member of the local had to verify the identity of the miner. Discovering the password, a reporter for the *Washington Evening Star* attempted to infiltrate a meeting by giving a fake local union number. As he approached the platform from which Keeney was about to talk, two miners grabbed him from behind and carried him toward the woods. A last minute shout to Keeney, whom he had interviewed before the march, saved the reporter from meeting the fate of Chafin's deputies who had tried to infiltrate the march. Keeney instructed the miners merely to escort the reporter out of the meeting grounds.[125]

The miners were prepared to fight; they had to be, for they not only sustained a week-long fight, but also defeated Chafin's army of over 2,000 men, who were equipped with machine guns and bombing planes. Blizzard was probably the generalissimo of the march. Approximately 2,000 army veterans were the field commanders, and they instructed the other miners in military tactics. A former member of the national rifle team of the U.S. Marine Corps and a former captain in the Italian army gave shooting lessons. Several former officers, including an ex-major, drilled the miners. After watching several hundred ex-servicemen drill the miners until they marched in "something like military order," a reporter walked to another area and heard an ex-serviceman tell a squad of miners how to fight machine guns: "Lie down, watch the bullets cut the trees, outflank 'em, get the snipers."[126]

In the field of battle the miners' military preparation and the sophistication of the plans of the march were even more apparent. The local at Blair, having been given prior instructions, had dug trenches in preparation for the marchers. An advanced patrol of 500 to 800 miners cut down the telephone and telegraph lines and cleared a sixty-five-mile area of Baldwin-Felts guards.[127]

At this point, the armed marchers were in complete control of the area from south of Charleston to the mountain range surrounding Logan and Mingo counties. All civil authority, the justices of the peace, deputy sheriffs, as well as Baldwin-Felts guards, were refugees. Company officials and their families fled the area. The miners were the law and order; as one coal operator later testified: "At any turn you were liable to butt into a colored man [miner] with a high-powered rifle." The operator continued, "I had no idea what terrorism could be until that anarchy came there without anybody to check it. . . . It is impossible to describe the terrorism that prevailed. The air was filled with the feeling that if you did not do as you were told to do that something would happen."[128]

But until the ascent up Blair Mountain, nothing happened. There was no slaughter of the innocents, no wanton pillage, and no destruction.

Looting did occur, but it was limited and selective. The marchers pilfered the company stores along the way for guns, ammunition, and provisions. At Sharples, black and white marchers broke into the coal company's racially segregated mess hall that was reserved for company officials and ate a meal. The company officials still at the mine did not try to enforce either class or racial policies.[129] This was still not a march of rage or vindictiveness. The miners were selective both in their consumption and destruction. Although they robbed the company stores at will, they did not harm the independent stores along the way. When they did obtain supplies from such stores, they paid for them. The miners did not molest or intimidate the people in the small, independent towns. The pastor of a church in Marmet reported: "I feel the public is due a statement as to their [the miners'] good deportment while they are here. They have treated the citizens of the town of Marmet with the utmost respect. Our church services go on as usual and we have no disturbances whatever. . . . There is not a single merchant in the town that has any complaint for these men buy and pay for what they get."[130]

When the miners arrived at the battle zone and began to fight, they still demonstrated preparedness and organization. Arriving on foot and in cars, trucks, and trains that they had commandeered, the miners, with cartridge belts and packs "rolled as neatly and smoothly as a stove pipe," responded to a bugler's sounding of assembly and their sergeants' shouts of "packs and guns, fall in."[131] A coal company official later testified that at Sharples:

> I noticed one company march up opposite the station, face to the front, and commanded by a man acting as a captain. He faced the men to the front, like the captain of an army would do. He called the roll and started with man No. 1. He said "Here," he got down to 39, 39 was absent. He called a man from the rear of the company probably a lieutenant, to know where the man was. Later he ordered the men to throw down their tin cups, he said they were going into action that night and he didn't want anything to rattle.[132]

Sentries were then posted along the mountain ridge. Sharpshooters with telescopic rifles were stationed at strategic locations to "clean out Chafin's machine gun nests." The other miners moved out in flank movements, each under the command of an army veteran. Harvey Dillon, a veteran of the Spanish-American War and World War I, commanded the Drawdey Creek division. Hewitt Creek division was led by a former army colonel, and another "high-ranking officer" led the charge up Blair Mountain. "It was carefully done," a coal operator later testified, "just as well as General Pershing had his [army] in France."[133]

The battle raged for over a week. A veteran of the Spanish-American War reported that he "heard about as much shooting at Blair as I did at Manila." Planes that the Logan County Coal Operators' Association had rented shelled the miners with gas and crudely made explosion bombs for three days. Both armies took prisoners, and each side was accused of torturing them.[134] And both sides killed. A message sent out from the headquarters of Chafin's army read:

> As I have returned from the front line trenches, thought I would try and tell you a little something about what has been going on over there. . . .We certainly have been doing some honest to God fighting the past few days. We lost three men yesterday; happened about 8 A.M. yesterday. Perhaps I will stop off and tell you all on my way home, if I don't get bumped off before I get away. Give my regards to the boys and kill all the red necks [miners] you can.[135]

A horrified nation sat back in disbelief as newspapers reported the largest armed conflict in American labor history. The headlines of an issue of the *New York Times* read:

> Fighting Continues in Mountains
> Planes Reported Bombing Miners
> Heavy Firing Unabated

An issue of the *Washington Evening Star* declared in bold letters:

> Martial Law Is Considered By Cabinet
> Fighting Now Ranging On Twenty-Mile Front

The *Charleston Gazette* headlined:

> Logan Now In A State of Siege[136]

The federal government moved to end the struggle that President Harding called a "civil war."[137] The U.S. War Department sent Brigadier General Henry Bandholtz to the battle front to meet with Keeney and Mooney and to order them to order the miners to disperse. On August 30, the president placed the entire state of West Virginia under martial law and issued a proclamation instructing the miners to cease fighting and to return home.[138]

Having been told by Bandholtz that they and District 17 would be held responsible for any further violence, Keeney and Mooney met with several thousand of the marchers in a ball park near Madison and asked them to return home. Keeney reportedly said:

> I've told you men, God knows how many times, that any time you want to do battle against Don Chafin and his thugs, I'll be right there

> in the front lines with you. I've been there before and you know it.
> But this time you've got more than Don Chafin against you. You've got more than the governor of West Virginia against you. You've got the government of the United States against you!
> Now I'm telling you for your own good and for the good of the cause and you've got to do it. Break up this march. Go home. Get back to your jobs. . . . You can fight the government of West Virginia, but, by God, you can't fight the government of the United States.[139]

Ignoring the president's proclamation and Keeney's reported request, the miners advanced steadily up the steep mountain ridge. When a position in the miners' approach weakened, a secret agent reported, hundreds of reinforcements appeared, and the advance continued.[140]

By the morning of September 1, the miners had captured one-half of the twenty-five-mile ridge and were ready to descend upon Logan and Mingo counties. A frantic coal operator in Logan wired a congressman from southern West Virginia, telling him to contact President Harding immediately and "say to him that unless he sends soldiers to Logan by midnight tonight that the town of Logan will be attacked by an army of four to eight thousand Reds and great loss of life and property sustained."[141] There was no need; the president had already issued orders for 2,500 federal troops, fourteen bombing planes, gas and percussion bombs, and machine guns to be sent into the area.[142] The armed march and the Mingo County strike were doomed; Chafin, the Baldwin-Felts mine-guard system, and the southern West Virginia coal establishment were saved.

## NOTES

1. Interview with Marion Preece, Delbarton, W.Va., summer 1975.
2. I am concerned with the misrepresentations in the monographic studies of these years, including Howard Lee, *Bloodletting in Appalachia* (Parsons, W.Va., 1969); Daniel Jordan, "The Mingo War," *Essays in Southern Labor History*, ed. Gary Fink and Merl Reed (Westport, Conn., 1977), 102–43; Cabell Phillips, "The West Virginia Mine War," *American Heritage* 25 (Aug. 1974), 58–61, 90–94. The greatest harm, however, has been done in the general studies of American labor history, for until we understand the most protracted and violent episodes in working-class history, we cannot understand the history of the American labor movement. For syntheses that misrepresent these years, see Jeremy Brecher, *Strike!* (San Francisco, 1972), 135–38, and Sidney Lens, *The Labor Wars* (New York, 1973), 265–66. Even worse is that some surveys of American labor history totally ignore these events. See, for example, Richard Boyer and Herbert Morais, *Labor's Untold Story* (New York, 1971).
3. Ed Jude, letter to the editor, *UMWJ*, Aug. 15, 1920, 14.

4. Keeney telegram to Cornwell, printed in *Charleston Gazette*, Nov. 1, 1919, and Thomas Epps to A. Mitchell Palmer, Jan. 26, 1920, Straight Numerical File, item 205194-50-135, Record Group 60, General Records of the Department of Justice, National Archives, Washington, D.C.

5. Winthrop Lane, *Civil War in West Virginia* (New York, 1921), chap. 14, especially 94-96; *UMWJ*, Dec. 27, 1917, 4; Frank Keeney, "Circular to the Officers and Members throughout District 17," Sept. 7, 1917, and Dec. 1918, District 17 Correspondence Files, UMWA Archives, UMWA Headquarters, Washington, D.C.; U.S. Congress, Senate Committee on Education and Labor, *West Virginia Coal Fields: Hearings . . . to Investigate the Recent Acts of Violence in the Coal Fields of West Virginia*, 67th Cong., 1st sess., 2 vols. (Washington, D.C., 1921-22), 1:197-98 (hereafter cited as *West Virginia Coal Fields*).

6. Keeney to John L. Lewis, Aug. 30, 1920, District 17 Correspondence Files; *Coal Age* 3 (Aug. 9, 1913): 213, 3 (Aug. 23, 1913): 285, 3 (Sept. 20, 1913): 433, 3 (Sept. 27, 1913): 469; *UMWJ*, Feb. 1, 1917, 7.

7. *New York Times*, July 6, 1920; *Coal Age* 18 (July 22, 1920): 185; *UMWJ*, July 15, 1920, 11-12.

8. James Doyle, letter to the editor, *UMWJ*, July 15, 1920, 11-12, and Ed Jude, letter to the editor, ibid., Aug. 15, 1920, 14.

9. Lane, *Civil War*, 99; *New York Times*, Nov. 13, 1919; testimony of A. H. Hester, *West Virginia Coal Fields*, 2:802-89; report of J. S. Martin, Nov. 11, 1919, and of A. E. Hayes, Nov. 1919, Bureau Section File, item 205194, Investigative File, Bureau of Investigation, Record Group 65, Records of the Federal Bureau of Investigation (FBI), National Archives.

10. C. F. Stoddard, "The Bituminous Coal Strike," *Monthly Labor Review* 9 (Dec. 1919): 1726-35; John Brophy, *A Miner's Life*, ed. John Hall (Madison, Wis., 1964), 138-41; C. A. Mauring to Lewis, Nov. 17, 1919, William Catlett to Lewis, Nov. 6, 1919, and Gypsy Local to Lewis, Oct. 27, 1919, District 17 Correspondence Files. For an analysis of the national coal strike of 1919 and the role of the federal government, see David Corbin, "The National Coal Strikes of 1919 and 1922, or John L. Lewis versus the U.S. Government," Historian's Office, U.S. Department of Labor, Washington, D.C.

11. *Charleston Gazette*, Oct. 28, 1919.

12. Keeney to Lewis, Oct. 31, 1919, District 17 Correspondence Files. Also see *Charleston Gazette*, Nov. 6, 7, 17, 1919.

13. *New York Times*, Nov. 6, 15, 16, 21, Dec. 10, 11, 1919. Keeney, in defiance of Palmer's injunction, secretly used District 17 funds to procure provisions for the strike. L. H. Kelly to Palmer, Nov. 6, 1919, Straight Numerical File, item 205194-57-1, Record Group 60, General Records of the Department of Justice.

14. Stoddard, "Bituminous Coal Strike," 1726-35; Brophy, *Miner's Life*, ed. Hall, 138-41; Corbin, "National Coal Strikes." For a scathing attack upon the international union's capitulation to Palmer and Wilson, see *Nation* 109 (Nov. 15, 1919): 630. For text of the agreement that should have ended the strike, see *UMWJ*, Mar. 15, 1920, 4. District 29 Correspondence Files, UMWA Archives, contain copies of notices the New River Coal companies posted, which told the miners the coal companies were not complying with the federal court order nor the wishes of the president. Copies are also in author's files.

15. Keeney to Lewis, Dec. 15 and 16, 1919, Lewis to Keeney, Dec. 17, 1919, Palmer to Keeney, Dec. 17, 1919, Straight Numerical File, item 205194-50-73, Record Group 60, General Records of the Department of Justice. Palmer did

publicly blast the New River operators for their noncompliance, but took no action. *New York Evening Post*, Dec. 20, 1919.

16. *UMWJ*, Feb. 1, 1919, 3 and Thomas Epps to Palmer, Jan. 26, 1920, Straight Numerical File, item 205194–50–135, Record Group 60, General Records of the Department of Justice. This file contains similar letters and petitions from rank-and-file miners as well as officials of Districts 17 and 29.

17. George Stone, president, Keeney's Creek Local, to Palmer, Jan. 14, 1920, Straight Numerical File, item 205194–50–116, and William Petry to Palmer, Nov. 13, 1919, Straight Numerical File, item 205194–50–16 5/6, Record Group 60, General Records of the Department of Justice. The file contains other letters that emphasized the strike was for "justice."

18. Affidavit of Tom Lethco, Feb. 12, 1920, Straight Numerical File, item 205194–50–16 5/6, and James Weir to L. H. Kelly, Jan. 21, 1920, Straight Numerical File, item 205194–50, Record Group 60, General Records of the Department of Justice; Non-Union Operators of Southern West Virginia, *Statement Made to the United States Coal Commission*, Report filed January 12, 1923 (n.p., n.d.), 10; *UMWJ*, May 1, 1919. For other fighting in Glen White, see *Charleston Gazette*, Apr. 8, 1921.

19. *UMWJ*, Apr. 5, 1921, 10–12, Oct. 1, 1922, 3–5; Non-Union Operators, *Statement Made to the U.S. Coal Commission*, 10.

20. *UMWJ*, Apr. 5, 1921, 10–12, Oct. 1, 1922, 3–5; Non-Union Operators, *Statement Made to the U.S. Coal Commission*, 10.

21. J. M. Daniel, letter to the editor, *UMWJ*, Jan. 15, 1921, 7.

22. Gompers is quoted in *Charleston Gazette*, Sept. 2, 1921.

23. Keeney is quoted in *UMWJ*, Nov. 15, 1919, 8.

24. J. R. Gilman to International Executive Board (IEB), Documents and Circulars File, Executive Board File, 1917–18, UMWA Archives, and resolution of Wickman Local, *UMWJ*, Sept. 15, 1919, 25. Keeney is quoted in Arthur Gleason, "Private Ownership of Public Officials," *Nation* 110 (May 29, 1920): 725.

25. Testimony of Frank Keeney, *West Virginia Coal Fields*, 1:170; *UMWJ*, Nov. 15, 1919, 1–3; Lane, *Civil War*, 106–8.

26. Testimony of Frank Keeney, *West Virginia Coal Fields*, 1:169–70; *UMWJ*, Nov. 15, 1919, 1–3; Lane, *Civil War*, 106–8.

27. Brecher, *Strike!*, 135–38.

28. John Spivak, *A Man in His Time* (New York, 1967), 69–70.

29. Report of John L. Lewis to 1921 UMWA Convention, *UMWJ*, Dec. 1, 1921, 13–15, and testimony of Fred Mooney, *West Virginia Coal Fields*, 1:15–16.

30. *UMWJ*, Dec. 1, 1921, 13–15; *West Virginia Coal Fields*, 1:115–24, 159; testimony of Fred Mooney, ibid., 16.

31. Keeney is quoted in *Coal Age* 18 (Aug. 19, 1920): 406–7. Mooney telegram to Palmer, May 8, 1920, Straight Numerical File, item 205194–50–182, Record Group 60, General Records of the Department of Justice.

32. Mooney telegram to Palmer, May 8, 1920, and C. B. Ames to Mooney, May 10, 1920, Straight Numerical File, item 205194–50–182, Record Group 60, General Records of the Department of Justice, and *Coal Age* 17 (May 27, 1920): 112.

33. *Charleston Gazette*, May 20, 21, 22, 1921; Fred Mooney, *Struggle in the Coal Fields*, ed. Fred Hess (Morgantown, W.Va., 1967), 72–74; *Coal Age* 18 (May 27, 1920): 1111–12; West Virginia State Federation of Labor, *Proceedings,*

*1921*, 131; O. F. McCutcheion, letter to the editor, *UMWJ*, June 15, 1920, 17; testimony of C. E. Lively, *West Virginia Coal Fields*, 1:363.

34. Keeney to Palmer, May 20, 1920, Straight Numerical File, item 205194-50-184, Record Group 60, General Records of the Department of Justice.

35. *UMWJ*, July 1, 1920, 8-9; *Coal Age* 18 (July 1, 1920): 79; *New York Times*, Jan. 8, 1921.

36. Green to Keeney, June 1, July 10, 1920, Keeney to Green, June 16, 1920, Keeney to Lewis, May 31, 1920, IEB Documents and Circulars File, 1913-14, District 17 Executive Board Meeting, June 1, 1920-June 1, 1922, UMWA Archives.

37. Martin Justice, letter to the editor, *UMWJ*, Nov. 1, 1920, 17, and T. H. Johnson, letter to the editor, ibid., Aug. 1, 1921, 7. Also see H. E. Phillips, letter to the editor, ibid., Feb. 1, 1921, 9.

38. *West Virginia Coal Fields*, 1:157-58, and C. P. Dibble, letter to the editor, *UMWJ*, Jan. 15, 1921, 10.

39. Assistant Chief of Staff of Military Intelligence to Commanding General, Fifth Army Corps, May 16, 1921, Secretary's Office, File 333.9, Record Group 159, Records of the Office of the Inspector General, National Archives; Bituminous Operators' Special Committee, *The United Mine Workers in West Virginia*, Statement Submitted to the U.S. Coal Commission, August 1923 (n.p., n.d.), 45-46; *New York Times*, July 5, 14, 23, Aug. 21, Nov. 18, 23, 30, Dec. 7, 12, 16, 1920.

40. Bituminous Operators' Special Comm., *UMW in West Virginia*, 45; *New York Times*, Nov. 30, 1920; W. H. Jenks, general manager of Norfolk and Western Railroad, to Secretary of War John Weeks, May 13, 1921, Secretary's Office, File 333.9, Record Group 159, Records of the Office of the Inspector General.

41. H. B. Fiske to Adjutant General, May 23, 1921, G. W. Read to Adjutant General, May 19 and 20, 1921, Secretary's Office, File 333.9, Record Group 159, Records of the Office of the Inspector General; *Coal Trade Bulletin*, June 1, 1921, 16; Bituminous Operators' Special Comm., *UMW in West Virginia*, 45-46, 51-56; *New York Times*, Oct. 17, Dec. 16, 1920; *Coal Age* 18 (Dec. 9, 1920): 1202.

42. Spivak, *Man in His Time*, 72, and Ed Jude, letter to the editor, *UMWJ*, Aug. 15, 1920, 14. The miners gunned down mine guards protecting strikebreakers, almost at will. *New York Times*, July 14, Aug. 22, Nov. 18, 1920; *New York Call*, Aug. 23, 1920.

43. Bituminous Operators' Special Comm., *UMW in West Virginia*, 51-56; *Coal Trade Bulletin*, June 1, 1921, 16; *Coal Age* 18 (Aug. 5, 1920): 312; testimony of J. R. Brockus, *West Virginia Coal Fields*, 1:324. Keeney is quoted, ibid., 181-82.

44. *Coal Age* 18 (July 29, 1920): 251, 18 (Sept. 9, 1920): 552, 18 (Dec. 2, 1920): 1157, 18 (Dec. 9, 1920): 1202; *Coal Trade Bulletin*, June 1, 1921, 16; *New York Times*, July 5, Aug. 21, 1920; J. L. Vaughan to Department of Justice, Aug. 30, 1920, Straight Numerical File, item 205194-50, Record Group 60, General Records of the Department of Justice.

45. James Doyle and Andrew Wilson to Lewis, *UMWJ*, July 15, 1920, 11-12, and an interview with John McCoy, Alum Creek, W.Va., summer 1975.

46. Report of William Austin, camp inspector, to Commanding General, Camp Sherman, Ohio, Sept. 21, 1920, Secretary's Office, File 333.9, Record Group 159, Records of the Office of the Inspector General.

47. Interview with John McCoy, Alum Creek, W.Va., summer 1975.

48. Edward Berman, *Labor Disputes and the President of the United States* (New York, 1924), 209; *New York Times*, Nov. 30, 1920; *Nation* 113 (Oct. 4, 1922): 334; Bituminous Operators' Special Comm., *UMW in West Virginia*, 51–55; *UMWJ*, Oct. 1, 1920, 15, and Oct. 15, 1920, 5. Also see *New York Times*, Sept. 24–25, 1920.

49. *UMWJ*, Oct. 1, 1920, 15, Oct. 15, 1920, 5; Josephus Daniels, *The Wilson Era: Years of War and After, 1917–1923* (Chapel Hill, N.C., 1946), 560. Also see *New York Times*, Sept. 24–25, 1920.

50. *Charleston Gazette*, Sept. 25, Nov. 10–25, 1920; Bituminous Operators' Special Comm., *UMW in West Virginia*, 51–55; Roy Hinds, "Last Stand of the Open Shop," *Coal Age* 18 (Nov. 18, 1920): 1037–39. A month later Cornwell again declared martial law and received federal troops, but not before he got a stern, if not condescending, lecture from Secretary of War Baker. Baker memo to Chiefs of Staff, Dec. 2, 1920, Secretary's Office, File 333.9, Record Group 159, Records of the Office of the Inspector General. Also, see Assistant Chief of Staff of Military Intelligence to Commanding General, Fifth Army Corps, n.d., in the same file.

51. *Charleston Daily Mail*, Apr. 5, 8, 21, 1921; E. F. Morgan to Secretary of War Weeks, May 14, 1921, and Gen. Read to Adjutant General, May 16, 1921, Secretary's Office, File 333.9, Record Group 159, Records of the Office of the Inspector General; *West Virginia Coal Fields*, 1:171; Bituminous Operators' Special Comm., *UMW in West Virginia*, 47–48; Winthrop Lane, "Senators Tour West Virginia," *Survey* 46 (Oct. 1, 1921): 23–24.

52. *Charleston Daily Mail*, Apr. 5, 8, 21, 1921; Morgan to Weeks, May 14, 1921, and Gen. Read to Adjutant General, May 16, 1921, Secretary's Office, File 333.9, Record Group 159, Records of the Office of the Inspector General; *West Virginia Coal Fields*, 1:171; Bituminous Operators' Special Comm., *UMW in West Virginia*, 47–48; Lane, "Senators Tour West Virginia," 23–24. Phil Murray, vice-president of the international union, justifiably declared: "Never in the history of the American labor movement has there been a more bitter conflict than the one in which our organization now finds itself in Mingo County." *UMWJ*, Nov. 1, 1921, 26–28.

53. Gen. Read to Adjutant General, May 16, 1921, and Assistant Chief of Staff of Military Intelligence to Commanding General, Fifth Army Corps, May 16, 1921, Secretary's Office, File 333.9, Record Group 159, Records of the Office of the Inspector General.

54. Assistant Chief of Staff of Military Intelligence to Commanding General, Fifth Army Corps, May 16, 1921, Secretary's Office, File 333.9, Record Group 159, Records of the Office of the Inspector General.

55. C. R. Wilson to Secretary of War Weeks, May 14, 1921, Secretary's Office, File 333.9, Record Group 159, Records of the Office of the Inspector General.

56. Harry Olmstead to Daugherty, May 17, 1921, Straight Numerical File, 205194–50–198, Record Group 60, General Records of the Department of Justice.

57. George Byerne to Gen. Read, May 14, 1921, Harry Kramen, commander of Williamson American Legion post, to President Harding, May 15, 1921, Howard Sutherland to Secretary of War Weeks, Sept. 10, 1921, Secretary's Office, File 333.9, Record Group 159, Records of the Office of the Inspector General.

58. Morgan to Weeks, May 12, 14, 16, 1921, Secretary's Office, File 333.9, Record Group 159, Records of the Office of the Inspector General.

59. Recorded telephone conversation between Gen. Jervey and Lt. Gen. J. C. Rhea on May 16, 1921. Also see Gen. Jervey Memo, May 16, 1921, Secretary's Office, File 333.9, Record Group 159, Records of the Office of the Inspector General.

60. Robert Rankin, *When the Civil Law Fails: Martial Law and Its Legal Basis in the United States* (Durham, N.C., 1939), 169; Berman, *Labor Disputes*, 210–11; Jordan, "Mingo War," in *Essays*, ed. Fink and Reed, 108–9.

61. Gen. Read to Adjutant General, May 14, 1921, Secretary's Office, File 333.9, Record Group 159, Records of the Office of the Inspector General, and Hinds, "Last Stand of the Open Shop," 1038.

62. Report of Major C. F. Thompson, major general, assistant chief of staff for Military Intelligence, May 23, 1921, Secretary's Office 333.9, Record Group 159, Records of the Office of the Inspector General; Arthur Warner, "Fighting Unionism with Martial Law," *Nation* 113 (Oct. 12, 1921): 395–96; Heber Blankenhorn, "Marchin' through West Virginia, ibid. (Sept. 14, 1921): 201; A. F. Hinrichs, *The United Mine Workers of America and the Non-Union Coal Fields* (New York, 1923): 152–53; *West Virginia Coal Fields*, 1: 230–36; testimony of J. R. Brockus, ibid., 339, 341–48; Bituminous Operators' Special Comm., *UMW in West Virginia*, 58–60; *Williamson* (W.Va.) *Daily News*, May 19, 1921; Winthrop Lane, "West Virginia: The Civil War in Its Coal Fields," *Survey* 47 (Oct. 29, 1921): 183.

63. Warner, "Fighting Unionism," 395–96; Lane, "West Virginia," 183; Hinrichs, *Non-Union Coal Fields*, 152–53.

64. James Baumgardner, letter to the editor, *UMWJ*, Aug. 1921, 5.

65. Davis is quoted in Hinrichs, *Non-Union Coal Fields*, 153.

66. Testimony of Bud Francis, *West Virginia Coal Fields*, 1:477, and testimony of William Ball, ibid., 482–83.

67. D. Robb, letter to the editor, *UMWJ*, Aug. 1, 1921, 7; affidavit of William Ball, *West Virginia Coal Fields*, 1:167; testimony of James Williams, ibid., 167; ibid., 309–10, 322–46, 557–601; Mooney, *Struggle*, ed. Hess., 77–78.

68. *New York Times*, July 9–11, 1921; *UMWJ*, July 15, 1921, 3, Aug. 1, 1921, 3, Oct. 15, 1921, 8.

69. Murray is quoted in *UMWJ*, Nov. 1, 1921, 16–18. Also see ibid., Oct. 15, 1921, 4, 5.

70. John Hutchinson, letter to the editor, ibid., July 15, 1921, 11.

71. Stanley Lipscomb to Harding, June 22, 1921, Secretary's Office, File 333.9, Record Group 159, Records of the Office of the Inspector General.

72. Lewis Farley and Frank Miley, letter to the editor, *UMWJ*, July 15, 1921, 10.

73. Ibid., Nov. 1, 1921, 9. Keeney is quoted in West Virginia State Federation of Labor, *Proceedings, 1921*, 196.

74. Keeney to Lewis, July 10, 1921, District 17 Correspondence Files.

75. *Charleston Gazette*, Aug. 2, 1921.

76. Murray is quoted in *UMWJ*, Sept. 22, 1910, 2. Historians who claim this was "the spark" include: Jordan, "Mingo War," in *Essays*, ed. Fink and Reed, 108–10, and Phillips, "West Virginia Mine War," 58–60.

77. William Petry, "A Statement on the Basic Causes of the Industrial Strife in the Coal Fields of West Virginia," Report to the international union, ca. 1921,

District 17 Correspondence Files. See also Elliot Northcutt to Attorney General Daugherty, Dec. 18, 1922, Straight Numerical File, item 205194–50–30, Record Group 60, General Records of the Department of Justice.

78. See, for example, the resolutions adopted by District 17, printed in *UMWJ*, Mar. 10, 1910, 3.

79. For Governor William Glasscock's position on the mine guards, see Harold West, "Civil War in the West Virginia Coal Mines," *Survey* 30 (Apr. 5, 1913): 46. For Governor Henry Hatfield's position on the mine guards, see David Corbin, "Betrayal in the West Virginia Coalfields: Eugene V. Debs and the Socialist Party of America, 1912–1913," *Journal of American History* 64 (Mar. 1978): 987–1010. For Governor Cornwell's position, see Lucy Lee Fisher, "John J. Cornwell, Governor of West Virginia, 1917–1921, Part II," *West Virginia History* 24 (July 1963): 55.

80. Keeney to Cornwell, Nov. 3, 1919, letter published in *Charleston Gazette*, Nov. 4, 1919, and *New York Times*, Nov. 8, 1919.

81. Gleason, "Private Ownership of Public Officials," 724; *Charleston Gazette*, Dec. 7, 1920; *UMWJ*, Dec. 15, 1920, 11. For the admission that mine guards were employed, see ibid., Aug. 1, 1921, 10–11; *New York World*, July 16, 1921; *New York Evening Post*, Oct. 1, 1921; testimonies of C. E. Lively and Harry Olmsted, *West Virginia Coal Fields*, 1:227–76, 354–92.

82. Gompers's editorial was published in *American Federationist* 28 (Oct. 1921): 852–53.

83. The quote from the *Memphis Press* was from an undated clipping in District 17 Correspondence Files; *Nation* 111 (Sept. 21, 1921): 92; *New York World*, Aug. 27, 1921; the quote from the *Baltimore Sun* was reprinted in the *UMWJ*, Nov. 1, 1921, 11.

84. *UMWJ*, Sept. 1, 1921, 8. Robb was accused of murdering Dan Whitt, but I have been unable to fiind out further details of Whitt's life or of the crime so far.

85. L. T. Belsher, letter to the editor, ibid., July 1, 1921, 16; Mooney to Cornwell, July 17, 1920, letter published ibid., Aug. 15, 1920, 7. Cornwell's speech was published, ibid., June 15, 1921, 6–7.

86. Ibid., Sept. 15, 1921, 9.

87. *Washington Evening Star*, Sept. 13, 1921; "Garyism in West Virginia," *New Republic* 28 (Sept. 21, 1921): 86–88; Charles Batley to Lewis, Aug. 31, 1921, District 17 Correspondence Files.

88. Turkey Knob, W.Va., Resolution, printed in *UMWJ*, Sept. 1, 1921, 5, and George Wolfe, Circular, to Winding Gulf Coal Operators, July 19, 1921, Justin Collins Papers, West Virginia University Library, Morgantown, W.Va.

89. Stanley Lipscomb, petition to President Harding, June 22, 1921, Secretary's Office, File 333.9, Record Group 159, Records of the Office of the Inspector General, and Petry, "Basic Causes of Industrial Strife."

90. *UMWJ*, Feb. 15, 1924, 15; Hoover to Judge Ames, Nov. 19, 1919, Straight Numerical File, item 205194–50, Record Group 60, General Records of the Department of Justice.

91. Mooney, *Struggle*, ed. Hess, 97; *West Virginia Coal Fields*, 1:220–21; *UMWJ*, Aug. 1, 1920, 11.

92. *Charleston Gazette*, Apr. 27, 1914; *Fairmont* (W.Va.) *Times*, Nov. 14, 1915; *Clarksburg* (W.Va.) *Daily Telegram*, Nov. 28, 1915; Evelyn Harris and Frank Krebs, *From Humble Beginnings: West Virginia State Federation of Labor 1903–1957* (Charleston, W.Va., 1960), 120–22. The materials, especially letters and petitions from the union, relating to the impeachment movement are con-

tained in File 50, Record Group 60, General Records of the Department of Justice.

93. Harris and Krebs, *From Humble Beginnings*, 120–22, and *Survey* 24 (Dec. 22, 1917): 348–49, and 39 (June 12, 1920): 376. For the legal implications of the case, see Felix Frankfurter and Nathan Greene, *The Labor Injunction* (Gloucester, Mass., 1963). For a brief history of the Hitchman case, which also emphasizes its West Virginia background, see Lee, *Bloodletting in Appalachia*, chap. 11.

94. *Nation* 114 (May 24, 1922): 612; *UMWJ*, Jan. 1, 1922, 3; Anna Rochester, *Labor and Coal* (New York, 1931), 208; Winthrop Lane, *The Denial of Civil Liberties in the Coal Fields* (New York, 1924), 26–28. For the background of *Swift vs. United States* and its relevence to the use of injunctions in West Virginia labor disputes, see Committee on Coal and Civil Liberties, "A Report to the U.S. Coal Commission," Aug. 11, 1923, 29–33, U.S. Department of Labor Library, Washington, D.C.

95. *Charleston Gazette*, Sept. 25, 1921.

96. Gompers is quoted in *UMWJ*, Mar. 15, 1923, 7; Mooney, *Struggle*, ed. Hess, 122. For the State Federation of Labor's denunciation of the state's courts, see its *Proceedings, 1922*, 114.

97. M. A. McCoy to Palmer, May 6, 1920, Straight Numerical File, item 205194–50–183, Record Group 60, General Records of the Department of Justice, and Mooney, *Struggle*, ed. Hess, 66.

98. *Charleston Gazette*, Sept. 25, Nov. 10–25, 1919, and Mooney, *Struggle*, ed. Hess, 63. Keeney and Mooney later participated in the state's gubernatorial campaigns of 1920 and 1924. They tried to direct the labor vote of central and northern West Virginia, where workers, as I pointed out in chap. 1, still retained political rights and some influence. In regard to southern West Virginia, District 17 officials, as this chapter shows, identified with the political stance of the miners. Similarly, Fred Barkey points out that the state's traditionally politically oriented Socialists became so frustrated with the state's corrupt politics that they turned to direct action. "The Socialist Party in West Virginia from 1898 to 1920" (Ph.D. diss., University of Pittsburgh, 1971), 197. Brecher's point is valid: "The relative disinterest of American workers in traditional forms of political action largely reflects the irrelevance of traditional politics to their daily problems; their militance at the workplace is their mode of political action." *Strike!*, 241.

99. William Page to Dr. O. J. Green, Sept. 23, 1902, in Shelden Harris, ed., "Letters from West Virginia: Management's Version of the 1902 Coal Strike," *Labor History* 9 (Spring 1969): 238; Wolfe to Justin Collins, Feb. 27, 1913, Collins Papers.

100. In southern West Virginia, workmen's compensation was a victory for the operators, not the miners. See chap. 1.

101. *Charleston Gazette*, Sept. 25, Nov. 10–25, 1919; Mooney, *Struggle*, ed. Hess, 63; *UMWJ*, Apr. 1, 1921, 13, and Apr. 15, 1921; clipping of *New York American* (n.d.), found in District 17 Correspondence Files; *Indianapolis Times*, Feb. 11, 1921.

102. John T. Harris, ed., *West Virginia Legislative Handbook and Manual, 1921* (Charleston, W.Va., 1921), 288, 322, and Russel Shelby, letter to the editor, *UMWJ*, June 1, 1921, 5.

103. Frank Keeney, William Petry, and Fred Mooney were the authors of the *Manifesto*. It was printed in West Virginia State Federation of Labor, *Proceedings, 1921*, 118–21.

104. Lee, *Bloodletting in Appalachia, 63*–64.

105. *UMWJ*, Aug. 15, 1921, 1–4; *Huntington* (W.Va) *Herald-Dispatch*, Aug. 3, 1921; *Charleston Gazette*, Aug. 1, 2, 4, 1921; Mooney, *Struggle*, ed. Hess, 88; Lee, *Bloodletting in Appalachia*, 65. The *UMWJ* received so many resolutions condemning these murders that it could not publish all of them. *UMWJ*, Sept. 1, 1921, 11.

106. *Charleston Gazette*, Aug. 8, 9, 1921, and *American Coal Miner*, Aug. 15, 1921, 10. Keeney denied having made this statement, but two people attending that meeting quoted him as having said it. See testimony of Earl Hager at the Blizzard murder trial, excerpted in Bituminous Operators' Special Comm., *UMW in West Virginia*, 66, and Lee, *Bloodletting in Appalachia*, 96.

107. Keeney is quoted in Bituminous Operators' Special Comm., *UMW in West Virginia*, 66–67.

108. Keeney's letter is excerpted and Holt is quoted, ibid., 68, 67.

109. Blankenhorn, "Marchin' through West Virginia," 201; interviews with Ike Peters, Eskdale, W.Va., and Frank Blizzard, Dry Branch, W.Va., summer 1975; Bituminous Operators' Special Comm., *UMW in West Virginia*, 72; testimony of Col. Stanley Ford, *West Virginia Coal Fields*, 2:1035.

110. Interview with Nimrod Workman, Chatteroy Hollow, W.Va., summer 1975.

111. Mooney, *Struggle*, ed. Hess, 90–99; interview with Ike Peters, Eskdale, W.Va., summer 1975; *Charleston Gazette*, Aug. 25, 1921. In attempting to persuade the marchers to return home, Jones read them a bogus telegram from President Harding that requested their dispersal. When Keeney followed Jones, he exposed the telegram as a fake. Several writers feel that this is the reason that the miners followed Keeney rather than Jones in this episode. I believe, however, that this is not the case; it seems too much, often erroneous, emphasis has been placed on that telegram. As in 1916 (see chap. 4), given a choice between following a national labor leader and their own, locally elected one, the miners chose the latter. Furthermore, during these years the local unions of District 17 were petitioning the international union to keep Jones out of southern West Virginia because, according to the miners at Whittaker, "we believe that the presence of Mother Jones in our District is detrimental to the interests of our District." Local unions at Nutter Fork and Cannelton also requested that Jones leave the district. An unidentified circular accused Mother Jones of working with Morgan and pointed out that "there was never any organization in West Virginia until the present administration was elected despite the twenty years in the state by Mother Jones." The circular further recalled the episode in 1916, "when the secession movement was on Mother Jones went on Cabin Creek and tried to get Keeney, Dwyer, and their crowd to lay down." It seems that the "boys" had grown up and left their "Mother." For the story of the exposure of the bogus telegram, see Mooney, *Struggle*, ed. Hess, 90–91. Writers who have emphasized the telegram as the turning point of the march include Dale Fetherling, *Mother Jones, the Miners' Angel* (Carbondale, Ill., 1974), 186, and Jordan, "Mingo War," in *Essays*, ed. Fink and Reed, 110. For the petitions and circulars denouncing Jones and demanding that she be kept out of southern West Virginia, see District 17 Correspondence Files. Copies are also in author's collection.

112. *Baltimore Evening Sun*, Aug. 30, 1921; testimony of Col. Jackson Arnold, *West Virginia Coal Fields*, 2:545; Bituminous Operators' Special Comm., *UMW in West Virginia*, 61. Logan County miners later filed affidavits claiming Chafin and his deputies threatened to kill them if they refused to fight. Interviews

with John Stowall and Walter Richardson, Sabine, W.Va., summer 1975.

113. For secrecy on the part of the miners, see Lane, *Civil War*, 106; William Wiley, *Facts about the Armed March* (Charleston, 1921), 24; *Washington Evening Star*, Sept. 8, 1921; J. S. Martin to William J. Burns, Sept. 1, Aug. 29, 1921, Bureau Section File, item 205194–50, Investigative File, Bureau of Investigation, Record Group 65, Records of the FBI. Reporters in Logan County covering Chafin's army also found it difficult to obtain and distribute information. The problem here was not secrecy, but what the editor of International News Service called a "censorship of terror." Chafin's deputies and the state police both verbally and physically intimidated reporters, exposed their camera film, refused to allow their dispatches to be sent out by telegraph, and even shot several of them. Boyden Sparks, the famed war correspondent for the *New York Tribune*, was shot in the head and in the calf. Chafin claimed his deputies mistook Sparks for a miner. Mildred Morris, a reporter for Universal Press, although she had a military press pass, was shot in the hand, arrested, and "for more than three hours subjected to indignities by State guards." (No doubt Chafin thought she was also a miner.) Her outraged supervisor, Marlen Pew, wrote that the "state guards of West Virginia . . . are of the belief that they can prevent the publication of the truth." *Charleston Gazette*, Sept. 4, 1921; *Washington Evening Star*, Sept. 8, 1921; *UMWJ*, Sept. 15, 1921, 2–5; *American Federationist* 28 (Oct. 1921): 866; Marlen Pew to J. W. Weeks, Sept. 3, 1921, Correspondence Files, item 000.77, Record Group 407, Records of the Adjutant General's Office, National Archives.

114. Mooney, *Struggle*, ed. Hess, 90, and Art Shields, "The Battle of Logan County," *Political Affairs* 50 (Sept. 1971): 27 (a reprint of an article that originally appeared in *The Liberator*, Oct. 1921). Keeney estimated the original size to be 10,000. *New York Times*, Aug. 24, 1921.

115. *Charleston Gazette*, Aug. 24, 1921; report of Lt. E. L. Brine, Intelligence Officer, Sept. 4, 1921, Bureau of Investigation, Record Group 65, Records of the FBI; Blankenhorn, "Marchin' through West Virginia," 288–89; Oscar Cartlidge, *Fifty Years of Coal Mining* (Charleston, 1936), 72; Shields, "Battle of Logan," 31; Bituminous Operators' Special Comm., *UMW in West Virginia*, 72; interviews with Oscar England, Mullins, W.Va., and Columbus Avery, Williamson, W.Va., summer 1975; *New York Times*, Aug. 26, 1921.

116. Interviews with Columbus Avery, Williamson, W.Va., Ted Walters, Welch, W.Va., and Oscar England, Mullins, W.Va., summer 1975.

117. E. F. Morgan telegram to Secretary of War Weeks, Aug. 30, 1921, Secretary's Office, File 333.9, Record Group 159, Records of the Office of the Inspector General. Also see his telegram to President Harding, Aug. 31, 1921, in the same file. Testimony of Col. Stanley Ford, *West Virginia Coal Fields*, 2:1031.

118. Blankenhorn, "Marchin' through West Virginia," 288–89, and an interview with Harry Hypes, Gauley Bridge, W.Va., summer 1975.

119. Mooney, *Struggle*, ed. Hess, 142–44; interviews with Ike Peters, Eskdale, W.Va., and Oscar Roebuck, Eskdale, summer 1975; testimony of Millard Halley, *West Virginia Coal Fields*, 1:567; testimony of J. H. Reed, ibid., 480.

120. U.S. Coal Commission, *Report of the United States Coal Commission*, Senate Document no. 195, 68th Cong., 2nd sess., 5 vols. (Washington, D.C., 1925); Phillips, "West Virginia Mine War," 58–59; Jordan, "Mingo War," in *Essays*, ed. Fink and Reed, 119–20.

121. *New York Times*, Aug. 29, 1921.

122. Hatfield is quoted, ibid., Aug. 31, 1921.

123. *Charleston Gazette*, Aug. 24, 25, 1921, and report of Lt. E. L. Brine, Intelligence Officer, Sept. 4, 1921, Straight Numerical File, item 205194-50, Record Group 60, General Records of the Department of Justice. Also see *Leslie's Weekly*, excerpted in *West Virginia Coal Fields*, 2:876-80.

124. Lane, *Civil War*, 106; Wiley, *Facts about the Armed March*, 23-24; Phillips, "West Virginia Mine War," 61; Blankenhorn, "Marchin' through West Virginia," 288-89; *Charleston Gazette*, Aug. 25, 1921.

125. Lane, *Civil War*, 106; Wiley, *Facts about the Armed March*, 24; *Washington Evening Star*, Sept. 8, 1921.

126. Testimony of Col. Jackson Arnold, *West Virginia Coal Fields*, 1:545, 550; Wiley, *Facts about the Armed March*, 24; Blankenhorn, "Marchin' through West Virginia," 200.

127. *New York Times*, Aug. 26, 1921; testimony of Col. Jackson Arnold, *West Virginia Coal Fields*, 1:547-50; Wiley, *Facts about the Armed March*, 19-20, 46-47.

128. *Charleston Daily Mail*, Sept. 2, 1921; report of Lt. E. L. Brine, Intelligence Officer, Sept. 4, 1921, and Agent Martin to Director Burns, Aug. 9, 1921, both in Bureau Section File, item 205194-50, Investigative File, Bureau of Investigation, Record Group 65, Records of the FBI; Wiley, *Facts about the Armed March*, 2, 19-20, 46-47; testimony of Col. Jackson Arnold, *West Virginia Coal Fields*, 1:547; Cartlidge, *Fifty Years*, 72; *Charleston Gazette*, Aug. 26, 1921; *New York Times*, Aug. 26, 1921; Mooney, *Struggle*, ed. Hess, 95; interview with Rogers Mitchell, Institute, W.Va., summer 1975.

129. Wiley, *Facts about the Armed March*, 23-25.

130. *Charleston Gazette*, Aug. 24, 25, 1921; *Leslie's Weekly*, reprinted in *West Virginia Coal Fields*, 2:876-80; *American Federationist* 28 (Oct. 1921): 865.

131. *Leslie's Weekly*, excerpted in *West Virginia Coal Fields*, 2:876-80; Mooney, *Struggle*, ed. Hess, 95; Lane, *Civil War*, 106; Wiley, *Facts about the Armed March*, 21; interview with Hubert Moss, Coalburg, W.Va., summer 1975.

132. Bituminous Operators' Special Comm., *UMW in West Virginia*, 64.

133. *Leslie's Weekly*, excerpted in *West Virginia Coal Fields*, 2:876-80; testimony of Quinn Martin, ibid., 1:554-55; testimony of Col. Jackson Arnold, ibid., 545-47; Shields, "Battle of Logan," 23; Mooney, ed., *Struggle*, ed. Hess, 93-94; Wiley, *Facts about the Armed March*, 24.

134. Bituminous Operators' Special Comm., *UMW in West Virginia*, 76; *National Labor Tribune*, Sept. 1, 1921; Shields, "Battle of Logan," 33; *New York Times*, Sept. 1-5, 1921; *Charleston Gazette*, Apr. 18, 1976; Wiley, *Facts about the Armed March*, 24; Martin to Burns, Aug. 29, Sept. 1, 1921, and Harold Nathan to Burns, Sept. 2, 1921, Bureau Section File, item 205194-50, Investigative File, Bureau of Investigation, Record Group 65, Records of the FBI.

135. G. C. B. to Mr. Blouins, n.d., found in District 17 Correspondence Files.

136. *Times*, Sept. 3, 1921; *Star*, Sept. 2, 1921; *Gazette*, Aug. 22, 1921.

137. Harding to Lewis, Aug. 29, 1921, District 17 Correspondence Files.

138. Berman, *Labor Disputes*, 211-12; Gen. Harris to Commanding General, Aug. 26, 1921, Secretary's Office, File 333.9, Record Group 159, Records of the Office of the Inspector General; *New York Times*, Aug. 31, 1921. Governor Morgan placed all state employees under the command of the U.S. Army. *Charleston Gazette*, Sept. 4, 1921.

139. Questions about Keeney's sincerity in asking the miners to return home

were immediately, and probably correctly, raised. See, for example, *Fairmont West Virginian*, Aug. 29, 30, 1921; *New York Times*, Apr. 28, 1922; "Investigation of Logan County," excerpted in *West Virginia Coal Fields*, 2:1019–28. Gen. Harris to Commanding General, Aug. 26, 1921, Secretary's Office, File 333.9, Record Group 159, Records of the Office of the Inspector General, and Phillips, "West Virginia Mine War," 90.

140. Report of Martin to Burns, Sept. 1, 1921, Bureau Section File, item 205194–50, Investigative File, Bureau of Investigation, Record Group 65, Records of the FBI. Also see Bituminous Operators' Special Comm., *UMW in West Virginia*, 65.

141. Shields, "Battle of Logan," 27, 32; Berman, *Labor Disputes*, 210–13; W. R. Thurmond to Wells Goodykoontz, Sept. 1, 1921, Secretary's Office, File 333.9, Record Group 159, Records of the Office of the Inspector General. The number of people killed during the fighting is as difficult to determine as are the other circumstances surrounding the march; both armies refused to make public their casualties. Mooney later recalled that the confirmed number of dead was only four. This extraordinarily low figure for a week of intense fighting between several thousand men is suspicious. A secret agent reported that he had seen ten miners killed and "five for the state forces." Chafin estimated that fifty people had been killed, thirty of them miners. A reporter at the battle estimated that about ten miners died while Chafin's army lost over 100 men; the reporter's estimate was largely based on the testimony of a black miner in Logan who claimed he had literally seen truckloads of police brought back from the battle front. Mooney, *Struggle*, ed. Hess, 98; Shields, "Battle of Logan," 33; *Baltimore Evening Sun*, Sept. 1, 1921.

142. "Mingo County, West Virginia," especially Major-General Charles T. Menoher to Commanding Officer Billy Mitchell, Sept. 1, 1921, Secretary's Office, File 370.6, Record Group 159, Records of the Office of the Inspector General, and Maurer Maurer and Calvin Senning, "Billy Mitchell, the Air Service and the Mingo Wars," *Airpower Historian* 12 (Apr. 1965): 37–43.

## CHAPTER IX

# "Land of the Free, Home of the Brave"

> The thing was not to talk, but to do; the thing was to get hold of others and arouse them, to organize them and prepare them for the fight.
>
> Upton Sinclair
> *The Jungle*

> What's the matter with those God Damn Wobblies? They sure talk a lot, but they don't do much.
>
> Brant Scott, District 17
> IEB representative

That the three years of regional resistance amounted to a revolutionary movement is difficult to argue, as most of the participants disclaimed any revolutionary intentions. The marchers on Logan had announced that their only goals were to hang the sheriff of Logan County, abolish the mine-guard system of southern West Virginia, overthrow the governor's proclamation of martial law in Mingo County, and unionize all of southern West Virginia. In attempting to accomplish these goals, they merely defied the state police, the governor, the courts, and the president of the United States, but not the federal troops—because they were not in rebellion "against constituted authority." Fred Mooney later reflected: "When the miners surrendered their arms to Brigadier General Bandholtz, they said to the grizzled veteran of many battles, 'General, we are not fighting our government. . . . ' Thus was established beyond question the fact that they were not in revolt against constituted authority, but had taken to arms because they believed there was no other way to correct the wrongs."[1]

The revolutionary implications of the armed march were further diminished since the marchers had the sympathy of the labor and liberal press. Liberal presses and associations were more concerned with the corruption in the West Virginia government and the tyranny of the coal companies than with the possible radical proclivities of the marchers. Representing the American Civil Liberties Union, Roger Baldwin, Rose Pastor Stokes, and Norman Thomas wrote President Warren G. Harding that while they "opposed all methods of violence," they recognized that the conflict was "due directly to the denial of civil rights to the miners by the employing interests and their officials whom they control."[2]

The marchers also had the moral support of organized labor, from the executive council of the American Federation of Labor to thousands of local unions throughout the country.[3] While careful not to associate the international union with the armed march, UMWA President John L. Lewis, both privately and publicly, lent the marchers his sympathy and support. During the armed march, Lewis told President Harding that the "bloodthirsty acts" of the "Hessians of Industry" were responsible "for the present outburst of indignation on the part of the West Virginia miners." Two weeks later, Lewis was still defending the armed march to the president; this time he blamed the West Virginia state government. "It is a matter of public knowledge," Lewis wrote, "that the state government of West Virginia has broken down and that the executive officers of the commonwealth are in league with the non-union coal interests." He helped to obtain defense attorneys for the trials that followed the march and then paid for the attorneys' fees and sent a special committee to see Harding to secure federal protection for Frank Keeney and Mooney, who were confined in the Logan County jail. And Lewis kept in constant communication with the incarcerated Keeney.[4]

The treason trials that followed the armed march also add to the confusion and misunderstanding about the struggle. The grand juries of Logan and Boone counties indicted around 550 of the participants in the march, including Bill Blizzard, Mooney, and Keeney, for murder, insurrection, and treason. The indictments charged that the marchers undertook to overthrow the governor's proclamation of martial law in Mingo County and had formed an army to "wage war against the state of West Virginia."[5]

The nation's presses scoffed at these charges. The *Nation* claimed that the indictments were trumped up charges to discredit, if not bankrupt, the UMWA. Most publications simply dismissed the charges of treason as inflated indictments. The *Daily News* of Cumberland, Maryland, declared: "We see no reason for charging these workers with any such offense. On the face of it the charge looks like a libel on them and a slander." The *New York Times* editorialized: "Whatever their offenses, the union miners and their leaders were not trying to subvert the Government of West Virginia in whole or in part. . . . In West Virginia indictments for treason seem to be thrown about as carelessly as if they were indictments for the larceny of a chicken."[6]

The disclosure that the coal companies were financing the treason trials and providing the prosecutors added to the nation's scorn for the "industrial feudalism" of West Virginia and probably precluded a clear analysis of the armed march.[7] The only time that the charges were taken seriously occurred when it was learned that the treason trials were to be conducted in the same courthouse in which John Brown was convicted of treason. Newspapers then compared the armed march with Brown's raid on Harpers Ferry. The *Nation* editorialized: "Public notice . . . is concentrated on the little hamlet of Charleston, West Virginia. There, sixty-three years ago, John Brown was hanged for his ideals and on April 24 last the trial of 500 persons was begun on charges of treason and conspiracy growing out of the miners' march over mountains last summer to rescue their comrades in Logan and Mingo Counties."[8]

To the coal operators of West Virginia, the indicted marchers were guilty. For a decade, they had been telling the world that union miners were murderers and that the UMWA intended to take over their mines and violently overthrow the state government.[9] Now they had the evidence to back up those charges. They were not merely attempting to discredit or bankrupt the UMWA with these charges. They wanted the men responsible for the uprising prosecuted and then punished as traitors. Referring to Lewis and Keeney, the *Williamson Daily News*, a paper owned and controlled by an operators' association, declared: "These men aided, abetted, and encouraged the insurrectionists in their mad march from the heart of Kanawha County to the edge of Logan, and each is liable to indictment for murder and larceny, if not for treason. . . . The juries of Logan and Boone Counties have the opportunity to send them to the penitentiary or to the gallows."[10]

Federal government officials were also concerned about the march and what it represented. When federal soldiers were ordered into Logan County to put down the uprising, Secretary of War John Weeks explained: "When railroads are closed up, and armed bands of men establish themselves in the country, marching to and fro overpowering resistance you may call it what you please. We call it insurrection."[11]

This view continued after the defeat of the armed march, and officials in both the War Department and the Department of Justice demanded that the persons responsible be prosecuted for insurrection—not against the state of West Virginia, but against the federal government. An assistant attorney general of the United States explained that

> the undisputed acts of the insurgents in remaining as an armed body in defiance of the [President's] proclamation after it had come to their attention constituted a violation of Section 4 of the Criminal Code of the United States in that it was an insurrection against the authority of the United States. Their refusal to disperse constituted

an opposition to an authority of the United States, viz., the authority of the President acting under a valid law of the United States, and since this opposition took the form of an armed, coherent body, it constituted an insurrection.[12]

In February 1922, acting Secretary of War J. M. Wainwright declared that it was "necessary that the law be vindicated." Wainwright explained to U.S. Attorney General Harry Daugherty that

in defiance of the President's proclamation the conflict went on between the insurgents and the West Virginia constabulary; firing was more or less continuous and resulted in a considerable number of casualties. . . . The insurgents willfully and deliberately defied the President's proclamation and actively challenged the power and authority of the United States until they were confronted by a force that made resistance dangerous and useless. . . . They had become public enemies whom the Army of the United States was authorized by law to shoot and kill.

Wainwright concluded that "in the interest of law and order and the security of the Federal Government and to vindicate the authority of the President duly exercised in accordance with the law, the prosecution of two or three of the leaders, by your department, is recommended."[13]

The federal government came close to prosecuting the miners. In a memo to the attorney general, Assistant Attorney General John Crum also demanded that the miners be tried, and noted, "Our machinery is ready, if this meets with your approval."[14]

Why the federal government did not prosecute is unclear, for officials of the Department of Justice never explained their reasons. When a West Virginia senator asked the attorney general what action the government planned to take, Daugherty replied, "At this time I can tell you that the Department has not yet determined, and may not for a little time, what action will be taken. . . . This is all I can say at present."[15] The federal government may have felt it was safer to allow the state of West Virginia to prosecute. Shortly after the indictments for treason, an assistant attorney general explained that "in view of the reported activity on the part of the State officials in prosecuting those responsible, we are of the opinion that no steps should be taken by the Federal Government at this time, for it may embarrass the State officials."[16]

This view seems verified by the statements of Governor Ephraim Morgan, who told J. Edgar Hoover in early October that the situation was "well in hand." Morgan later explained to a secret agent that federal prosecution would cause "undue anxiety" and asked that the state be allowed to prosecute. A week later, however, Daugherty reported to Harding that Morgan "urges that there should be prosecution in the Federal Courts against certain leaders of the mob in West Virginia."[17] However, by March 1922, Harding opposed federal prosecution, and other federal officials were willing to drop the charges against the miners. This does not mean the federal government viewed the armed march with less concern, *only* that officials did not choose to prosecute. Crum wrote Daugherty: "While I am disposed to feel that there is adequate law for this prosecution, nevertheless, many of our insurrections, particularly that of the Civil War, were not followed by prosecution."[18] Although the federal government did not prosecute, its officials, like the coal operators, considered the armed march an insurrection against constituted authority.

The immensity and intensity of the struggle in southern West Virginia from 1919 to 1921 suggest that the miners were more than "dimly conscious" that they were fighting for something larger than themselves and the selfish enhancement of their own particular positions. The southern West Virginia miners were not revolutionaries or radicals by traditional standards. They did not speak in a socialist vocabulary, nor did they call for the establishment of a "socialist commonwealth"—they were not informed by the intellectual traditions behind that concept. Mooney had read widely—Winthrop Lane claimed under other circumstances that he "would have been a student"—but, although he had read the works of Plato, Voltaire, Victor Hugo, and Thomas Carlyle, Mooney confessed, "Marx I could not absorb. His work was too difficult for I had only reached the sixth reader in public school."[19] Considering that the other miners in southern West Virginia were probably less well read and less educated than Mooney, it is doubtful if they understood theoretical socialism. Some of the miners in Kanawha and Fayette counties had joined the Socialist party during the Paint Creek–Cabin Creek strike and once or twice had voted the Socialist ticket. But there is little indication that they understood socialism as a way of life and government and as a means of social salvation. Keeney had once been a member of the Socialist party of America, but by 1919 he was declaring that "it is the duty of the United Mine Workers of District 17 to eliminate from their organization any group that is preaching their different 'isms.'"[20]

Just as traditional politics had little meaning to the miners, radical political parties, like the International Workers of the World (IWW), the Socialist party of America, and the Communist party, were of little interest. Denied access to the ballot box, why should they have listened to the political ideas of any radical organization? Furthermore, the coal operators also denied them access to information from which to formulate and articulate traditional radical philosophies.[21] Even when opportunities were presented, the miners were not any more interested; during the Paint Creek-Cabin Creek strike, Ralph Chaplin discovered that "there was little use in proclaiming the virtues of the I.W.W. to the striking coal miners. . . . The miners not only had a union already . . ., but being in the midst of a two year strike, they were more interested in remaining alive than in listening to arguments in favor of a dual organization."[22] A secret agent covering the armed march noted the absence of outside radical organizations in southern West Virginia, but pointed out that "miners' agitations are of extremely radical type. . . . Outside agitators who come into this section of West Virginia are absorbed into this class industrial struggle."[23]

The miners believed in action, not rhetoric. During the march Brant Scott, District 17 representative to the International Executive Board, asked a reporter who had returned from the Northwest what action the Wobblies had taken against the men who lynched Frank Little. When the journalist answered that no action had been taken, Scott exclaimed: "What's the matter with those God Damn Wobblies? They sure talk a lot, but they don't do much."[24]

Without the education or leisure to formulate an ideology, and, except for violence, denied any means to express it, it is not surprising that the miners articulated their values and their visions of a workers' utopia in the language they knew best—Americanism. That the southern West Virginia miners expressed their intentions for a new social, political, and economic order in terms of Americanism may have represented what Leon Trotsky called "a chronic lag of ideas and relations behind new objective conditions" and what Karl Marx described as a "superstitious regard for the past" ("the traditions of the dead generations weigh like a nightmare on the minds of the living"), but it does not render their intentions and actions any less revolutionary. Marx himself recognized the potency and creativity involved in using time-honored values. Writing about the 1848 revolution in France, Marx explained that "the resurrection of the dead served to exalt the new struggles, rather than parody the old, to exaggerate the given task in the imagination, rather than to flee from solving it in reality, and to recover the spirit of the revolution, rather than to set its ghost walking again."[25]

The Americanism that fired the southern West Virginia miners was not a chauvinistic or imperialistic nationalism, nor one of national honor and power. The Americanism that the miners espoused was one that promised liberty, equality, and dignity to all people, regardless of race, religion, or current condition of servitude. It promised the miners freedom from the absolutism of coal companies, as it guaranteed them fundamental rights supposedly inalienable, but conspicuously absent in the company towns. Miners demanded the right to be safe and secure in their homes against company eviction and unreasonable search and seizure. They wanted freedom of religion; following a coal company's banishment of a minister from its town for delivering a pro-union sermon, the local miners passed a resolution that declared: "Whereas, Article I of the Constitution of the United States reads as follows: Congress should make no law respecting an establishment of religion . . . we question the right of any corporation in denying the same and condemn the act of the superintendent and appeal to all liberty loving citizens to help and assist us in restoring our rights guaranteed under the Constitution of the United States."[26]

Americanism promised the miners "their Constitutional rights of free speech and free assemblage," the denial of which, according to District 17 Vice-President William Petry, was largely responsible for "all the trouble around here." Petry explained that the miners "cannot meet in halls or in homes; to suggest organizing is to invite assault or worse at the hands of the so-called Deputy Sheriffs, who are actually the employees of the operators."[27]

Consequently, Americanism also promised the miners freedom from the tyranny of the Baldwin-Felts guards. "Imagine 23,000 coal miners under the 'iron heel' of the gunmen system for twenty years," wrote Mooney. "Think of the crushed hopes, smothered ambitions, cracked heads and the thousands of insults to which they had been subjected at the hands of the bludgeonites." If the mine guards "cease[d] cracking heads," Mooney explained, the miners would "have their freedom guaranteed by the American Constitution" and thus, "for the first time in their lives enjoy the civil liberties of which they had been deprived for two or more decades."[28]

Americanism promised them freedom from arbitrary discharge and blacklists, economic rights, such as living and working where they pleased, and the right to belong to the UMWA. "I am barred from work in this field [New River field]," a miner wrote to President Woodrow Wilson. "I am applying to you as chief executive of this nation for what legal steps to take to secure my rights as an American citizen."[29] "The coal operators," wrote another, this time to President Harding, "are depriving the coal miners of the right to belong to the labor organization which is their inherent right given to all citizens of the United States."[30]

Unionism was seen by the coal diggers as the means by which they could, and would, Americanize the coal fields. In authorizing union organizers to be sent into Logan and Mingo counties, a joint convention of Districts 17 and 29 declared that "in the event the company thugs prevent them from exercising their state lawful rights, in such event we petition our district officials to call a joint Special Convention for the purpose of adopting some further action . . . in order that the coal miners of this state may enjoy and exercise their state lawful rights." The strike for the union in Mingo County, a miner explained, was in reality a "struggle for freedom and liberty."[31]

The miners recognized that their rights as American citizens could not be obtained without a fight. "The oppressed of all time," a miner wrote, "when goaded too far have risen up in their might and thrown off the yoke of the oppressor. Unfortunately these uprisings have been accompanied with bloodshed and riot." Covering a strike in southern West Virginia, a reporter observed that "the talk of the miners and their wives forsooth, is of guns and shooting . . . [because] freedom has never been won without bloodshed."[32]

Therefore, the miners emphasized the constitutional right to bear arms. "We have a constitutional right to bear arms—about the only right left to us," Keeney once declared. "I've a high-powered rifle, three pistols, and a thousand rounds of ammunition at home. I'd like to see anyone take away that gun—except smoking."[33] The miners' obsession with this constitutional right did not stem from a frontier tradition of "gun totin' " and violence. After the southern West Virginia miners took control of District 17, Keeney and Mooney impeached district organizer Ed Scott for the "wild, and reckless use of a gun" while he was on an organizing trip.[34] The miners viewed guns as their protectors and guarantors and, eventually, the provider of their rights. As Keeney stated, "The only way you can obtain your rights [in this state] is with a high-powered rifle." Similarly, the local union at Ramage passed a resolution that declared, "As a final arbiter of the rights of public assembly, free speech, and a free and uncensored press we will not for a single moment hesitate to meet our enemies upon the battle fields, and there amid the roar of the cannon and the groans of the dying and the crash of the systems purchase again our birthright of blood-bought freedom."[35]

Contemporaries, U.S. senate investigators, and historians correctly wrote that the Mingo County strike was a strike for the union,[36] but what they failed to understand and appreciate was that the UMWA was not an end in itself. To these miners, as to the miners in the New River and Kanawha fields and elsewhere in southern West Virginia, the UMWA meant social and economic justice, and especially a means to insure their fundamental rights as Americans. And the union meant prestige and working-class power. A poem written by a Mingo County miner in 1921 declared:

Awake! the coal digger has found his might,
The union boys have put the foe to flight;
All loyal men are marching on each day,
To certain victory that will crown our way,

. . . .

While now our wives and children in white tents lay,
Yet we are nearer victory day by day;
For we will WIN—the end is almost here—
Peace and freedom for Tug River men is near.[37]

Similarly, another Mingo County miner wrote the *UMWJ*: "The miners of Mingo will be free, and with the God of Justice on their side and the help of the UMW, the shot that was fired at Lexington will re-echo in these mountain fastnesses where

Liberty lies bleeding at the feet of tyranny
A craven is that man
Resigned to live in slavery,
But glory comes to those who dare
For love of Home and Liberty."[38]

"The members of this local are standing pat. Everyday there is more unionism than ever in this local," wrote a disabled veteran of World War I working at Sprigg. "I truly hope the Lord is with the United Mine Workers on Tug River and that they will gain their liberty." When asked by a U.S. senator why the Mingo miners had struck, Keeney replied that they "are trying to establish principles whereby the men could be protected in the lawful rights, and we could not do it in any other way."[39]

Obscured in the miners' use of Americanism was an ideology, containing values, beliefs, principles, and goals, as coherent, radical, and understanding of an exploitative and oppressive system as any ideology announced by Socialists, Communists, or Wobblies. E. P. Thompson cogently pointed out the overlooked obvious when he wrote that "the exploitative relationship is more than the sum of grievances and mutual antagonisms. It is a relationship that can be seen to take distinct forms in different historical contexts, forms which are related to corresponding forms of ownership and State power."[40] The grievances and antagonisms of the southern West Virginia miners must be found in the politics and company towns of southern West Virginia, not in New York, Chicago, or Europe, where owners of the coal fields lived. Their intentions and goals were peculiar to them, and the radicalism that they espoused was peculiar to their situation.

Hidden in their firm belief in Americanism was a language common to all social revolutionaries. Although the miners did not overtly attack capitalism, neither were their economic grievances expressed in terms of bread-and-butter issues or living wages; they were expressed in terms of social justice. "Employers want to measure wages in cold dollars and cents," explained Keeney, "while the miners insist that they be measured in human values, not by what they will actually purchase of the necessities of life. The miners take the position that low wages deprive children of education, of good food, destroy self-respect, and drive men to degradation." A Mingo County miner declared, "The operators have banked large sums of money the coal loader should have." A miner from Lochgelly wrote that "many children are in need of shoes and of clothes which is very often overlooked by the ones who make prices of things soar beyond the reach of a man's earnings who has to toil for a mere existence. This should be looked into more than the wage question."[41] Although they were not calling for its abolition, the miners were certainly questioning the profit-making system; they believed that it was unjust and that the pursuit of profit should be subordinated to human needs. They realized that abuse and antagonism were intrinsic in the relations between coal operators and miners in southern West Virginia.

Without attacking private property, the miners recognized and assailed class injustice and the class rule that it promulgated. "These men of wealth," wrote Mooney, "ground their employees into profits in order that they might fare sumptuously and clothe their offspring in purple and fine linen. . . . Each dollar . . . [they spent] represented anguish, pain, misery, blighted hopes, crushed aspirations, stifled ideals . . . every dollar . . . represented the groans of overworked and underfed employees, the pitiful supplications of undernourished women, and the heart-rending whines of starving babies."[42]

Class injustice was political as well as economic. "The coal opperations they do the way they please, Because NoBody looks after them," wrote a Raleigh County miner during the height of warfare in his coal field. During the Mingo County war, a miner from Mont Coal declared, "There is only one class of people in West Virginia that is allowed to talk and that is the class that turns its eyes upward and sings 'Pass me not, O golden dollar.' That is the operators."[43]

The explanation for the miners' drive for unionization and the revolt that it prompted, then, is found outside the sphere of strictly economic conditions. Indeed, when the Mingo miners struck and fought for their union, they were receiving more than the union wage rate.[44] The explanation is found in terms of class consciousness. When the miners spoke of their exploitation and oppression, they spoke not of individual operators, not of good and bad operators, not of benevolent and abusive operators, but of *the operators*. "The operators have banked large sums . . ." ; "The coal operators they do the way they please"; only one class of people was allowed to talk and that was "the operators." The operators conspired not in this or that incident alone, but in all essential exploitative and oppressive relationships. As one miner explained, "The operators will have to account for the way they have treated the women and children of the miners." A poem written by a Mingo County miner read:

> *We* who have furnished all *your* golden flood,
> And sealed *your* comfort with *our* heart's red blood;
> Will starve and freeze through winter's snows and rain,
> Before *we'll* ever be *your* slaves again.
>
> *You* seemed to think slaves should not go to bed,
> But toil for *you* until the east was red;
> *You* were well satisfied when *we* served *you* thus
> And gayly traveled on *your* way of thievery and lust.
>
> Why should *we* toil while bright earth greets the skies,
> And let every sneaking thief rob *our* guts who tries?
> *We* learned *our* lesson—and *we* learned *your* lies,
> Before *we* had the brains to O-R-G-A-N-I-Z-E.[45]

The development of class consciousness was a historical process, not the product of monetary circumstances brought about by the postwar recession. A resolution by the District 29 convention in 1921 began with words that emphasize this point: "Whereas the men working in and around the coal mines of West Virginia for the past thirty years, have not been allowed to exercise their state and constitutional rights, and have been forced to work under the most brutal mine guard system, and the most horrible working conditions." Mooney explained that the "struggle now going on in Mingo County . . . is the continuance of the struggle begun in West Virginia twenty-three years ago and extending through that period."[46] And that struggle stemmed from antagonistic class outlooks; the operators had shared attitudes about power, profit, paternalism, and unionism. The miners responded with a thirty-year experience of shared styles of life and work, brotherhood, religion, love, and a common perception of not only what was wrong, but the way things should be.

Intending to Americanize the southern West Virginia coal fields, the miners planned to replace an order that had existed for over a generation. Americanism meant the abolition of company stores, company churches and company ministers, company schools and company teachers, company houses and evictions, yellow-dog contracts, blacklists, scrip, and mine guards.

The miners' resistance stemmed not from the brutalities and discipline demanded by industrialization, but from the exploitation and oppression of the class system that developed in the coal fields. That they proposed to remove the evils of this system by bringing the ideas and ideals of the U.S. Constitution and the Declaration of Independence to the coal fields did not mean that they were looking backward or were intending to return to a pre-industrial bliss of sharecropping and subsistence farming. Motivated by the ideas and ideals that standard radical political parties expressed, the miners demanded and fought for a more ethical and humane industrial-economic order, based on freedom, equality, a just distribution of wealth, and political liberty—the democratic right to rule their own lives.

The means by which they would institute the new order was the UMWA. "We are waiting for the time to come when all places can say what we are saying today, 100 per cent organization," wrote William Vandal of Sanderson, "and then we can all live in peace and harmony."[47] Hubert Kirk of Miami, Cabin Creek, declared: "I hope the time will come when there will not be such a thing as a scab or a guard in the country, and then we will be free people."[48]

This was the promise of Americanism. A miner from Longacre began a letter with a simple introduction: "We hold these truths to be self-evident that all men are created equal; that they are endowed by their Creator with certain inalienable rights, that among these are liberty."[49]

To unionize and Americanize the southern West Virginia coal fields would have certainly changed, indeed revolutionized, the way of life and work in southern West Virginia. "Let us change the system of master and slave," wrote a miner from Poca, "and then the people of the United States will be able to say for once that this is 'The Land of the Free and the Home of the Brave.'"[50]

## NOTES

1. Fred Mooney, *Struggle in the Coal Fields*, ed. Fred Hess (Morgantown, W.Va., 1967), 164–65.

2. American Civil Liberties Union to President Harding, Oct. 19, 1921, Straight Numerical File, item 205194–50–281, Record Group 60, General Records of the Department of Justice, National Archives, Washington, D.C.; U.S. Congress, Senate Committee on Education and Labor, *West Virginia Coal Fields: Hearings to Investigate the Recent Acts of Violence in the Coal Fields of West Virginia*, 67th Cong., 1st sess., 2 vols. (Washington, D.C., 1921–22), 1:557–58 (hereafter cited as *West Virginia Coal Fields*). For favorable editorials, see *New York World*, Aug. 27, 1921, and *Nation* 114 (Sept. 14, 1921): 1.

3. For letters and petitions of associations and local unions throughout the country that sympathized or in other ways supported the marchers, see Straight

Numerical File, items 205194-50-246 to 205194-50-86, Record Group 60, General Records of the Department of Justice. Samuel Gompers, "Peace Reigns in (Mingo) Warsaw," *American Federationist* 28 (Oct. 1921): 852-54, and "AF of L Executive Council Denounces Feudal Action in West Virginia," *UMWJ*, Oct. 15, 1922, 7.

4. Lewis to Harding, Aug. 26 and Sept. 14, 1921, District 17 Correspondence Files, UMWA Archives, UMWA Headquarters, Washington, D.C. Lewis made similar statements publicly. See *New York Times*, Aug. 31, 1921; *UMWJ*, Oct. 15, 1921, 7; Mooney, *Struggle*, ed. Hess, 102-3; Lewis to Samuel Montgomery, Sept. 21, 1921, Lewis to Keeney, Sept. 21 and Oct. 1, 1921, Keeney to Lewis, Sept. 22, 1921, also Lewis to Keeney, Sept. 19 and 22, 1921, Lewis to Harding, Sept. 14, Oct. 4, Oct. 7, and Oct. 11, 1921, District 17 Correspondence Files. A myth persists that Lewis was angry with Keeney for "trying to shoot the union into West Virginia" and eventually dismissed him as District 17 president because of it. See Howard Lee, *Bloodletting in Appalachia* (Parsons, W.Va., 1969), 115. Other writers who stressed this point of view include Dale Fetherling, *Mother Jones, the Miners' Angel* (Carbondale, Ill., 1974), 190, and Cabell Phillips, "The West Virginia Mine War," *American Heritage* 25 (Aug. 1974): 93. There is, I believe, no evidence to support this myth. In addition to arranging and financing the defense and obtaining protection for Keeney and Mooney, Lewis gave Keeney great moral support following the march. During the UMWA convention Lewis telegrammed Keeney that "every delegate is in deep sympathy with yourself and Secretary Mooney and associates of our organization who are being victimized by the mine guard government of West Virginia. We trust that your courage will be maintained for a continuance of the fight for recognition of the principles of freedom and democracy." Keeney replied that "it is gratifying to note the position taken by yourself in the name of the mine workers of the United States in our behalf." Lewis to Keeney, Sept. 21, 1921, and Keeney to Lewis, Sept. 22, 1921, District 17 Correspondence Files. Lewis's dismissal of Keeney came in 1924, when Keeney and Mooney both resigned over differences with Lewis on how union contracts should be negotiated. In a forthcoming book I explore the post-1922 period in the history of the southern West Virginia miners.

5. *UMWJ*, Feb. 1, 1922, 21, Apr. 15, 1924, 5, July 1, 1924, 7; Charles Ambler and Festus Summers, *West Virginia: The Mountain State* (Englewood Cliffs, N.J., 1958), 457-59. There is confusion over how many miners were indicted. Lee (*Bloodletting in Appalachia*, 104) claimed there were 525; Ambler and Summers (*West Virginia*, 458-59) reported 543.

6. *Nation* 114 (May 24, 1922): 612; *Cumberland Daily News* is quoted in *UMWJ*, May 15, 1922, 6; *New York Times*, May 3, 1922.

7. *New York Call*, n.d., clipping found in District 17 Correspondence Files; testimony of Walter Thurmond, *West Virginia Coal Fields*, 1:539; *New York Times*, Oct. 25, 1922; *UMWJ*, Nov. 15, 1922, 10.

8. *Nation* 114 (May 3, 1922): 509, and *New York Call*, n.d., clipping found in District 17 Correspondence Files.

9. See chap. 5.

10. *Williamson* (W.Va.) *Daily News*, excerpted in *West Virginia Coal Fields*, 2:748. The revolutionary implications of the armed march were also diminished, no doubt, since Keeney, Blizzard, and the other leaders of the march were acquitted of treason in the state trials. While several of the marchers were convicted of murder, only one, Walter Allen, was found guilty of treason. Allen was convicted

for having led "hundreds" of miners from Dry Branch, in Raleigh County, to join the march. He was sentenced to ten years in prison but "jumped his bail" while the case was pending in the state supreme court of appeals. He was never apprehended. James Randall, "Miners and the Law of Treason," *North American Review* 216 (Sept. 1922), 312–33.

11. *Illinois State Journal*, Sept. 3, 1921, found in District 17 Correspondence Files.

12. Attorney Herron memo to Assistant Attorney General John Crum, Nov. 16, 1921, Straight Numerical File, item 205194-50-194, Record Group 60, General Records of the Department of Justice.

13. Wainwright to Daugherty, Feb. 28, 1922, Straight Numerical File, item 205194-50-298, Record Group 60, General Records of the Department of Justice.

14. Crum memo to Daugherty. Nov. 22, 1921, Straight Numerical File, item 205194-50-295, Record Group 60, General Records of the Department of Justice.

15. Daugherty to Howard Sutherland, Oct. 3, 1921, Straight Numerical File, item 205194-50-263, Record Group 60, General Records of the Department of Justice.

16. Crum to L. H. Kelly, Sept. 26, 1921, Straight Numerical File, item 205194-50-236, Record Group 60, General Records of the Department of Justice. For comments of other officials of the Department of Justice who maintained that the armed march constituted an insurrection against the United States government and that the federal government should prosecute, see Kelly to Daugherty, Sept. 5, 1921, H. S. Ridgely to Crum, Sept. 19, 1921, W. A. Bethel to Judge Advocate General, Nov. 15, 1921, Straight Numerical File, item 205194-50-236, Record Group 60, General Records of the Department of Justice.

17. Calvin Weakley to Director of Bureau of Investigation, Oct. 7, 1921, Hoover to Daugherty, Oct. 8, 1921, Daugherty to Harding, Oct. 22, 1921, Straight Numerical File, item 205194-50-274, Record Group 60, General Records of the Department of Justice. The federal government may have decided not to prosecute because of a lengthy report made by secret agents detailing the political corruption in southern West Virginia. See Elliot Northcutt to Daugherty, Dec. 8, 1922, in the same file.

18. Crum memo to Daugherty, Mar. 2, 1922, and Daugherty memo to Crum, Mar. 3, 1922, Straight Numerical File, item 205194-50-298, Record Group 60, General Records of the Department of Justice.

19. Mooney, *Struggle*, ed. Hess, 48, and Winthrop Lane, *Civil War in West Virginia* (New York, 1921), 88.

20. *UMWJ*, Apr. 1, 1919, 13. See also Brant Scott to Lewis, Nov. 4, 1922, and James McCleary, "Report of the Revolving Charter of Local Union Located at Pursglove," Oct. 28, 1922, District 17 Files. For opposition of local unions to national radical organizations, see *Charleston Gazette*, Nov. 4, 1919. For a similar position stated by Mooney, see *UMWJ*, Dec. 1, 1923, 15. Fred Barkey ("The Socialist Party in West Virginia from 1898 to 1920" [Ph.D. diss., University of Pittsburgh, 1971], 216–18) argued that fear of the state's 1919 "Red Flag" and Constabulary laws led to these public repudiations of radicalism. (The "Red Flag" law made it illegal to "speak, print, publish, or communicate . . . ideals, institutions, or forms of government hostile, inimical or antagonistic to . . . the consti-

tution and laws of this state or the United States." Evelyn Harris and Frank Krebs, *From Humble Beginnings: West Vriginia State Federation of Labor, 1903–1957* [Charleston, W.Va., 1960], 131–32. The Constabulary law created the West Virginia state police.) Barkey's argument may be true for the state's Socialists and northern labor leaders, but not for District 17 officials nor the southern West Virginia miners. During this period the miners and their officials were already defying presidential proclamations and were engaging in open warfare with the coal establishment and governmental authorities (see chap. 8). It seems unlikely that they would be intimidated by a government that they held in contempt and that they were fighting. Indeed, the passage of the laws only served to increase the miners' contempt for the state government. *Charleston Gazette*, Apr. 6, 1923, and *West Virginia Federationist*, May 10, 1923, 10.

21. See chaps. 3 and 5.

22. Ralph Chaplin, *Wobbly, the Rough and Tumbling Life of an American Radical* (Chicago, 1948), 128.

23. Report of Harold Nathan to Director William J. Burns, Sept. 5, 1921, Bureau Section File, 205194–50, Investigative File, Bureau of Investigation, Record Group 65, Records of the Federal Bureau of Investigation (FBI), National Archives.

24. Interview with Art Shields, Garrett Park, Md., Nov. 20, 1977. Frank Little, an organizer and member of the general executive board of the IWW, was lynched from a railroad trestle during a copper strike in Butte, Mont., in August 1917. Selig Perlman and Phillip Taft, *History of Labor in the United States*, 4 (New York, 1966), 401–2. Gabriel Kolko and William Preston made a relevant point to the miners' belief in violence, rather than rhetoric or voting, as a means of political expression: "Radicalism had to move 'beyond politics,' not because it had no political ideas but because . . . it had no political means." William Preston, "Shall This Be All? U.S. Historians versus William D. Haywood et al.," *Labor History* 11 (Summer 1971): 452.

25. Karl Marx, *The Eighteenth Brumaire of Louis Bonaparte*, in *Surveys from Exile*, ed. David Fernback, Political Writings II (New York, 1974), 148. Trotsky is quoted in Jeremy Brecher, *Strike!* (San Francisco, 1972), 249.

26. Mike O'Leary et al., letter to the editor, *UMWJ*, July 15, 1922, 9.

27. Petry, "A Statement on the Basic Causes of the Industrial Strife in the Coal Fields of West Virginia," District 17 Correspondence Files.

28. Mooney, *Struggle*, ed. Hess, 59–60.

29. H. Allen to Wilson, Dec. 20, 1919, Straight Numerical File, item 205194–50–86, Record Group 60, General Records of the Department of Justice.

30. Stanley Lipscomb, petition to Harding, June 22, 1921, Secretary's Office, item 333.9, Record Group 159, Records of the Office of the Inspector General, National Archives.

31. Wickham Local, letter to the editor, *UMWJ*, Sept. 15, 1919, 25, and J. B. Wiggens, letter to the editor, ibid., June 15, 1921, 9.

32. John Hutchinson, letter to the editor, ibid., July 15, 1921, 11, and Winifred Chappell, "Embattled Miners," *Christian Century* 48 (Aug. 19, 1931): 1044.

33. Herber Blankenhorn, "Marchin' through West Virginia," *Nation* 113 (Sept. 14, 1921): 289. Also see Keeney to Sheriff Pinson (Mingo County), n.d., printed in *West Virginia Coal Fields*, 1:110–11.

34. For writers who emphasized that the "gun totin'" traditions of the hill-

billies provoked the violence, see Daniel Jordan, "The Mingo War," in *Essays in Southern Labor History*, ed. Gary Fink and Merl Reed (Westport, Conn., 1977), 119–20, and Phillips, "West Virginia Mine War," 58–59. The impeachment of Scott is covered in Mooney, *Struggle*, ed. Hess, 54–55. This also seems to be the view of Melvyn Dubofsky and Warren Van Tine in *John L. Lewis* (New York, 1977), 77.

35. Keeney is quoted in Bituminous Operators' Special Committee, *The United Mine Workers in West Virginia*, Statement Submitted to U.S. Coal Commission, August 1923 (n.p., n.d.), 66; Lee, *Bloodletting in Appalachia*, 96; Ramage Resolution, printed in *West Virginia Coal Fields*, 1:190.

36. Winthrop Lane, "West Virginia: The Civil War in Its Coal Fields," *Survey* 54 (Oct. 29, 1921): 179; Phillips, "West Virginia Mine War," 58; Jordan, "The Mingo War," in *Essays*, ed. Fink and Reed, 113.

37. E. H. Burgett, "Reveries of a Tug River Miner," cited in West Virginia State Federation of Labor, *Proceedings, 1921*, 153–54.

38. D. Robb, letter to the editor, *UMWJ*, Feb. 1, 1921, 4–5.

39. Andrew Summer, letter to the editor, ibid., Aug. 15, 1921, 17, and testimony of Frank Keeney, *West Virginia Coal Fields*, 1:180–82.

40. E. P. Thompson, *The Making of the English Working Class* (New York, 1963), 203.

41. *Charleston Gazette*, Feb. 11, 1925; Andrew Summer, letter to the editor, *UMWJ*, Aug. 15, 1921, 17; William Catlett et al. to Lewis, Nov. 6, 1919, District 17 Correspondence Files.

42. Mooney, *Struggle*, ed. Hess, 23.

43. Sam Greco to unnamed "Sirs," June 4, 1920, Straight Numerical File, item 205194-50-191, Record Group 60, General Records of the Department of Justice, and Sanford Alexander, letter to the editor, *UMWJ*, Aug. 15, 1921, 17.

44. Roy Hinds, "The Last Stand of the Open Shop," *Coal Age* 18 (Nov. 18, 1920): 1038; *West Virginia Coal Fields*, 1:159; John Owens, "Gunmen in West Virginia," *New Republic* 28 (Sept. 21, 1921): 90–91. Keeney also acknowledged that wages were never an issue in the Mingo County strike. *Coal Age* 18 (July 22, 1920): 175.

45. T. H. Seals, letter to the editor, *UMWJ*, Mar. 1, 1921, 11, and Burgett, "Reveries of a Tug River Miner," 153–54. The miners often defined class boundaries in terms of Americans. A Mingo miner who had been kidnapped and questioned by mine guards wrote the *UMWJ*, explaining: "A good American does not like the kind of company I was forced to mix with that night." A meeting of union miners at Williamson praised the pro-union city officials of Matewan as possessing "an unusual true Americanism." Sanford Alexander, letter to the editor, *UMWJ*, Aug. 15, 1921, 17, Aug. 15, 1920, 7–8. Similarly, see L. M. Denares, letter to the editor, ibid., Sept. 5, 1922, 4.

46. Resolution, Third Biennial Convention of District 29, printed in *UMWJ*, Nov. 1, 1921, 9. Mooney is quoted, ibid., Dec. 1, 1920, 14–15, and *West Virginia Coal Fields*, 1:10. The miners' recognition of their situation as a historical phenomenon, based upon the industrial-capitalistic development of southern West Virginia is significant. It demolishes the idea that the labor violence of these years stemmed from the feuding, "gun-totin'" traditions of the native mountaineers. Furthermore, it reveals the fruition of the miners' sense of class consciousness. Jeremy Brecher writes: "Class consciousness involves more than an individual sense of oppression. It requires the sense that one's being part of an oppressed

group." Jeremy Brecher, *Strike!*, 247.

47. William Vandal, letter to the editor, *UMWJ*, Apr. 15, 1921, 16.

48. Letter to the editor, ibid., Mar. 2, 1913, 2.

49. William Edwards to William Green, June 3, 1916, Alphabetical File B, International Executive Board (IEB) Meeting File, UMWA Archives.

50. B. F. Cochran, letter to the editor, *UMWJ*, Jan. 1, 1921, 10.

# Bibliographical Essay on Sources

The principal difficulty in writing grass-roots sociocultural labor history is source material. This difficulty has led historians to label rank-and-file workers condescendingly as "the inarticulate" and has caused them to concentrate upon the workers' institutions and elite. Selig Perlman, for example, once justified the institutional approach by writing: "The individual workman leaves no historical record, but the labor movement does. That is the closest we can get to the experimental method in dealing with social movements."*

In contrast to this view, rank-and-file workers, at least in southern West Virginia, were articulate. They expressed their beliefs, anger, and aspirations not only in songs and actions, but also in written records, left not in one or two or three depositories, but in many places throughout the country. The difficulty in researching rank-and-file workers is not the absence of written documents, but finding them.

This bibliographical essay cannot list all of the sources and depositories that I examined for this study. Dozens of collections, such as the National Urban League Papers at the Library of Congress, yielded little in quantity, but much in quality. To list all of these collections would make this essay longer than the manuscript. Only the most important and more extensive sources are cited.

My study of the life, work, and protest movement of the southern West Virginia coal diggers stemmed from four main sources. First was my own collection of documents and interviews. I interviewed over 300 elderly coal miners and their families who lived and worked in southern West Virginia during the period studied. Their stories provided insight and information that could never have been gleaned from traditional sources. Many of them also gave me, or let me copy, their personal papers and documents (e.g., pay slips, housing contracts, blacklists, and company notices), which tell of the arrogance and tyranny of the coal companies.

---

*Selig Perlman, "The Basic Philosophy of the American Labor Movement," *Annals of the American Academy of Political and Social Science* 254 (Mar. 1951): 60.

The UMWA Archives, at the UMWA Headquarters, Washington, D.C., provided invaluable information. The archives contained a wealth of material that provided an inside story of the local union experience (both institutional and social) in southern West Virginia.

Two of the most extensive and insightful, but neglected, sources of rank-and-file opinions and values upon which I relied heavily were the letters to the editor in the *United Mine Workers' Journal* and the testimonies of rank-and-file miners to the three U.S. Senate committees that came to southern West Virginia to investigate the labor violence:

U.S. Senate, Committee on Education and Labor, *Conditions in the Paint Creek District, West Virginia*, 63rd Cong., 1st sess., 3 vols. (Washington, D.C.: Government Printing Office, 1913)

U.S. Senate, Committee on Education and Labor, *West Virginia Coal Fields: Hearings . . . to Investigate the Recent Acts of Violence in the Coal Fields of West Virginia and Adjacent Territory and the Causes Which Led to the Conditions Which Now Exist in Said Territory,* 67th Cong., 1st sess., 2 vols. (Washington, D.C.: Government Printing Office, 1921–22)

U.S. Senate, Committee on Interstate Commerce, *Conditions in the Coal Fields of Pennsylvania, West Virginia, and Ohio. Hearings . . .,* 70th Cong., 1st sess., 2 vols. (Washington, D.C.: Government Printing Office, 1928).

These letters and testimonies revealed that the miners could articulate quite well their accomplishments, their goals, and their motivations.

Manuscript collections have supplied pertinent information about the southern West Virginia coal miners. The Letterbooks of Samuel Gompers at the Library of Congress and the John Mitchell Papers and the Mother Mary Jones Papers at the Catholic University of America, Washington, D.C., contained significant information about the successes and failures of national organizers in southern West Virginia and their encounters with the miners, the coal operators, and the mine guards. The Katherine Pollak Ellickson Papers at Wayne State University, Detroit, Mich., furnished much excellent material. The Socialist Party Papers at Duke University, Durham, N.C., were disappointing in the lack of material on the Paint Creek–Cabin Creek strike, which occurred when the Socialist party had reached its peak in southern West Virginia. The papers of John J. Cornwell, governor of West Virginia from 1916 to 1920, at the West Virginia Department of Archives and History, Charleston, included letters of rank-and-file miners and resolutions of local unions during World War I that related information about the miners' ideas on state government, the coal operators, and the war.

Other key sources were the reports, records, and archives of U.S. government agencies. The papers of the U.S. Department of Labor (Record Group 174) at the National Archives included letters from rank-and-file miners in southern West Virginia during World War I and the New Deal. Reports of agencies of the Department of Labor that investigated conditions in southern West Virginia were filled with social analyses. Especially informative were:

U.S. Department of Labor, Women's Bureau, *Home Environment and Employment Opportunities of Women in Coal-Mine Workers' Families*, Bulletin no. 45 (Washington, D.C.: Government Printing Office, 1925)

Nettie McGill, *Welfare of Children in the Bituminous Coal Communities in West Virginia*, Bulletin no. 117 (Washington, D.C.: Government Printing Office, 1923).

Material in the records of the Chief o` Ordnance (Record Group 156) in the National Archives documented military operations in southern West Virginia during the Mingo County strike and showed that the federal government, like the miners and the operators, viewed the strike as a war.

Also helpful were the records of the U.S. Coal Commission, 1922 (Record Group 68) at the National Archives. The papers dealt with living conditions and living standards in the coal fields and the life-styles of the miners and their families.

The Committee on Coal and Civil Liberties, "A Report to the U.S. Coal Commission," August 11, 1923, U.S. Department of Labor Library, Washington, D.C., Winthrop Lane, *The Denial of Civil Liberties in the Coal Fields* (New York: George H. Doran, 1924), and the U.S. Coal Commission, *Report of the United States Coal Commission*, Senate Document 195, 68th Cong., 2nd sess., 5 vols. (Washington, D.C.: Government Printing Office, 1925) were all important sources.

The records of the Department of Justice (Record Group 60) and the Federal Bureau of Investigation (Record Group 65) were invaluable in documenting the industrial warfare in West Virginia between 1919 and 1921. On the other hand, the manuscript census at the National Archives was of little value, since company towns were not enumerated.

Rich in the social history of coal mining in southern West Virginia was Fred Mooney's autobiography, *Struggle in the Coal Fields*, ed. J. W. Hess (Morgantown: West Virginia University Library, 1967). Mooney worked as a miner on Paint Creek and served as secretary-treasurer of District 17 under Frank Keeney. Mooney was exceptional as a miner; he was well read and self-educated. After an interview with him, Winthrop Lane wrote that "under other circumstances Mooney would have been a student." The autobiography was a rich, personal story, and Mooney expressed strongly his views about all aspects of life (religion, family, the union, the coal operators) in southern West Virginia. Of somewhat lesser importance, but helpful, were: Mary "Mother" Jones, *Autobiography of Mother Jones*, ed. Mary Field Parton (Chicago: Charles H. Kerr and Co., 1925), and Ralph Chaplin, *Wobbly, the Rough and Tumbling Life of an American Radical* (Chicago: University of Chicago Press, 1948). Chaplin was working on a local Socialist newspaper during the Paint Creek–Cabin Creek strike. His writings and insights were beneficial in interpreting this labor-management conflict.

The viewpoints, policies, and actions of the southern West Virginia coal operators have been somewhat easier to ascertain, as they were prolific in praising and defending their efforts and exploits. They wrote articles, especially in *Coal Age* and *West Virginia Review*, and published numerous pamphlets telling of their accomplishments, describing life in the company towns, and giving their views on the UMWA. Two of the operators wrote autobiographies: W. P. Tams, *The Smokeless Coal Fields of West Virginia* (Morgantown: West Virginia University Library, 1963), and Walter Thurmond, *The Logan Coal Field of West Virginia* (Morgantown: West Virginia University Library, 1964). Nearly all of the major coal operators in the area testified before the U.S. Senate committees previously mentioned. Also the coal operators less-than-patriotic conduct during World War I is revealed in the Cornwell Papers, the Secretary of Navy files (Record Group 80) at the National Archives, and *Coal Age*. The most extensive and insightful information on the everyday thinking and actions of the operators and their officials came from the Justin Collins Papers at the West Virginia University Library, Morgantown.

The annual reports of various agencies and departments of the West Virginia state government have supplied pertinent information. Of particular benefit have been the reports of the West Virginia Bureau of Negro Welfare and Statistics. This source was used cautiously because of the pro-operator biases of its supervisor, Edward Hill, who was also editor of the admittedly anti-union black newspaper, the *McDowell Times*. Nevertheless, the reports contained valuable information on the black migration into the coal fields and the conditions and behavior of the black miners after settlement. The reports of the West Virginia State Supervisor of Free Schools revealed the contributions that the coal companies made to educational development in southern West Virginia, the social change prompted by industrialization, and the operators' attempts to use education as a form of social control. The annual reports of the West Virginia State Department of Mines were disappointing in their lack of information on strikes and the nature of company towns (probably because most of the directors during this period were coal operators), but they provided useful statistics concerning the rise of the coal industry and the racial and ethnic composition of the state's mining work force.

Journals and periodicals, especially the *United Mine Workers' Journal* and *Coal Age*, have furnished significant factual and descriptive information. Pregnant with social history were the liberal, muckraking periodicals during the period (e.g., *Nation, New Republic, Survey*). Intent on exposing the corruption and brutality of the southern West Virginia coal operators, they constantly carried articles about events and conditions in the area; in doing so, they provided excellent depictions of life in the company towns. While I disagree with their explanations for the violence, the articles were useful for their stories, facts, and interviews with southern West Virginia miners. Similarly, the *American Federationist* was quite helpful. Although it had relatively little material on the nonunion, southern West Virginia coal fields, the journal carried several lengthy, caustic, but factual, articles (usually written by Samuel Gompers) blasting the state government, the courts, the coal operators, and the mine guards in the state. Particular journals have been of great assistance for specific events in the area. For instance, the *International Socialist Review* provided thorough coverage of the Paint Creek–Cabin Creek strike from the rank-and-file miner's viewpoint.

Finally, I used local newspapers heavily throughout the study. The local labor and socialist presses, the Huntington *Socialist and Labor Star, Wheeling Majority*, and *Charleston Labor Argus*, have been most valuable. Also useful were the local commercial presses, the *Charleston Gazette*, the *Charleston Daily Mail, Huntington Herald–Dispatch*, and *Logan Banner*. The black newspaper, the *McDowell Times*, although admittedly pro-operator, provided insight into the life and work of the black miners in southern West Virginia.

Three secondary works deserve comment. Anyone familiar with E. P. Thompson's *The Making of the English Working Class* (New York: Random House, 1963) and Eugene Genovese, *Roll, Jordan, Roll: The World the Slaves Made* (New York: Random House, 1974) will immediately recognize my dependence upon these books for ideas, structure, and guidance. I have a few strong disagreements with Fred Barkey's "The Socialist Party in West Virginia from 1898–1928" (Ph.D. dissertation, University of Pittsburgh, 1971); however, these points of difference do not diminish my respect for this study nor its author; both were of great help.

# Selected Bibliography

## PRIMARY SOURCES

### *Manuscript Collections*

Author's personal collection of taped interviews, payroll books, company notices, and personal correspondence. College Park, Md.
Justin Collins Papers. West Virginia University Library, Morgantown.
John J. Cornwell Papers. West Virginia Department of Archives and History, Charleston.
Katherine Pollack Ellickson Papers. Wayne State University, Detroit, Mich.
A. B. Fleming Papers. West Virginia University Library, Morgantown.
Letterbooks of Samuel Gompers. Library of Congress, Washington, D.C.
Mother Jones Papers. Catholic University of America, Washington, D.C.
John Mitchell Papers. Catholic University of America, Washington, D.C.
National Archives, Washington, D.C.:
Adjutant General, Office of the (1780s–1917). Record Group 94.
Adjutant General, Office of the (1917– ). Record Group 407.
Chief of Ordnance. Record Group 156.
Children's Bureau. Record Group 102.
Federal Bureau of Investigation. Record Group 65.
Inspector General, Office of the. Record Group 159.
Justice, Department of. Record Group 60.
Labor, Department of. Record Group 174.
Mines, Bureau of. Record Group 70.
National Defense, Council of. Record Group 62.
Navy, Department of the. Record Group 80.
U.S. Coal Commission. Record Group 68.
National Urban League Papers. Library of Congress, Washington, D.C.
Socialist Party Papers. Duke University, Durham, N.C.
United Mine Workers' Archives. UMWA Headquarters, Washington, D.C.
*United Mine Workers' Journal.* Letters to the editor from miners in southern West Virginia, 1895–1940.

### *U.S. Government Documents*

Committee on Coal and Civil Liberties. "A Report to the U.S. Coal Commission," Aug. 11, 1923. U.S. Department of Labor Library, Washington, D.C.

U.S. Bituminous Coal Commission. *Majority and Minority Reports, 1920.* Washington, D.C.: Government Printing Office, 1920.

U.S. Coal Commission. *Report of the United States Coal Commission.* Senate Document 195, 68th Cong., 2d sess., 5 vols. Washington, D.C.: Government Printing Office, 1925.

U.S. Senate, Committee on Education and Labor. *Conditions in the Paint Creek District, West Virginia.* 63rd Cong., 1st sess., 3 vols. Washington, D.C.: Government Printing Office, 1913.

______. *Digest of Report on Investigation of Paint Creek Coal Fields of West Virginia.* 63rd Cong., 2d sess. Washington, D.C.: Government Printing Office, 1914.

______. *West Virginia Coal Fields: Hearings . . . to Investigate the Recent Acts of Violence. . . .*, 67th Cong., 1st sess., 2 vols. Washington, D.C.: Government Printing Office, 1921–22.

U.S. Senate, Committee on Interstate Commerce. *Conditions in the Coal Fields of Pennsylvania, West Virginia, and Ohio. Hearings . . . .*, 2 vols., 70th Cong., 1st sess. Washington, D.C.: Government Printing Office, 1928.

U.S. Department of Agriculture. *Economics and Social Problems and Conditions in the Southern Appalachians.* Bulletin no. 205. Washington, D.C.: Government Printing Office, 1935.

U.S. Department of Labor, Bureau of Labor Statistics. *Hours and Earnings in Anthracite and Bituminous Coal Mining.* Bulletin no. 316. Washington, D.C.: Government Printing Office, 1922.

______. *Housing by Employers.* Bulletin no. 263. Washington, D.C.: Government Printing Office, 1920.

______. *Labor Relations in the Fairmont, West Virginia Bituminous Coal Field*, by Boris Emmet. Bulletin no. 361. Washington, D.C.: Government Printing Office, 1924.

______. *Union Movement among Coal Mine Workers*, by Frank Warne. Bulletin no. 51. Washington, D.C.: Government Printing Office, 1904.

______. *Wages and Hours of Labor in Bituminous-Coal Mining: 1933.* Bulletin no. 601. Washington, D.C.: Government Printing Office, 1934.

U.S. Department of Labor, Children's Bureau. *Welfare of Children in the Bituminous Coal Communities in West Virginia*, by Nettie McGill. Bulletin no. 117. Washington, D.C.: Government Printing Office, 1923.

U.S. Department of Labor, Women's Bureau. *Home Environment and Employment Opportunities of Women in Coal-Mine Workers' Families.* Bulletin no. 45. Washington, D.C.: Government Printing Office, 1925.

U.S. Immigration Commission. *Immigrants in Industries.* 2 vols. Washington, D.C.: Government Printing Office, 1911.

### *West Virginia State Documents*

Barnes, J. Walter. "Federal Fuel Administration." In *West Virginia Legislative Handbook and Manual, 1918*, edited by John Harris. Charleston: Tribune Printing Co., 1918.

*The (Mining) Code of West Virginia, 1906*. St. Paul, Minn.: West Virginia Publishing Co., 1906.

West Virginia. *Legislative Handbook and Manual* (1916–30). Charleston: Tribune Publishing Co.

West Virginia Bureau of Negro Welfare and Statistics. *Negro Housing Survey of Charleston, Keystone, Kimball, Wheeling, and Williamson*. Charleston: Jarrett Printing Co., 1938.

______. *Annual Reports* (1914–30).

West Virginia House of Delegates. *Journal* (1897).

West Virginia Mining Investigation Commission. *Report*. Charleston: Tribune Printing Co., 1912.

West Virginia State Board of Agriculture. *Biennial Reports* (1899–1900).

West Virginia State Department of Mines. *Annual Reports* (1900–40).

West Virginia State Legislature. *Acts* (1899).

West Virginia Supervisor of Free Schools. *Biennial and Annual Reports* (1890–1940).

*Books*

Brophy, John. *A Miner's Life*. Madison: University of Wisconsin Press, 1964.

Cartlidge, Oscar. *Fifty Years of Coal Mining*. Charleston, W.Va.: Rose City Press, 1936.

Chaplin, Ralph. *Wobbly, the Rough and Tumbling Life of an American Radical*. Chicago: University of Chicago Press, 1948.

Conway, Alan, ed. *The Welsh in America: Letters from Immigrants*. Minneapolis: University of Minnesota Press, 1961.

Daniels, Josephus. *The Wilson Era: Years of War and After, 1917–1923*. Chapel Hill: University of North Carolina Press, 1946.

DeCaux, Len. *Labor Radical: From Wobblies to CIO*. Boston: Beacon Press, 1970.

Gompers, Samuel. *Seventy Years of Life and Labor: An Autobiography*. 2 vols. New York: E. P. Dutton, 1925.

Jones, Mary "Mother." *The Autobiography of Mother Jones*. Edited by Mary Parton. Chicago: Charles H. Kerr and Co., 1972.

MacCorkle, William A. *The Recollection of Fifty Years of West Virginia*. New York: G. P. Putnam's Sons, 1928.

Mooney, Fred. *Struggle in the Coal Fields: The Autobiography of Fred Mooney*. Edited by J. W. Hess. Morgantown: West Virginia University Library, 1967.

Samson, Leon. *Toward a United Front: A Philosophy for American Workers*. New York: Farrar & Rinehart, 1933.

Tams, W. P. *The Smokeless Coal Fields of West Virginia*. Morgantown: West Virginia University Library, 1963.

Thurmond, Walter. *The Logan Coal Fields of West Virginia*. Morgantown: West Virginia University Library, 1964.

*Periodicals*

"Activities in a Modern Mining Town." *West Virginia Review*, June 1925, 329.
Anderson, L. C. "Mine Labor Conditions in West Virginia." *Outlook* 82 (1906): 861-62.
"Appeal for the Miner." *American Federationist* 4 (Aug. 1897): 119.
Blankenhorn, Heber. "Marchin' through West Virginia." *Nation* 113 (Sept. 14, 1921): 288-89.
"Boycott of a State." *Engineering and Mining Journal*, Oct. 12, 1907, 698.
Brown, Phil. "Negro Labor Moves Northward." *Opportunity* 1 (May 1923): 5-6.
Cabell, C. A. "Building a Model Mining Community." *West Virginia Review*, Apr. 1927, 210.
"The Cabin Creek YMCA, Decota." *Coal Age* 4 (Nov. 15, 1913): 741.
Chaplin, Ralph. "Violence in West Virginia." *International Socialist Review* 13 (Apr. 1913): 731-32.
Chappell, Winifred. "Embattled Miners." *Christian Century* 48 (Aug. 19, 1931): 1043.
Chilton, Isabella. "Self-Improvement for a Company Town." *Survey* 41 (Mar. 15, 1919): 873.
Coleman, McAllister. "A Week in West Virginia." *Survey* 53 (Feb. 1, 1925): 532-34.
Collins, Justin. "My Experiences in the Smokeless Coal Fields of West Virginia." *West Virginia Review*, June 1926, 354-55.
Collis, Edgar. "The Coal Miner: His Health, Diseases and General Welfare." *Journal of Industrial Hygiene* 7 (May 1925): 235-36.
"Conference with Governor Atkinson." *American Federationist* 4 (Aug. 1897): 122.
Conley, Phil. "An Enterprise That Costs the State $30,000 Annually." *West Virginia Review*, Sept. 1929, 450-51.
______. "McDowell County." *West Virginia Review*, June 1924, 18.
"The Constitution in a Labor War." *Literary Digest*, Apr. 5, 1913, 756-58.
Coxe, Edward. "The Competition of West Virginia Coal with Ohio Coal." *Engineering and Mining Journal*, Oct. 8, 1898, 424.
Davis, Jerome. "Human Rights and Coal." *Journal of Social Forces* 3 (Nov. 1924): 102-6.
Davis, P. O. "Negro Exodus and Southern Agriculture." *American Review of Reviews* 68 (Oct. 1923): 401-7.
DuBois, W. E. B. "Does the Negro Need Separate Schools?" *Journal of Negro History* 10 (July 1925): 328-35.
"Environment and Opportunities for Women of Coal Miners' Families." *Monthly Labor Review* 21 (1925): 333-34.
Evans, Chris. "The Miners' Strike." *American Federationist* 4 (Dec. 1897): 241.
"Fireco—A New Mining Town in West Virginia." *Coal Age* 12 (Sept. 12, 1918): 1217.
Fowler, George. "Social and Industrial Conditions in the Pocahontas Coal Fields." *Mining Magazine* 27 (June 1904): 383.
"Free Speech in Logan." *Survey* 50 (Apr. 15, 1923): 79-81.

"Garyism in West Virginia." *New Republic* 28 (Sept. 21, 1921): 86–88.
Gist, F. W. "Migratory Habits of the Negro under Past and Present Conditions." *Manufacturers Record*, Mar. 13, 1924, 77–79.
Gleason, Arthur. "Company-Owned Americans." *Nation* 110 (June 12, 1920): 794–95.
______. "Private Ownership of Public Officials." *Nation* 110 (May 29, 1920): 724–25.
Gompers, Samuel. "Russianized West Virginia." *American Federationist* 20 (Oct. 1913): 825–35.
Hall, Dawson. "Have Mining Engineers Accepted All the Developments in Machinery for Handling Coal Imply?" *Coal Age* 12 (July 7, 1921): 13.
Hamill, R. H. "Design of Buildings in Mining Towns." Address to West Virginia Mining Institute. *Coal Age* 11 (June 16, 1917): 1045–48.
Haring, H. A. "Three Classes of Labor to Avoid." *Industrial Management*, Dec. 1921, 370.
Higgins, S. C. "The New River Coal Fields." *West Virginia Review*, Oct. 1927, 26–28.
"High Prices Not So High for the Miners." *Literary Digest*, Dec. 20, 1919, 69–74.
Hinds, Roy. "The Last Stand of the Open Shop." *Coal Age* 18 (Nov. 18, 1920): 1037–40.
Hutchinson, Helen, "What the YMCA Is Doing in West Virginia." *UMWJ*, Feb. 1, 1921, 16.
"Industrial Relations and Labor Conditions: Economic Condition of the Negro in West Virginia." *Monthly Labor Review* 18 (1922): 713–14.
Jones, Mary "Mother." "A Picture of Freedom in West Virginia." *International Socialist Review* 3 (Sept. 1902): 177–79.
Jordan, Margaret. "A Plea for the West Virginia Miner." *Coal Age* 6 (Dec. 5, 1914): 914–15.
Keeley, Josiah. "After the War." *Coal Age* 13 (Oct. 10, 1918): 868–69.
______. "The Cabin Creek Consolidated Coal Company." *West Virginia Review*, June 1926, 348.
______. "Successful Wives in Coal Camps." *Coal Age* 11 (Apr. 7, 1917): 591–92.
Kennedy, D. C. "Kanawha Coal Field." *West Virginia Review*, June 1925, 334–35.
Kirchway, Fred. "Miners' Wives in the Coal Strikes." *Century* 105 (Nov. 1922): 83–90.
Kirchway, Fred. "Mountaineers Shall Always Be Free." *Nation* 124 (July 12, 1922): 38.
Kneeland, Frank. "The Moving Picture in Coal Mining." *Coal Age* 5 (June 27, 1914): 1037.
______. "Patriotic Demonstration at Gary, W.V." *Coal Age* 11 (May 12, 1917): 826–27.
Lane, Winthrop. "Black Avalanche." *Survey* 47 (Mar. 25, 1922): 1044.
______. "The Conflict on the Tug." *Survey* 46 (June 18, 1921): 398–99.
______. "Labor Spy in West Virginia." *Survey* 46 (Oct. 22, 1921): 110–12.

______. "Senators Tour West Virginia." *Survey* 46 (Oct. 1, 1921): 23–24.

______. "West Virginia: The Civil War in its Coal Fields." *Survey* 47 (Oct. 29, 1921): 177–84.

"McDowell County, Coal Mining Being Almost Sole Industry, Provides Free Dentistry." *Coal Age* 18 (Oct. 7, 1920): 743–44.

Marcy, Leslie. "Hatfield's Challenge to the Socialist Party." *International Socialist Review* 13 (June 1913): 882.

Marcy, Mary. "Unions Repudiate Debs' Escort Haggerty." *International Socialist Review* 14 (July 1913): 22–23.

Mead, C. H. "The Winding Gulf Coal Fields." *West Virginia Review*, Apr. 1927, 211–12.

"Moonshine's Lively Part in Mingo Troubles." *Literary Digest*, Sept. 17, 1921, 37–39.

Myers, James. "Close-ups in the Coal Fields." *New Republic* 68 (Sept. 16, 1931): 118–20.

Norton, Helen. "Feudalism in West Virginia." *Nation* 133 (Aug. 12, 1931): 154–55.

Older, Cora. "Last Day of the Paint Creek Court Martial." *Independent* 74 (May 15, 1913): 1085–88.

Owens, John. "Gunmen in West Virginia." *Nation* 111 (Sept. 21, 1921): 90–92.

Parsons, Floyd. "Coal Mining in Southern West Virginia." *Engineering and Mining Journal*, Nov. 9, 1907, 881–84.

______. "Mining Coal on the Virginian Railroad." *Coal Age* 1 (May 18, 1912): 1039–43.

"Payday Drinking." *Coal Age* 4 (Oct. 4, 1913): 478–79.

Price, William. "Steel Corporation Mines at Gary." *Colliery Engineer* 34 (Mar. 1914): 464–72.

"Prohibition and Labor in West Virginia." *Coal Age* 13 (July 25, 1918): 181.

"Prohibition in West Virginia." *Coal Age* 5 (Mar. 7, 1914): 416.

"Religious Work in Mining Towns." *West Virginia Review*, Feb. 1925, inside cover.

Reynolds, Sam, and W. H. Reynolds. "Human Element in the Coal Industry." *Coal Age* 4 (July 12, 1913): 1.

"The Rise of a Call Boy." *West Virginia Review*, Mar. 1929, 206–7.

Rodgers, Jack. "I Remember that Mining Town." *West Virginia Review*, Apr. 1938, 203–5.

Rosenhelm, W. S. "Billion Dollar Coal Field." *West Virginia Review*, Oct. 1924, 20–21.

Shaw, Ira. "Welfare Work among Miners." Paper presented to the West Virginia Coal Mining Institute. *Coal Age* 1 (July 5, 1913): 21.

Shepherd, William. "Big Black Spot." *Colliers*, Sept. 19, 1931, 12–13.

Shields, Art. "The Battle of Logan County." *Political Affairs* 50 (Sept. 1971): 27–38.

"Shortage of West Virginia Coal." *Coal Age* 13 (Aug. 8, 1918): 274.

"Sifting West Virginia's Wrongs." *Literary Digest*, June 7, 1913, 1259–60.

"Sociological Conditions in West Virginia." *Coal Age* 2 (Nov. 23, 1912): 733.

Speranze, Gino. "Forced Labor in West Virginia." *Outlook* 74 (June 13, 1903): 407–10.

"Story of West Virginia's Famous Smokeless Coal Fields." *West Virginia Review*, June 1926, 290–99.

Stroup, Thomas. "Cause and Growth of Unionism among Coal Miners." *Mining and Metallurgy*, Sept. 1928, 468.

Sullivan, Jesse. "West Virginia's Greatest Industry." *West Virginia Review*, Apr. 1927, 233–34.

"Theater for Miners." *Dramatic Mirror* 53 (Aug. 7, 1919): 1227.

"This Week." *New Republic* 28 (Sept. 7, 1921): 1.

Thompson, Wyatt. "How Victory Was Turned into Settlement." *International Socialist Review* 14 (July 1913): 12–17.

______. "Strike Settlement in West Virginia." *International Socialist Review* 14 (Aug. 1913): 87–89.

Thurmond, Walter R. "The Town of Thurmond, 1884–1961." *West Virginia History* 22 (July 1961): 240–54.

Townson, Charles R. "Industrial Program of the Young Men's Christian Association." *Annals of the American Academy of Political and Social Sciences* 103 (Sept. 1922): 134–37.

"Unionism and Coal Production." *Harvard Business Review* 4 (Apr. 1926): 334–40.

"U.S. Senate Decides to Investigate West Virginia." *Current Opinion* 55 (July 1913): 3–4.

Warner, Arthur. "Fighting Unionism with Martial Law." *Nation* 113 (Oct. 12, 1921): 395–96.

______. "West Virginia—Industrialism Gone Mad." *Nation* 113 (Oct. 5, 1921): 372–73.

West, Harold. "Civil War in the West Virginia Coal Mines." *Survey* 30 (Apr. 5, 1913): 37–50.

"West Virginia Bribery Scandal." *Literary Digest*, Mar. 1913, 441–42.

West Virginia Coal Operators' Association. "Community Work in Mining Towns." *West Virginia Review*, June 1924, cover.

______. "Fuel for Miners' Families." *West Virginia Review*, Aug. 1925, 1.

______. "Gardens in Mining Towns." *West Virginia Review*, Oct. 1924, inside cover.

______. "Hard Work and Skill in the Coal Business." *West Virginia Review*, Oct. 1928, 1.

______. "Playgrounds in Mining Communities." *West Virginia Review*, Dec. 1924, 1.

______. "Stores in Mining Towns." *West Virginia Review*, Jan. 1925, 1.

______. "The Water Supply in Mining Towns." *West Virginia Review*, Nov. 1924, 1.

______. "West Virginia Coal and the Nation." *West Virginia Review*, Feb. 1929, 29.

"The West Virginia Strike." *Coal Age* 3 (Apr. 26, 1913): 650.

Wieck, Agnes. "Ku Kluxing in the Miner's Country." *New Republic* 38 (Mar. 26, 1924): 122–24.

Wilson, Isabella. "Welfare Work in a Mining Town." *Journal of Home Economics* 11 (Jan. 1919): 21–23.

Wolfe, George. "Winding Gulf District." *West Virginia Review*, June 1925, 326–27.

"Women and Child Labor." *Monthly Labor Review* 21 (Aug. 1925): 333–34.

### *Bulletins and Proceedings*

Downing, Thomas. "Where to Build Our Mining Towns and What to Build." *West Virginia Coal Mining Institute Proceedings*. Charleston, W.Va., 1923.

Lovejoy, Owen. "Child Labor in the Soft Coal Mines." *Proceedings of the Third Annual Meeting of the National Child Labor Committee*. New York, 1907.

National Coal Association. *Bulletins*. Washington, D.C., 1922–40.

United Mine Workers of American National Convention. *Proceedings*. Indianapolis, Ind., 1902–24.

West Virginia Chief Inspector of Mines. *Bulletins*. Charleston, 1920–30.

West Virginia Mine Workers Union. *Strike Bulletins*, 1931–32.

West Virginia State Federation of Labor. *Proceedings of Annual Conventions*. Charleston, 1906–30.

### *Pamphlets*

Bituminous Operators' Special Committee. *Comparative Efficiency of Labor in the Bituminous Industry under Union and Non-Union Operations*. Report submitted to the U.S. Coal Commission, Sept. 10, 1923. N.p., nd.

______. *The United Mine Workers in West Virginia*. Statement submitted to the U.S. Coal Commission, Aug. 1923. N.p., n.d.

Brown, John W. *Constitutional Government Overthrown in West Virginia*. Wheeling, W.Va.: Majority Co., 1913.

Clopper, E. N. *Child Labor in West Virginia*. National Child Labor Committee, pamphlet no. 86. New York: National Child Labor Committee, 1902.

______. *Child Labor in West Virginia in 1910*. National Child Labor Committee, pamphlet no. 142. New York: National Child Labor Committee, 1910.

Coolidge, William. *Brief in Behalf of Island Creek Coal Company*. Boston: n.p., 1921.

Dawson, J. W. *The Greatest Crisis that Ever Confronted the State of West Virginia*. Address before the West Virginia Board of Trade. N.p., n.d.

Logan County Coal Operators' Association. *The Issue in the Coal Fields of Southern West Virginia*. Statement to President Harding. N.p., n.d.

Logan District Mines Information Bureau. "Company Stores Pocket Mine Workers' Pocket Books." *Coal Facts* no. 7, Dec. 17, 1921.

______. "Why Coal Producers Aid West Virginia Schools." *Coal Facts*, no. 4, Nov. 4, 1921.

______. "Why Coal Production from the Non-Union Fields Constitutes a National Safeguard." *Coal Facts* no. 5, Dec. 30, 1921.

______. "Why Miners Do Not Own Their Own Homes." *Coal Facts* no. 3, n.d.

Lovejoy, Owen. *Child Labor in the Soft Coal Mines.* National Child Labor Committee, pamphlet no. 44. New York: National Child Labor Committee, n.d.

McKinley, J. C. *The Coal Crisis.* Address before the West Virginia Board of Trade. N.p., n.d.

Matthews, William G. *Martial Law in West Virginia.* Address before the West Virginia Bar Association, July 16, 17, 1913. Washington, D.C.: Government Printing Office, 1913.

Murray, Philip. *The Case of the West Virginia Coal Mine Workers.* Report to the Committee on Education and Labor, U.S. Senate. Washington, D.C.: UMWA, 1921.

Non-Union Operators of Southern West Virginia. *Statement to the United States Coal Commission.* Report filed Jan. 12, 1923. N.p., n.d.

Olmsted, Harry. *The Issue in the Coal Fields of Southern West Virginia.* N.p., n.d.

Operators' Association of the Williamson Field. *Statement Made before the Subcommittee on Education and Labor.* Report filed July 14, 1921. N.p., n.d.

______. *Statement to President Harding.* N.p., n.d.

Robinson, Neil. *West Virginia on the Brink of a Labor Struggle.* Charleston: West Virginia Mining Association, 1912.

White, I. C. *The Coal Resources along the Virginian Railway and Its Tributary Regions in West Virginia.* Morgantown, W.Va.: n.p., 1912.

Vinson, Z. T. *Advocating Co-operation, and Organization of West Virginia Coal Operators.* Address before the West Virginia Coal Mining Institute. Huntington, W.Va.: Standard Printing Co., 1913.

Wiley, William. *Facts about the Armed March.* Charleston, W.Va.: Charleston Printing Co., 1921.

### *Newspapers*

WEST VIRGINIA

*Argus Star* (Charleston)
*Beckley Herald-Post*
*Charleston Daily Mail*
*Charleston Gazette*
*Charleston Labor Argus*
*Fairmont Times*
*Fairmont West Virginian*
*Huntington Advertiser*
*Huntington Herald-Dispatch*
*Labor Argus* (Charleston)
*Logan Banner*
*McDowell Times*
*Martins Ferry Evening Journal*

*Montgomery Miners' Herald*
*Pineville Independent Herald*
*Socialist and Labor Star* (Huntington)
*Wheeling Majority*
*Williamson Daily News*

OTHERS

*Appeal to Reason*, Girard, Kans.
*Baltimore Sun*
*Indianapolis Times*
*Life and Labor Bulletin*, Washington, D.C.
*Milwaukee Leader*
*National Labor Tribune,* Pittsburgh, Pa.
*New York Call*
*New York Evening Post*
*New York Journal*
*New York Press*
*New York Times*
*New York World*
*Washington Evening Star*

## SECONDARY SOURCES

### *Books*

Allport, Gordon. *The Nature of Prejudice.* Reading, Mass.: Addison-Wesley Co., 1954.

Ambler, Charles. *History of Education in West Virginia.* Huntington, W.Va.: Standard Printing Co., 1951.

______. *Sectionalism in Virginia.* Chicago: University of Chicago Press, 1910.

Ambler, Charles H., and Festus P. Summers. *West Virginia: The Mountain State.* Englewood Cliffs, N.J.: Prentice-Hall, 1958.

Aronowitz, Stanley. *False Promises: The Shaping of American Working Class Consciousness.* New York: McGraw-Hill, 1973.

Baratz, Morton. *The Union and the Coal Industry.* New Haven, Conn.: Yale University Press, 1954.

Barnum, Darold. *The Negro in the Bituminous Coal Mining Industry.* Philadelphia: University of Pennsylvania Press, 1970.

Berman, Edward. *Labor Disputes and the President of the United States.* New York: Columbia University Press, 1924.

Bernstein, Irving. *The Lean Years: A History of the American Worker, 1920–1933.* Boston: Houghton Mifflin Co., 1960.

Blackwell, James. *The Black Community: Diversity and Unity.* New York: Dodd, Mead & Co., 1975.

Bloch, Louis. *The Coal Miners' Insecurity.* New York: Russell Sage Foundation, 1922.

Bonnett, Clarence. *History of Employers' Associations in the United States.* New York: Vantage Press, 1956.

Brandes, Stuart. *American Welfare Capitalism, 1880–1940.* Chicago: University of Chicago Press, 1976.

Brecher, Jeremy. *Strike!* San Francisco: Straight Arrow Books, 1972.

Brody, David. *Steelworkers in America, the Nonunion Era.* New York: Harper & Row, 1969.

______. "The Rise and Decline of Welfare Capitalism." In *Change and Continuity in Twentieth Century America: The Twenties,* edited by John Braeman, Robert Bremner, and Everett Walters, pp. 221–62. Columbus: Ohio State University Press, 1968.

Brown, Robert McAfee. *The Spirit of American Protestantism.* London: Oxford University Press, 1974.

Brunner, Edmund. *Industrial Village Churches.* New York: Institute of Social and Religious Research, 1930.

Carnoy, Martin. *Education as Cultural Imperialism.* New York: David McKay Co., 1974.

Cole, G. D. H. *Labour in the Coal Mining Industry.* London: Oxford, Clarendon Press, 1923.

Coleman, McAlister. *Men and Coal.* New York: Farrar & Rinehart, 1943.

Conley, Phil. *History of the Coal Industry of West Virginia.* Charleston, W.Va.: Educational Foundation, 1960.

______. *Life in a West Virginia Coal Field.* Charleston, W.Va.: American Constitutional Association, 1923.

Corbin, David. "'Frank Keeney Is Our Leader.'" In *Essays in Southern Labor History,* edited by Gary Fink and Merl Reed, pp. 144–56. Westport, Conn.: Greenwood Press, 1976.

Crawford, T. H. *An American Vendetta, a Story of Barbarism in the U.S.* New York: Delford, Clarke & Co., 1889.

Deutscher, Isaac. *The Prophet Armed, Trotsky, 1879–1921.* London: Oxford University Press, 1954.

Dick, William M. *Labor and Socialism in America: The Gompers Era.* Port Washington, N.Y.: Kennikat Press, 1972.

Dubofsky, Melvin. *We Shall Be All: A History of the Industrial Workers of the World.* Chicago: Quadrangle, 1969.

______, and Warren Van Tine. *John L. Lewis, a Biography.* New York: Quadrangle/The New York Times Book Co., 1977.

DuBois, W. E. B. *Souls of the Black Folk.* New York: Fawcett Publications, 1961.

Erikson, Erik. *Childhood and Society.* New York: W. W. Norton & Co., 1963.

______. *Identity: Youth and Crisis.* New York: W. W. Norton & Co., 1968.

Fetherling, Dale. *Mother Jones, the Miners' Angel.* Carbondale: Southern Illinois University Press, 1974.

Fowke, Edith, and Joe Glazer, eds. *Songs of Work and Protest.* New York: Dover, 1973.

Fowler, Charles. *Collective Bargaining in the Bituminous Coal Industry.* New

York: Prentice-Hall, 1927.

Franklin, John Hope. *From Slavery to Freedom*. New York: Random House, 1967.

Frazier, E. Franklin. *The Negro Family*. Chicago: University of Chicago Press, 1969.

Genovese, Eugene. *Roll, Jordan, Roll: The World the Slaves Made*. New York: Random House, 1974.

Geertz, Clifford. "Religion as a Cultural System." In *The Religious Situation*, edited by Donald Cutler, pp. 639–88. Boston: Beacon Press, 1968.

Gluck, Elsie. *John Mitchell, Miner*. New York: John Day Co., 1929.

Goodrich, Carter. *Migration and Economic Opportunities*. Philadelphia: University of Pennsylvania Press, 1936.

______. *The Miners' Freedom*. Boston: Marshall Jones, 1925.

Graebner, William. *Coal-Mining Safety in the Progressive Period*. Lexington: University of Kentucky Press, 1976.

Green, Archie. *Only a Miner: Studies in Recorded Coal Mining Songs*. Urbana: University of Illinois Press, 1972.

Gutman, Herbert. "The Negro and the United Mine Workers." In *The Negro and the American Labor Movement*, edited by Julius Jacobson, pp. 49–127. Garden City, N.Y.: Doubleday & Co., 1968.

______. "The Workers' Search for Power: Labor in the Gilded Age." In *The Gilded Age: A Reappraisal*, edited by H. Wayne Morgan, pp. 31–54. Syracuse, N.Y.: Syracuse University Press, 1963.

Hamilton, Charles. *The Black Preacher*. New York: William Morrow & Co., 1972.

Handlin, Oscar. *The Uprooted*. New York: Grosset & Dunlap, 1951.

Harrington, Michael. *Socialism*. New York: Bantam Books, 1972.

Harris, Evelyn, and Frank Krebs. *From Humble Beginnings: West Virginia State Federation of Labor, 1903–1957*. Charleston: West Virginia Labor History Publishing Fund, 1960.

Harrison, J. F. C. *Quest for a New Moral World: Robert Owen and the Owenites in Britain and America*. New York: Charles Scribner's Sons, 1969.

Hinrichs, A. F. *The United Mine Workers of America and the Non-Union Coal Fields*. New York: Columbia University Press, 1923.

Holiday, Robert. *Politics in Fayette County*. Montgomery, W.Va.: Montgomery Herald, 1958.

Hourwich, Isaac A. *Immigrants and Labor*. New York: G. P. Putnam's Sons, 1912; rev. ed., New York: AMS Press, 1972.

Hopkins, C. Howard. *History of the YMCA in North America*. New York: Association Press, 1951.

Hudson, Charles. "The Structure of a Fundamentalist Christian Belief System." In *Religion and the Solid South*, edited by Samuel Hill, pp. 122–42. Nashvill, Tenn.: Abingdon Press, 1972.

Hunt, Edward, F. G. Tryon, and Joseph Willits. *What the Coal Commission Found*. Baltimore: Williams & Williams Co., 1925.

Jackson, Kenneth. *The Ku Klux Klan in the City, 1915*–1930. New York: Oxford University Press, 1967.

Jones, Virgil. *The Hatfields and the McCoys*. Chapel Hill: University of North Carolina Press, 1948.

Jordan, Daniel. "The Mingo War: Labor Violence in the Southern West Virginia Coal Fields, 1919–1922." In *Essays in Southern Labor History*, edited by Gary Fink and Merl Reed, pp. 102–43. Westport, Conn.: Greenwood Press, 1976.

Katz, Michael. *Class, Bureaucracy, and Schools: The Illusion of Educational Change in America*. New York: Praeger, 1971.

______. *The Irony of Early School Reform*. Cambridge, Mass.: Harvard University Press, 1968.

Kaufman, Stuart B. *Samuel Gompers and the Origins of the American Federation of Labor*. Westport, Conn.: Greenwood Press, 1973.

Kirkland, Edward. *Industry Comes of Age*. New York: Holt, Rinehart & Winston, 1961.

Korson, George. *Coal Dust on the Fiddle: Songs and Stories of the Bituminous Coal Industry*. Philadelphia: University of Pennsylvania Press, 1943.

Lambie, Joseph. *From Mine to Market: The History of Coal Transportation on the Norfolk and Western Railroad*. New York: New York University Press, 1954.

Lane, Winthrop. *Civil War in West Virginia*. New York: Huebsch, 1921.

______. *The Denial of Civil Liberties in the Coal Fields*. New York: George H. Doran, 1924.

Lee, Howard. *Bloodletting in Appalachia*. Parsons, W.Va.: McClain Printing Co., 1969.

Lens, Sydney. *The Labor Wars*. Garden City, N.Y.: Doubleday & Co., 1973.

Levinson, Edward. "*I Break Strikes*." New York: n.p., 1935; reprinted in *American Labor from Conspiracy to Collective Bargaining*, edited by Leon Stein and Philip Taft. New York: Arno Press, 1969.

Lipset, Seymour Martin. *Political Man: The Social Basis of Politics*. New York: Doubleday & Co., 1960.

Lipset, Seymour, and Reinhart Bendix. *Social Mobility in Industrial Society*. Berkeley: University of California Press, 1963.

Lloyd, A. L. *Come All Ye Bold Miners*. London: Lawrence & Wishart, 1952.

McDonald, David, and Edward Lynch. *Coal and Unionism*. Silver Spring, Md.: Cornelius Printing Co., 1930.

Marx, Karl, and Friedrich Engels. *On Religion*. Moscow: Foreign Languages Publishing House, 1957.

Maurer, James. "Has the Church Betrayed Labor." In *Labor Speaks for Itself on Religion*, edited by Jerome Davis, pp. 19–28. New York: Macmillan Co., 1929.

Mead, Sydney. *The Lively Experiment*. New York: Harper & Row, 1963.

Mintz, Sidney. Foreward to *Afro-American Anthropology*, edited by Norman E. Whitten and John Szwed. New York: Free Press, 1970.

Morris, Homer. *The Plight of the Bituminous Coal Miner*. Philadelphia: University of Pennsylvania Press, 1934.

Musick, Ruth Ann, ed. *The Telltale Lilac Bush and Other West Virginia Ghost Tales*. Lexington: University of Kentucky Press, 1965.

Nevins, Allan. *Abram S. Hewitt with Some Account of Peter Cooper.* New York: Octagon Books, 1935.

Northrup, Herbert. *Organized Labor and the Negro.* New York: Harper & Brothers, 1944.

Osofosky, Gilbert. *Harlem: The Making of a Ghetto.* New York: Harper & Row, 1966.

Perlman, Selig. *A Theory of the Labor Movement.* New York: Macmillan, 1949.

Perlman, Selig, and Philip Taft. *History of Labor in the United States: 1896–1932.* Vol. 4. New York: Augustus M. Kelly, 1966.

Peters, J. T., and H. B. Carden. *History of Fayette County.* Fayetteville, W.Va.: Fayette County Historical Society, 1926.

Pope, Liston. *Millhands and Preachers.* New Haven, Conn.: Yale University Press, 1942.

Radosh, Ronald, ed. *Debs.* Englewood Cliffs, N.J.: Prentice-Hall, 1971.

Rankin, Robert. *When Civil Law Fails: Martial Law and Its Legal Basis in the United States.* Durham, N.C.: Duke University Press, 1939.

Rayback, Joseph. *A History of American Labor.* New York: Collier-MacMillan, 1966.

Reed, Merl, and Gary Fink, eds. *Essays in Southern Labor History.* Westport, Conn.: Greenwood Press, 1977.

Reid, H. *The Virginian Railway.* Milwaukee, Wis.: Kalmbach, 1961.

Rice, Otis. *The Allegheny Frontier.* Lexington: University of Kentucky Press, 1970.

Rich, Mark. *Some Churches in Coal Mining Communities of West Virginia.* Charleston: West Virginia Council of Churches, 1951.

Rochester, Anna. *Labor and Coal.* New York: International Publishers, 1931.

Ross, Malcomb. *Machine Age in the Hills.* New York: Macmillan, 1933.

Roy, Andrew. *History of the Coal Miners of the United States.* Columbus, Ohio: Trouger Printing Co., 1907.

Rudé, George. *The Crowd in the French Revolution.* New York: Oxford University Press, 1972.

Selvin, David. *The Thundering Voice of John L. Lewis.* New York: Lothrop, Lee & Shepard, 1969.

Shurick, Adam. *The Coal Industry.* Boston: Little, Brown & Co., 1924.

Spear, Allan. *Black Chicago, the Making of a Negro Ghetto.* Chicago: University of Chicago Press, 1974.

Spence, Robert. *The Land of the Guyandot.* Detroit: Harlo Press, 1976.

Spero, Sterling, and Abram Harris. *The Black Worker.* New York: Atheneum Press, 1974.

Spivak, John. *A Man in His Time.* New York: Horizon Press, 1967.

Spring, Joel. *Education and the Rise of the Corporate State.* Boston: Beacon Press, 1972.

Suffern, Arthur. *The Coal Miners' Struggle for Industrial Status.* New York: Macmillan Co., 1926.

______. *Conciliation and Arbitration in the Coal Industry of America.* Boston: Houghton Mifflin Co., 1915.

Summers, Festus. *Johnson Newlon Camden: A Study in Individualism.* New York: G. P. Putnam's sons, 1937.

Taylor, Philip. *The Distant Magnet.* New York: Harper & Row, 1971.

Teleky, Ludwig. *History of Factory and Mine Hygiene.* New York: Columbia University Press, 1948.

Thernstrom, Stephan. *The Other Bostonians.* Cambridge, Mass.: Harvard University Press, 1973.

Thompson, E. P. *The Making of the English Working Class.* New York: Vintage Books, 1963.

Troeltsch, Ernst. *The Social Teaching of the Christian Churches.* Vol. 1. London: George Allen & Unwin, 1930.

Turner, Charles. *Chessies Railroad.* Richmond, Va.: Garrett and Massie, 1956.

Valentine, Charles. *Culture and Poverty: Critique and Counter-Proposals.* Chicago: University of Chicago Press, 1968.

Ward, Robert D., and William Rodgers. *Labor Revolt in Alabama.* Tuscaloosa: University of Alabama Press, 1965.

Watkins, Gordon, and Paul Dodd. *Labor Problems.* New York: Thomas Y. Crowell, 1940.

Weinstein, James. *The Decline of American Socialism.* New York: Vintage Books, 1969.

Weller, Jack. *Yesterday's People.* Lexington: University of Kentucky Press, 1966.

Wiebe, Robert. *The Search for Order: 1877–1920.* The Making of America Series, edited by David Donald. New York: Hill & Wang, 1967.

Williams, John Alexander. *West Virginia: A Bicentennial History.* New York: W. W. Norton, 1976.

Williams, Robin. *The Reduction of Inter-Group Tensions.* New York: Social Science Research Council, 1947.

Wilson Edmund. *The American Jitters.* Freeport, N.Y.: Books for Libraries Press, 1968.

Wright, Helen. *Coal's Worst Year.* Boston: Richard Badger, 1925.

### *Periodicals*

"Ansted-Ganley Mountain Coal Company and the Development of Ansted." *West Virginia Review*, Sept. 1930, 408.

Bailey, Kenneth. "Hawk's Nest Coal Company Strike, 1880." *West Virginia History* 30 (July 1969): 625–34.

______. "Tell the Boys to Fall in Line." *West Virginia History* 32 (July 1971): 224–37.

Bauman, Zygmunt. "Marxism and the Contemporary Theory of Culture." *Co-Existence* 5 (July 1968): 161–71.

Berthoff, Rowland. "The Social Order of the Anthracite Region, 1825–1902." *Pennsylvania Magazine of History and Biography* 89 (July 1965): 261–91.

Brammel, Bernard. "Eugene V. Debs: Blue-Denim Spokesman." *North Dakota Quarterly* 43 (Spring 1972): 12–28.

Cantrell, Betty, Grace Phillips, and Helen Reed. "Widen: The Town J. G. Bradley Built." *Goldenseal* 3 (Jan.-Mar. 1977): 2-6.

Carter, Charles. "Murder to Maintain Coal Monopoly." *Current History* 15 (Jan. 1922): 597-603.

______. "The West Virginia Coal Insurrection." *North American Review* 198 (Oct. 1913): 457-69.

Clignet, Remi. "Damned if You Do, Damned if You Don't: The Dilemmas of the Colonizer-Colonized Relations." *Comparative Education Review* 15 (Oct. 1971): 300.

Cohen, David, and Marvin Lazerson. "Education and the Corporate Order." *Socialist Revolution* 8 (Mar.-Apr. 1972): 47-72.

Corbin, David. "The 'Ups and Downs' of a British Miner in the West Virginia Coal Fields." *Goldenseal* 2 (Oct.-Dec. 1976): 35-49.

Duncan, Ross. "Case Studies in Emigration: Cornwall, Gloucestershire and New South Wales, 1877-1886." *Economic History Review* 16 (June 1963): 272-89.

Field, Lewis, Reed Ewing, and David Wayne. "Observations on the Relation of Psychosocial Factors to Psychiatric Illness among Coal Miners." *International Journal of Social Psychiatry* 3 (Autumn 1957): 133-45.

Foster, George. "What Is Folk Culture?" *American Anthropologist* 55 (Apr.-June 1953): 159-73.

Fox-Genovese, Elizabeth. "The Many Fates of Moral Economy." *Past and Present* 58 (Feb. 1973): 161-68.

Frankel, Emil. "Occupational Classes among Negroes in Cities." *American Journal of Sociology* 35 (Mar. 1930): 718-38.

Genovese, Eugene. "On Antonio Gramsci." *Studies on the Left* 7 (Mar.-Apr. 1967): 301.

Goodrich, Carter. "Machine and the Miner." *Harper* 154 (1927): 649-54.

______. "Nothing but a Coal Factory: Machinery in Coal Mining." *New Republic* 44 (Sept. 1925): 91-93.

Green, James. "Behaviorism and Class Analysis." *Labor History* 13 (Winter 1972): 89-106.

______. "The Brotherhood of Timber Workers, 1910-1913: A Radical Response to Industrial Capitalism in the Southern U.S.A." *Past and Present* 40 (Aug. 1973): 161-200.

Grimsted, David. "Rioting in Its Jacksonian Setting." *American Historical Review* 77 (Apr. 1972): 361-97.

Gutman, Herbert. "Protestantism and the American Labor Movement: The Christian Spirit in the Gilded Age." *American Historical Review* 71 (Oct. 1966): 74-101.

______. "Work, Culture, and Society in Industrial America, 1815-1919." *American Historical Review* 78 (June 1973): 531-88.

Harris, Abram. "The Negro in the Coal Mining Industry." *Opportunity,* Feb. 1926, 45-48.

______. "Negro Migration to the North." *Current History* 20 (Sept. 1924): 921-25.

Harris, Sheldon. "Letters from West Virginia: Management's Version of the 1902 Coal Strike." *Labor History* 9 (Spring 1969): 228-40.

Haynes, George. "Negro Migration." *Opportunity*, Oct. 1924, 303–6.
Hobsbawn, E. J. "Methodism and the Threat of Revolution in Britain." *History Today* 7 (Feb. 1957): 115–24.
Johnson, Guy. "Negro Migration and Its Consequences." *Social Forces* 2 (Mar. 1924): 404–8.
Johnson, James. "The Wilsonians as War Managers: Coal and the 1917–18 Winter Crisis." *Prologue* 9 (Winter 1977): 193–208.
Karsh, Bernard, and Jack London. "The Coal Miners: A Study of Union Control." *Quarterly Journal of Economics* 68 (1954): 415–36.
Laing, James. "Negro Migration to the Mining Fields of West Virginia." *Proceedings of the West Virginia Academy of Science* 10 (1936).
______. "The Negro Miner in West Virginia." *Social Forces* 36 (Mar. 1936): 416–22.
Leamer, Laurence. "Twilight of a Baron." *Playboy*, May 1973, 114, 120, 168–76.
Lynch, Lawrence. "The West Virginia Coal Strike." *Political Science Quarterly* 29 (Dec. 1914): 626–63.
Mack, Raymond. "Riot, Revolt or Responsible Revolution: Of Reference Groups and Racism." *Sociological Quarterly* 10 (Spring 1969): 147–56.
Massey, Glen. "Legislators, Lobbyists and Loopholes, 1875–1901." *West Virginia History* 32 (Apr. 1971): 135–70.
Maurer, Maurer, and Calvin Senning. "Billy Mitchell, the Air Service, and the Mingo Way." *Airpower Historian* 12 (Apr. 1965): 37–43.
Miller, Kelly. "Education of the Negro in the North." *Educational Review* 62 (Oct. 1921): 232–38.
Minard, Ralph. "Race Relations in the Pocahontas Coal Fields." *Journal of Social Issues* 8 (1952): 37.
Montgomery, David. "The 'New Unionism' and the Transformation of Workers Consciousness in America, 1909–1922." *Journal of Social History* 7 (Summer 1974): 509–29.
______. "The Shuttle and the Cross: Weavers and Artisans in the Kensington Riots of 1844." *Journal of Social History* 5 (Summer 1972): 411–46.
______. "Spontaneity and Organization: Some Comments." *Radical America* 7 (Nov.-Dec. 1973): 70–80.
______. "Workers' Control of Machine Production in the Nineteenth Century." *Labor History* 17 (Fall 1976): 485–508.
Newby, Robert, and David Tyack. "Victims without Crimes: Some Historical Perspectives on Black Education." *Journal of Negro Education* (Fall 1971): 192–206.
Northern, E. E. "The Religions of the Mountaineer." *West Virginia Review*, Aug. 1933, 317–36.
Olson, Mancure. "Rapid Economic Growth as a Destabilizing Force." *Journal of Economic History* 23 (1963): 529–52.
Perlman, Selig. "The Basic Philosophy of the American Labor Movement." *Annals of the American Academy of Political and Social Sciences* 254 (Mar. 1951): 57–63.
Phillips, Cabell. "The West Virginia Mine War." *American Heritage* 25 (Aug.

1974): 58–61.

Pollard, Sidney. "Factory Discipline in the Industrial Revolution." *Economic History Review* 16 (Dec. 1963): 254–71.

______. "Investment, Consumption, and the Industrial Revolution." *Economic History Review* 11 (1958): 215–18.

Posey, Thomas. "Unemployment Compensation and the Coal Industry in West Virginia." *Southern Economic Journal* 7 (1947): 347–61.

Preston, William. "Shall This Be All? U.S. Historians versus William D. Haywood et al." *Labor History* 12 (Summer 1971): 435–53.

Randall, James. "Miners and the Law of Treason." *North American Review* 216 (Sept. 1922): 312–22.

Rice, Otis. "Coal Mining in the Kanawha Valley to 1861." *Journal of Southern History* 31 (Nov. 1965): 293–315.

Ross, Mike. "Life Style of the Coal Miner." *Appalachian Medicine*, Mar. 1971, 6–7.

______. "Life Style of the Coal Miner." *United Mine Workers' Journal,* July 1, 1970, 12.

Ryan, Frederick. "The Development of the Coal Operators' Associations in the Southwest." *Southwestern Social Science Quarterly* 14 (Sept. 1933): 133–44.

Sherrill, Robert. "West Virginia Miracle: The Black Lung Rebellion." *Nation* 218 (Apr. 28, 1969): 531.

Simmons, Charles, John Rankin, and U. G. Carter. "Negro Coal Miners in West Virginia, 1875–1925." *Midwest Journal* 60 (Spring 1954): 60–69.

Stoddard, C. F. "The Bituminous Coal Strike." *Monthly Labor Review*, Dec. 1919, 61–78, 1725–42.

Surface, G. T. "Negro Mine Laborers: Central Appalachian Coal Field." *Annals of the American Academy of Political Science* 33 (Mar. 1909): 116–17.

Thompson, E. P. "The Moral Economy of the English Crowd in the Eighteenth Century." *Past and Present* 50 (Feb. 1971): 82–96.

______. "Patricial Society, Plebian Culture." *Journal of Social History* 7 (Summer 1974): 382–405.

______. "Time, Work-Discipline, and Industrial Capitalism." *Past and Present* 38 (Dec. 1967): 56–97.

Thompson, Edgar. "Mines and Plantations and the Movements of Peoples." *American Journal of Sociology* 37 (Jan. 1932): 603–11.

Tyron, F. G. "The Irregular Operation of the Bituminous Coal Industry." *American Economic Review* 11 (Mar. 1921): 57–73.

Wiebe, Robert. "The Anthracite Strike of 1902: A Record of Confusion." *Mississippi Valley Historical Review* 48 (Sept. 1961): 229–51.

Williams, John Alexander. "The New Dominion and the Old." *West Virginia History* 33 (July 1972): 317–407.

Woods, Roy. "History of the Hatfield-McCoy Feud with Special Attention to the Effects of Education on It." *West Virginia History* 21 (Oct. 1960): 27–37.

*Theses and Dissertations*

Anson, Charles. "A History of the Labor Movement in West Virginia." Ph.D. dissertation, University of North Carolina, 1940.

Barb, John M. "Strikes in the Southern West Virginia Coal Fields, 1912–1922." M.A. thesis, West Virginia University, 1949.

Barkey, Fred. "The Socialist Party in West Virginia from 1898 to 1920." Ph.D. dissertation, University of Pittsburgh, 1971.

Cubby, Edwin. "The Transformation of the Tug and Guyandot Valleys: Economic and Social Change in West Virginia, 1888–1921." Ph.D. dissertation, Syracuse University, 1962.

Gillenwater, Mack. "Cultural and Historical Geography of Mining Settlements in the Pocahontas Coal Fields of Southern West Virginia." Ph.D. dissertation, University of Tennessee, 1972.

Goodall, Elizabeth J. "History of the Charleston Industrial Area." M.A. thesis, West Virginia University, 1937.

Harvey, Helen B. "From Frontier to Mining Town in Logan County, West Virginia." M.A. thesis, University of Kentucky, 1942.

Hubbard, Danny. "Rodgers and His Railroad: A History of the Virginian Railway from Conception to Merger." M.A. thesis, Marshall University, 1974.

Hurst, Mary. "Social History of Logan County, West Virginia, 1765–1923." M.A. thesis, Columbia University, 1924.

Livingston, William. "Coal Miners and Religion: A Study of Logan." Th.D. dissertation, Union Theological Seminary, 1951.

Merrill, William. "Economics of the Southern Smokeless Coals." Ph.D. dissertation, University of Illinois, 1953.

Morris, Thomas. "The Coal Camp: A Pattern of Limited Community Life." M.A. thesis, West Virginia University, 1950.

Posey, Thomas. "The Labor Movement in West Virginia, 1900–1948." Ph.D. dissertation, University of Wisconsin, 1948.

Thomas, Jerry Bruce. "Coal County: The Rise of the Southern Smokeless Coal Industry and its Effect on Area Development, 1872–1910." Ph.D. dissertation, University of North Carolina, 1971.

White, Elizabeth. "Development of the Bituminous Coal Mining Industry in Logan County, West Virginia." M.A. thesis, Marshall University, 1956.

*Pamphlets*

Barkus, Gary. "The West Virginia Tax Structure, the People and Coal." *Report.* Appalachian Research and Defense Fund, Nov. 30, 1971.

Corbin, David. *The Socialist and Labor Star.* Huntington, W.Va.: Appalachian Movement Press, 1972.

Dix, Keith. *Work Relations in the Coal Industry: The Hand Loading Era, 1880–1930.* West Virginia University Bulletin, Series 78, no. 7-2. Morgantown:

West Virginia University Institute for Labor Studies, 1977.
Miller, Tom. *Who Owns West Virginia*? Huntington, W.Va.: Huntington Herald-Advertiser and Herald-Dispatch, 1974.
Thompson, James. *Significant Trends in the West Virginia Coal Industry, 1900–1957*. Vol. 6, no. 1. Morgantown: West Virginia University Business and Economic Studies, 1958.

*Miscellaneous*

Corbin, David. "The National Coal Strikes of 1919 and 1922, or Jòhn L. Lewis versus the U.S. Government." Historian's Office, U.S. Department of Labor, Washington, D.C.
Eller, Ronald. "Industrialization and Social Change in Appalachia, 1880–1930." Paper presented at the Southern Historical Association Convention, Atlanta, Ga., Nov. 10–13, 1976.
Lawrence, Randall. "Here Today, Gone Tomorrow: Coal Miners in Appalachia, 1880–1940." Paper presented at the Organization of American Historians Convention, Detroit, Mich., Apr. 1–4, 1981.

# Index

## *A Note on the Author*

David Alan Corbin was born in Charleston, West Virginia, in 1947. He received his bachelor's and master's degrees from Marshall University, Huntington, West Virginia (1969 and 1972, respectively) and his Ph.D. from the University of Maryland, College Park (1978). He is presently writing a sequel on the southern West Virginia coal miners from 1922 to 1980, under a research grant from the National Endowment for the Humanities.